Getting Started with Pro/ENGINEER® Wildfire

Third Edition

Robert Rizza

Department of Mechanical Engineering
Milwaukee School of Engineering
Milwaukee, Wisconsin

PEARSON

Prentice Hall

Upper Saddle River, New Jersey
Columbus, Ohio

Library of Congress Cataloging-in-Publication Data

Rizza, Robert 1965–.
     Getting started with Pro/Engineer Wildfire / Robert Rizza. — 3rd ed.
          p. cm.
     Rev. ed. of: Getting started with Pro/Engineer. 2nd ed. Prentice Hall, c2002.
     Includes bibliographical references and indexes.
     ISBN 0-13-146474-4
     1. Engineering design–Data processing. 2. Pro/ENGINEER. 3. Computer-aided design.
     I. Rizza, Robert, 1965–. Getting started with Pro/Engineer. II. Title.

TA174.R555 2005
620'.0042'0285536—dc22   2004053433

Executive Editor: *Debbie Yarnell*
Managing Editor: *Judith Casillo*
Production Editor: *Louise N. Sette*
Production Coordination: *Karen Fortgang, bookworks*
Design Coordinator: *Diane Ernsberger*
Cover Designer: *Terry Rohrbach*
Cover art: *Jeff Anderson, Mike Feldner, David H. Zanon, Jr., Matt Wubben, Mark Gales, Randy Hoffman, Andrew Brahmstedt, and Joseph A. King.*
Production Manager: *Deidra Schwartz*
Marketing Manager: *Jimmy Stephens*

This book was set by *The GTS Companies*/York, PA Campus. It was printed and bound by Courier Kendallville, Inc. The cover was printed by Phoenix Color Corp.

The author and publisher of this book have used their best efforts in preparing this book. These efforts include the development, research, and testing of the theories and programs to determine their effectiveness. The author and publisher make no warranty of any kind, expressed or implied, with regard to these programs or the documentation contained in this book. The author and publisher shall not be liable in any event for incidental or consequential damages in connection with, or arising out of, the furnishing, performance, or use of these programs.

PTC and Pro/ENGINEER are trademarks or registered trademarks of Parametric Technology Corporation or its subsidiaries in the United States and in other countries.

Pearson Education Ltd.   Pearson Education Australia Pty. Limited
Pearson Education Singapore Pte. Ltd.   Pearson Education North Asia Ltd.
Pearson Education Canada, Ltd.   Pearson Educación de Mexico, S.A. de C.V.
Pearson Education—Japan   Pearson Education Malaysia Pte. Ltd.

10 9 8 7 6 5 4 3

CONTENTS

PREFACE

This book originated from an introductory engineering graphics and computer-aided design (CAD) course taught at North Dakota State University (NDSU). The purpose of the course was to introduce the engineering student to sketching, visualization, and parametric modeling. The course emphasized that CAD is the cornerstone of the engineering program and a way for the engineer to communicate design ideas graphically. This book was developed with this philosophy.

This edition details the use of Pro/ENGINEER Wildfire and contains all new high-resolution screen captures. Additional models have been added. These are the parts used in the ButterFly Valve, Vise, and HoldDown Clamp assemblies.

The layout of the book is similar to the previous editions. However, the chapters on drawings, sections, and tolerancing now appear earlier in the book. The *Redefine* option has been moved to Chapter 2 right after the introduction of Sketcher. This change was made in order to show the student some techniques for dealing with errors introduced during the construction of a feature.

New tutorials have been added, including one on using points in the Sketcher, as well as some additional tutorials on dealing with failed features. Furthermore, in order to increase the effectiveness of the tutorials, the objectives of each tutorial are listed at the beginning of the tutorial.

Earlier editions of this text used short-term and long-term projects. This edition follows the same approach. In order to show how the tutorials and hence the projects are related to one another, a list of parts and assemblies is given. The reader should consult this list when planning which tutorial to work through.

Throughout the book, Pro/E options and commands are given in *italic* face. Snapshots of the icons have been added in line with the text to allow the reader to readily find the feature on the software interface.

Additional exercises at the end of each chapter further reinforce the options and concepts presented in that chapter. The models in these exercises were selected to challenge the student in the use of the options under consideration. In many of the chapters, new assignments have been added. Many of the exercises from the previous editions have been reordered in terms of complexity and/or new figures produced.

Additional parts and formats are available with this book and are needed to complete some of the tutorials. These parts and formats may be downloaded from the Prentice Hall website at www.prenhall.com/rizza. They are also available from the author's website at http://people.msoe.edu/~rizza.

ACKNOWLEDGMENTS

Any work, whatever its magnitude, is impossible without the support and encouragement of family, friends, and colleagues. I thank Eric Svendsen, my editor at Prentice Hall, for his support at getting this project on the right track and completed. Special thanks go to Keven Mueller for inquiring about the manuscript in the first place.

Gratitude must go to the faculty of the Mechanical Engineering and Applied Mechanics (MEAM) Department at NDSU for their support during the initial work on this project. In particular, Mohammed Mahinfalah and James Stone deserve many heartfelt thanks for their continual support.

Shaikh Rahman and Ken Jones, at the time students in the MEAM department, completed many of the original models for the homework assignments. Their hard work and timeliness are greatly appreciated.

I also wish to thank Greg Gessel, currently at John Deere, for his useful comments on the first edition of this book and for his continual support. Debra L. Smith of Johnson Controls made many suggestions regarding the *Resolve* options. Thanks, Deb.

Though Solidworks and not Pro/ENGINEER users, Joe Musto and William "Ed" Howard of the Milwaukee School of Engineering have always been willing to listen to my ideas on CAD. The discussions I had with them gave me new perspectives on many of the topics in the text. Thanks, to both of you.

I would like to acknowledge the reviewers of this text: Dr. Mustapha S. Fofana, Worcester Polytechnic Institute; Dr. David N. Rocheleau, University of South Carolina; and Dr. Robert V. Pieri, North Dakota State University.

Lastly, but in no way in the least, thanks go to my wife Barbara and son Paul for their loving encouragement.

Robert Rizza
Milwaukee School of Engineering

LIST OF PARTS AND ASSEMBLIES

Part or assembly name	Tutorial
2DFrame	13.3, 13.6, 15.5
AdjustmentScrew	5.4, 7.5, 15.4
AngleBracket	2.4, 4.1, 7.1, 8.10, 20.5
BaseOrient*	1.2, 1.3, 1.4, 1.5, 1.6, 2.6
BlockAssembly	19.1
Blower	16.2
ButterFlyValve_Plate	13.5
BushingT3	3.1, 4.2
CoffeeMug	14.2, 15.1
Collar	13.1
ConnectingArm	6.2, 7.2, 8.7, 11.4
CoverPlate	17.2, 18.2
EmergencyLightBase	5.5, 9.1, 18.1
EmergencyLightHolder	6.1, 8.11, 9.5, 9.6
EmergencyLightReflector	14.3
Gasket	15.2
Handle	2.1
HeatTakeoff	16.1
HoldDownClamp_Arm	2.4
KingPostTruss	13.4, 13.7, 15.6
Nut	8.3, 17.1
PlateWithCounterboreHoles	4.4, 8.1, 8.12, 19.2
PipeFlange	4.5, 7.3, 7.4, 8.2, 11.1, 11.2, 11.3
Propeller	3.2, 4.3, 5.3, 8.4, 8.5
PulleyWheel	14.1, 19.2
PulleyAssembly	19.2, 19.3, 19.4, 19.5, 19.6, 19.7, 19.8, 20.3, 20.4
RemoteControlCoverLower	18.3
RodSupport	3.7, 4.6, 10.4, 13.2, 20.1, 20.2
ShifterFork	3.1, 3.3, 3.4, 3.5, 3.6, 9.2, 9.3, 9.4
Spring	15.3
SquarePlate	8.13
SquarePlatewithHoles*	10.2, 10.3, 12.1, 12.2
TensionPlate	2.2
UBracket	5.1, 6.3
VibrationPad	8.6, 8.8, 8.9
WasherT52	5.2
Wing	13.8, 18.4

* This part is available to download from www.prenhall.com/rizza or http://people.msoe.edu/~rizza.

CHAPTER 1

GETTING ACQUAINTED WITH THE PRO/E INTERFACE

INTRODUCTION AND OBJECTIVES

Pro/Engineer (Pro/E) is a software package developed and sold by Parametric Technology Corporation for the Computer Aided Design and Drawing (CADD) market. It uses parametric or constraint-based modeling.

Parametric modeling allows the user to represent the dimensions of the object with parameters. For example, in traditional CADD packages, entering the given dimensions would construct the block shown in Figure 1.1. If the model dimensions were later changed, the model would have to be redrawn. This is an inefficient approach. With a parametric modeler, the dimensions of the object are represented in the computer as parameters, as shown in Figure 1.2(a). Then, numbers are assigned to the parameters. If the

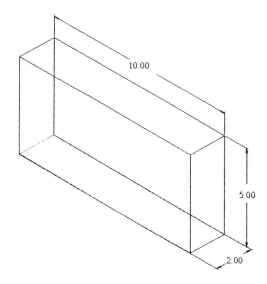

FIGURE 1.1 *Block under consideration.*

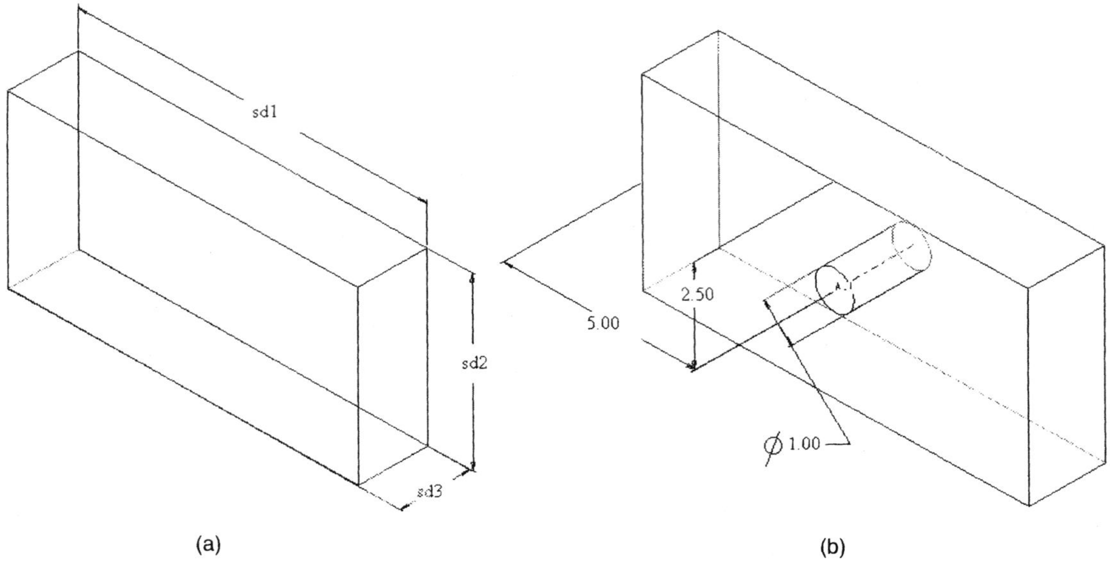

(a)   (b)

FIGURE 1.2 *An object may be represented by parameters.*

dimensions of the object are changed, the model can be easily modified, since only the numerical values of the parameters need to be changed.

Pro/E is also a feature-based modeler. Feature-based modeling is a way to combine the commands needed to produce a common manufactured feature. For example, in traditional CADD packages, a hole in a part would be produced by creating two circles and at least one line connecting the circles with the prescribed dimensions (see Figure 1.2(b)). In a software package such as Pro/E, the information to create the hole is grouped together in one feature. The circles and line(s) making up the hole are represented and identified by Pro/E as a single feature, and the relationships among the individual parts making up the feature are maintained. Thus, if any changes are made to any part of the elements making up the hole, the entire feature is easily updated.

Nonconstraint-based modelers are called primitive modelers. Such modelers use a finite set of geometric primitives to construct a three-dimensional model (see Figure 1.3). A three-dimensional primitive modeler constructs a model by combining these primitives, as in Figure 1.4.

Constraint-based modeling begins by creating a two-dimensional sketch. This sketch does not have to be accurate, as it will be constrained by adding dimensions. When the generation process is performed, the dimensions constrain the sketch and the geometry is redefined using parametric equations. A three-dimensional model is generated from the sketch by extruding (as in Figure 1.4), revolving the sketch about an axis of revolution or sweeping along a trajectory. This process often mimics the manufacturing process.

Constraint-based modelers are more intuitive because they capture the design intent and the dimensions are related to the geometry. A change in one of the dimensions will cause the model to change immediately. Any design flaw will be readily visible, because the model is automatically updated.

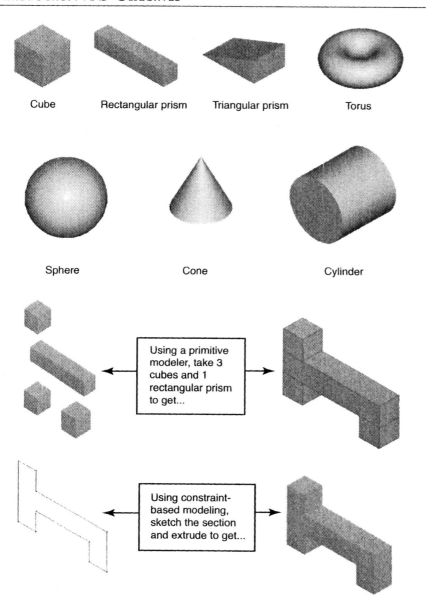

FIGURE 1.3 *Primitive modelers use a set of geometric shapes to construct models.*

Cube   Rectangular prism   Triangular prism   Torus

Sphere   Cone   Cylinder

FIGURE 1.4 *The difference between a primitive and constraint-based modeler.*

Using a primitive modeler, take 3 cubes and 1 rectangular prism to get...

Using constraint-based modeling, sketch the section and extrude to get...

The approach in this book follows the method used in constraint-based modeling. That is, we begin with the options of Pro/E that are used to generate a sketch and proceed to the options that are used to make the two-dimensional model three-dimensional.

However, in this chapter, we concentrate on the Pro/E interface. The interface is the display that is visible after the software has been loaded. It is our intent to discuss the pertinent options that are used to create, retrieve, and print files. Therefore, the objectives of this chapter are for the user to

1. Learn about the Pro/E interface
2. Develop an ability to retrieve a part file, reorient the part, and print a hard copy of the model through the tutorials
3. Attain some experience in customizing the display

1.1 ENGINEERING GRAPHICS AND PRO/ENGINEER

Before diving into the software, let us review some basic engineering graphics. Recall from your engineering graphics course that different types of lines are used to visualize an object. These include visible, dashed, centerlines and cutting plane lines. Visible lines are used to indicate edges and other visible geometric entities. Dashed lines are used to represent hidden edges. If a model is shown as a wireframe, then all edges are represented using visible lines. As we will see shortly, Pro/E allows the user to represent a model as a wireframe, hidden line, no hidden line, or shaded model. In Figure 1.5, the same model is shown in each of the four different display modes. Note that on the screen hidden lines are shown in gray, not dashed. Dashed lines are drawn by the software when drawing layouts are produced. In Figure 1.5, note the centerline through the hole.

The mode used to represent a model depends on personal preference and the task at hand. In this text, we will use the formats as needed.

From your engineering graphics course, you may recall the relationship between an orthographic view and a three-dimensional pictorial of a model. The representations in Figure 1.5 are all isometric pictorials. Two-dimensional orthographic views may be constructed from an isometric. In fact, this is the power of a modern CAD system. Three-dimensional models are constructed and then two-dimensional views of the model are obtained by looking along different directions of the three-dimensional model (see Figure 1.6).

In Figure 1.6, all six orthographic views are shown. In general, most models may be represented in layouts using the front, top, and either the right-hand or left-hand view. Some models, especially those exhibiting axisymmetry, may be represented using just two.

In this text, two-dimensional views will be used whenever they best represent the information that needs to be conveyed. In addition, two-dimensional views will be used to generate features by drawing on a sketching plane. We will see how this works in Chapter 2 when we begin to use the Sketcher.

FIGURE 1.5 *A model may be represented in several different modes.*

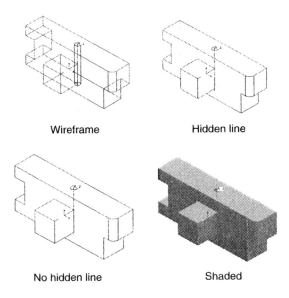

Wireframe   Hidden line

No hidden line   Shaded

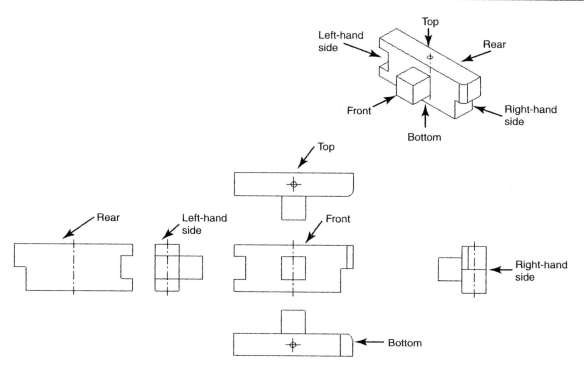

FIGURE 1.8 *In general, six orthographic views may be obtained from a three-dimensional model.*

1.2 THE PRO/E INTERFACE

The actual loading and running of the software is system dependent. Before proceeding any further with this book, you should check the exact procedure for loading and running the software on your system.

The Pro/E interface is illustrated in Figure 1.7. For the sake of clarity, we have changed the default background color of the drawing window from gray to white. The interface is characterized by three separate panels: the window explorer, the browser, and the main drawing window. Notice that each panel may be collapsed or enlarged to change its relative size.

The purpose of the window explorer is to allow easy access to data files. The user may scroll through the folders and find the desired part, assembly, or drawing and then preview the object in the browser, as in Figure 1.8. If the computer is connected to the Web, the browser may be used to obtain data and information. In particular, the latest help and tutorial files may be downloaded directly from the PTC website.

Below the main window is a smaller window. This is the message window. All messages from Pro/E to the user are displayed in this window.

Across the top is a Microsoft Windows–style toolbar containing various buttons (see Figure 1.7). When the software is first loaded, the first two buttons are active and may be used to create a new file or open an existing file, respectively. Pro/E uses pull-down menus and icons for navigation. That is, you will activate various options by mouse clicks and pull-down menus. Depending on the options, pop-up menus and dialogue boxes may also appear. These interfaces appear in order to enter all the information required

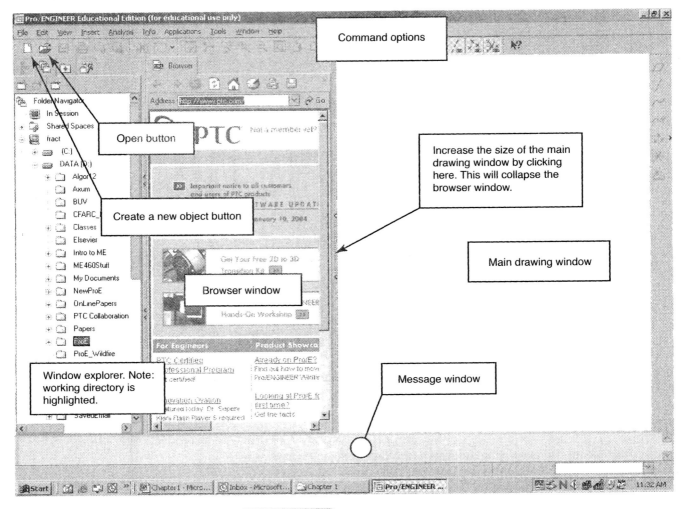

FIGURE 1.7 *The Pro/E interface.*

to complete the requested task and then disappear when all the appropriate information has been provided by the user.

In the toolbar there are ten command options: *File, Edit, View, Insert, Analysis, Info, Applications, Tools, Window,* and *Help.* The options are used in the following manner:

1. *File.* The pull-down menu contains options for creating, retrieving, printing, or saving a file in various configurations.

2. *Edit.* This option modifies existing features or can undo or redo features.

3. *View.* This option allows the user to change the display of the model.

4. *Insert.* The Insert option allows the user to add new features to the current model.

5. *Analysis.* This option performs geometric, feasibility, and optimization analysis.

6. *Info.* This option allows the user to obtain information concerning the model in memory

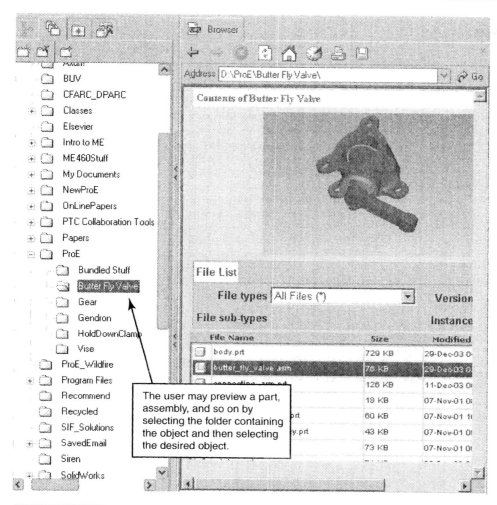

The user may preview a part, assembly, and so on by selecting the folder containing the object and then selecting the desired object.

FIGURE 1.8 *The Pro/E interface has a built-in browser that may be used to preview an object.*

7. *Applications.* For users with a license to applications, such as Pro/Mechanica, this command allows the user to quickly interface with the application.

8. *Tools.* The Tools command is a potpourri of options, which include editing and loading the configuration file, defining and loading *Mapkeys*, and setting preferences.

9. *Window.* This standard windows command allows the user to toggle between active windows.

10. *Help.* The last command allows the user to activate Pro/E's online help system. The information is displayed in a Web browser–like environment.

The tutorials that follow are designed to illustrate the use of these options. In particular, we show how to use some options to change the way the software looks and operates. This warrants a word of caution. While we have changed the drawing window background to white, we have left the colors used for the various entities in their default setting. In order to avoid any confusion caused by differences in the appearance of your screen and the figures in this book, we suggest that beginning Pro/E users do the same.

Furthermore, in Tutorial 1.1, we show how to make changes to the "config. pro" file. Changes to this file may be made available for all subsequent use of the software by placing this file in the directory from which Pro/E loads. Your system administrator can help you find this directory.

1.2.1 TUTORIAL 1.1: BECOMING FAMILIAR WITH THE INTERFACE

This tutorial illustrates:

- Setting a working directory
- Turning various display items on and off
- Customizing toolbars

1. Initiate the software using the specific commands for your system.
2. By comparing your desktop with Figure 1.7, identify the main window and the message window.
3. Click on *File* and then *Set Working Directory*. The Select *Working Directory* box, as shown in Figure 1.9, will appear.
4. Scroll down and find your assigned folder. Then select *OK*.
5. Select *Tools*. You will see several options listed in the pull-down menu. Among these are *Environment*, *MapKeys*, and *Options*.
6. Select *Environment* (Figure 1.10).
7. Try turning the display of the datum planes off. Click on this toggle and notice the message in the message window. You should see the message "Datum planes will NOT be displayed."
8. Select the toggle again. Pro/E should respond with "Datum planes will be displayed."

FIGURE 1.9 *The* Select Working Directory *box is used to find and set the working directory.*

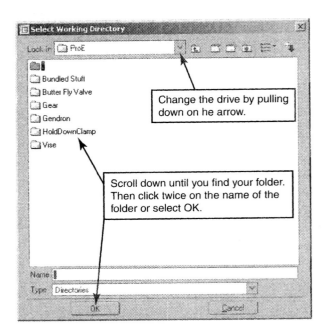

9. Change the orientation from *Trimetric* to *Isometric*, which is preferred by many users, by clicking on the pull-down next to Trimetric and selecting Isometric. Pro/E will display the message: "The default view is Isometric."
10. Notice the remaining options in the *Environment* dialog box (Figure 1.10). Take some time and turn these options on and off. As you do so, take note of the messages sent by the software. After doing so, select *OK*.
11. Now practice customizing the configuration file. Select *Tools* and *Options*.
12. Make sure that "Current Session" is listed in the *Showing* field. Then, unselect the "Show only options loaded from file." This will list all the options.
13. Scroll down the list of options until "allow_anatomic_features" is found.
14. Select the option with your mouse. In the "Value" field, change the "no" to "yes." Press the *Add/Change* button . Select *Close*.
15. You may wish to customize your display by adding additional icons to the toolbar or loading other toolbars that are not currently displayed. Select *Tools* and *Customize Screen*. The *Customize* box (shown in Figure 1.11) will appear.
16. For illustrative purposes, we will add a button to the toolbar that will orient the model to standard orientation. Click on *View* in the *Categories* list.
17. Several grayed icons will appear. They are gray because a part is not active. However, one or more may be dragged onto the

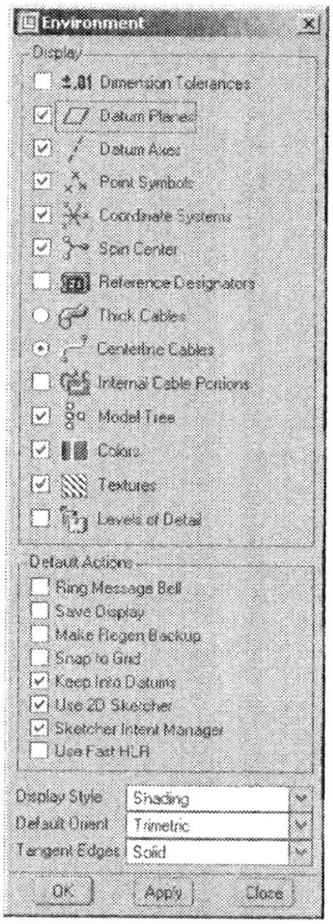

FIGURE 1.10 *The* Environment *dialog box contains toggles for turning the display of certain features on and off. It also contains the field for changing the display orientation and style.*

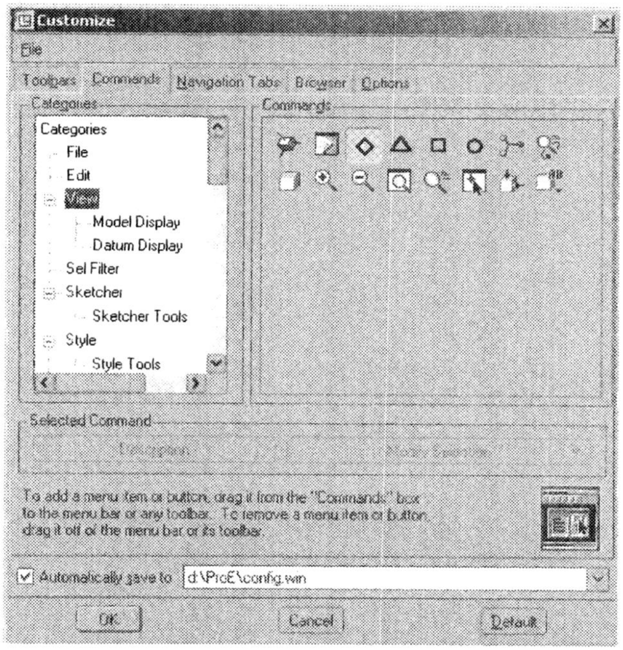

FIGURE 1.11 *The commands available to the user may be customized by activating various built-in toolbars or by dragging desired icons to the toolbar currently displayed.*

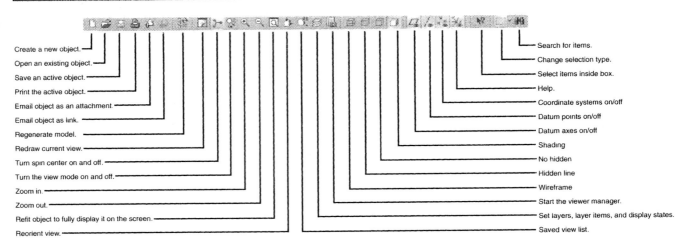

Create a new object.
Open an existing object.
Save an active object.
Print the active object.
Email object as an attachment.
Email object as link.
Regenerate model.
Redraw current view.
Turn spin center on and off.
Turn the view mode on and off.
Zoom in.
Zoom out.
Refit object to fully display it on the screen.
Reorient view.

Search for items.
Change selection type.
Select items inside box.
Help.
Coordinate systems on/off
Datum points on/off
Datum axes on/off
Shading
No hidden
Hidden line
Wireframe
Start the viewer manager.
Set layers, layer items, and display states.
Saved view list.

FIGURE 1.12 *Several icons are available in the default toolbar.*

toolbar. The *Default Orientation* icon is the first icon from the left. Grab this icon and drag it onto the toolbar. Then select *OK*.

18. If you wish, you may drag additional icons onto the toolbar using this approach.

19. The default icons are reproduced in Figure 1.12, along with a short description of their use. The description is also available by placing your mouse pointer over the icon. Try placing your mouse pointer over ⬜. In the message window, you should see: "Create a new object."

20. Now, let us see how we can change the various color schemes. Choose *View*, *Display Settings*, and *System Colors* (Figure 1.13).

21. The colors for the various entities may be changed by selecting the desired tab. For example, by selecting *Datum*, the user may change the datum color scheme. In this book will use the default scheme, so no changes will be made to the entities at this point.

22. However, we will change the background window color. Select *Scheme* and then *Black on White* and then *OK*.

23. Additional schemes are available such as *White on Black*. Try selecting these additional schemes until you find one that you prefer.

1.3 PART FILES AND PRO/E

In Pro/E, part files are saved with the extension ".prt." Multiple versions of a part are allowed. The software will save each version with a version number appended to the end of the file name extension. For example "BaseOrient.prt.1" is the first version of the part "BaseOrient." In order to delete all but the current version from the disk use the path:

1. Select *File*, *Delete*, *Old Versions*.
2. Then, enter the name of the part.

FIGURE 1.13 *The color scheme for the system and the displayed entities may be modified by using the options found in the* System Colors *dialog box.*

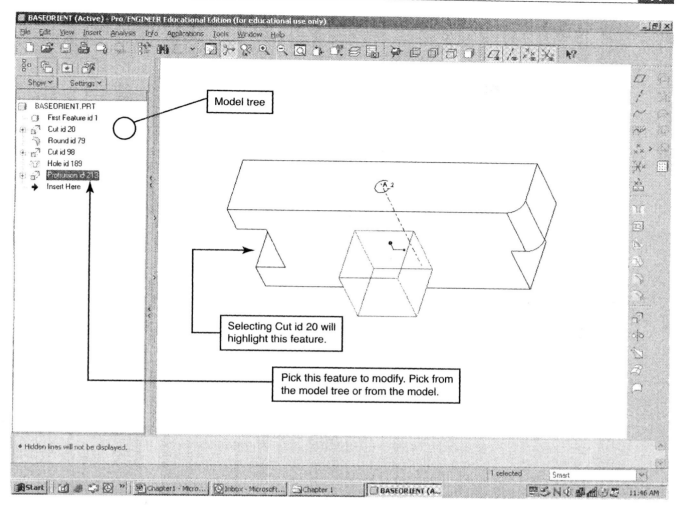

FIGURE 1.14 *The part "BaseOrient" loaded in Pro/E. Note the window with the* Model Tree *containing a list of the features in the model.*

The software will also delete *all* versions of a file on the disk. In this case, the path is:

1. Choose *File, Delete, All Versions*.
2. Then, enter the name of the part.

Pro/E loads all models into memory. Thus, it is possible that you will exhaust your computer's memory. It is a good idea to erase a model from memory after you are through working on it. In order to remove a model from memory, use the sequence *File, Erase,* and *Current*. If you wish to erase all the models, except the currently displayed model, use the path: *File, Erase,* and *Not Displayed*.

Every part has a *Model Tree* (Figure 1.14). The model tree lists all the features in the model and the order in which they were created. The model tree can be used to select features on the model by clicking the element in the model tree list. The name of the feature may be changed in the model tree by double-clicking the current name and entering a new one.

Because Pro/E is a parametric modeler—that is, the models are constructed using parameters—existing features may be easily modified. In particular, the dimensions defining the feature may be simply changed with a click of the mouse.

In order to change the value of a dimension, first select the feature in the model tree. This will highlight the feature. Then, right-click on the feature name. This will display a pop-up menu with several options. One of these options is *Edit*. Selecting *Edit* will display the dimensions of the feature. The value of any of the displayed dimensions may then be simply changed by clicking on the value and entering a new value.

Suppose we wanted to save this new version of a model, but under a different name. We can easily do this by using the *Save a Copy* option. However, Pro/E does not automatically update the model in memory to the new name.

The part called "BaseOrient.prt" has been developed to illustrated several key aspects of working with part files. Before proceeding further, you need to make sure that you have access to this part file. It may be downloaded from the Prentice Hall website (www.prenhall.com/rizza).

1.3.1 TUTORIAL 1.2: RETRIEVING "BASEORIENT"

This tutorial illustrates:

- Opening a part file
- Saving a copy
- Modifying a part

1. Select *File* and *Working Directory*.
2. Scroll down the list of available folders or change the connection to the drive until you find the appropriate folder.
3. Double-click on the name of the folder or click on the *OK* button.
4. Select *File* and *Open*, or use the *Open* icon.
5. Scroll to find the file called "BaseOrient." In order to open the file, double-click or highlight the name and select the *Open* button.
6. With your mouse, select the feature in the model tree list called "Cut id 20." Pro/E will highlight the feature (the default color is red), as shown in Figure 1.14.
7. Now, select "Protrusion id 213" from the model tree or click on the boxlike feature on the model as shown in Figure 1.14.
8. Click on your right mouse button.
9. Select *Edit* from the pop-up menu.
10. Pro/E will display the dimensions of the feature as shown in Figure 1.15. Select the 6.00 dimension. Enter a value of 8.00. Hit the enter or return key.
11. Then rebuild the model by selecting *Edit* and *Regenerate*. Upon regeneration, the software will update the model. The updated model is shown in Figure 1.16.
12. Save the modified model. Select *File* and *Save a Copy*, or the *Save As* icon.
13. In the space provided, enter the new name: "BaseOrient2." Choose *OK*.

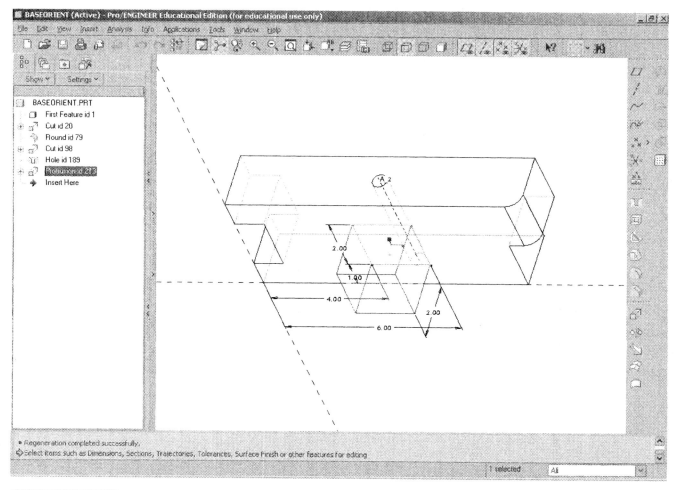

FIGURE 1.15 *The main window with the "BaseOrient" part. The dimensions of the boxlike feature ("Protrusion id 213") are displayed.*

14. In order to see if the model was properly saved, erase "BaseOrient" from memory by selecting *File*, *Erase*, and *Current*.

15. Select ⬚.

16. Scroll down and find "BaseOrient2."

17. Double-click on the file or hit *OK* after clicking once. Does this part contain the modified protrusion? It should.

1.4 CHANGING THE ORIENTATION OF A MODEL

The orientation of a model can be changed using one of two methods. If you have a three-button mouse, the orientation can be changed by using the Control (CTRL) or Shift key and the middle mouse button. Figure 1.17 shows the various key and button combinations necessary to achieve the desired task.

Furthermore, some of the icons shown in Figure 1.12 may also be used to change the orientation of the model.

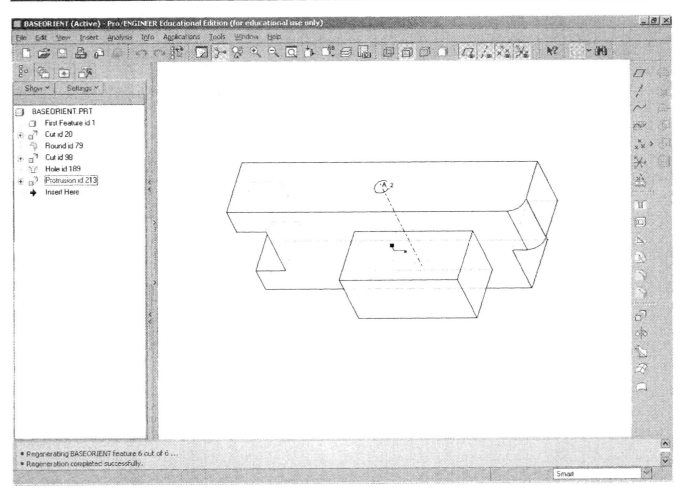

FIGURE 1.16 *After modifying the 6.00 dimension to 8.00, you will see the model change in appearance.*

1. The *Zoom in* icon will allow the user to zoom in to a portion of the model. The user must select the region to zoom. This is done by dragging a window across the region of interest.

2. The *Zoom out* icon performs the reverse operation of the *Zoom in* icon. All the user needs to do is click on the icon. Every time the icon is selected, the zoom out action is performed.

3. The *Refit* icon may be used to resize the model so that it fits into the window.

4. Finally, the *Reorient view* icon will launch the *Orientation* dialogue box, which may be used to reorient the model by defining different views. This approach will form the basis of the next tutorial.

The model can also be reoriented by defining orthographic views. In defining such views, two orthogonal surfaces are needed.

The part "BaseOrient" contains several planar surfaces. In the following tutorial, we will use these planes to reorient the model.

Action	Key	Example
SPIN	None	
PAN	Shift	
ZOOM	Control	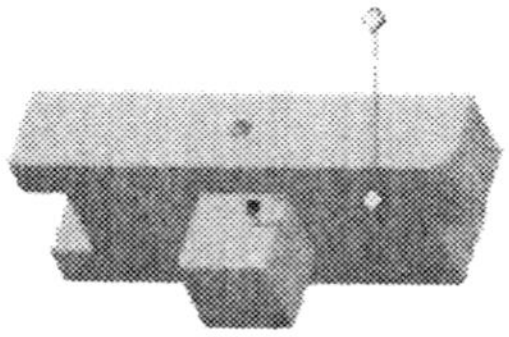
TURN	Control	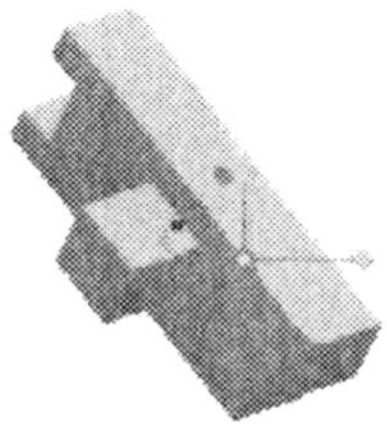

FIGURE 1.17 *Key and button combinations.*

This approach makes use of the *View* and *Reorient* commands (or the *Reorient* icon). Consider the actions described in Figure 1.18. According to the figure, by properly defining the *Top* and *Front* the model can be reoriented. Notice that there are other possibilities for obtaining the same orientation. This is because the model has multiple planar surfaces.

Reorientation requires the user to select a surface. A hidden surface may be accessed by using a query select. In Wildfire, this is accomplished by pressing the right mouse button and selecting *Next* and *Previous* as needed. When the hidden surface is highlighted, then it may be selected by clicking with the left mouse button.

1.4.1 TUTORIAL 1.3: CHANGING THE ORIENTATION OF "BASEORIENT"

This tutorial illustrates:

• Changing the orientation of a part

1. Use ![icon]. Retrieve the part "BaseOrient."
2. Press the Control (CTRL) key and middle mouse button. Select anywhere on the model with your mouse pointer.

FIGURE 1.18 *The model can be reoriented by using appropriate selections from the Orientation dialog box.*

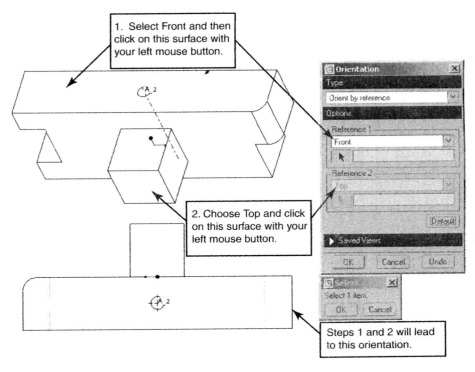

1. Select Front and then click on this surface with your left mouse button.

2. Choose Top and click on this surface with your left mouse button.

Steps 1 and 2 will lead to this orientation.

3. The model should grow larger as you pull back and smaller as you push forward.
4. Now, let go of the control key and depress the middle mouse button. Again, move the mouse. This should rotate the model.
5. Finally, release the control key; depress and hold the shift key and the middle mouse button.
6. Move the mouse. This will move the model in the same direction.
7. Push the mouse away from you; the model should move up.
8. Move the mouse toward you; the model should move in the downward direction.
9. Select 🔳 and drag the mouse across the model.
10. Release the mouse button; the software will zoom in on the chosen region of the model.
11. Try clicking on 🔳 several times.
12. After you are satisfied with the operation of these icons, choose 🔍. This will reorient the model to fit within the drawing window.
13. Reorient the model to its default orientation by selecting 🔳 and *Default Orientation*.
14. Now click on the *Reorient* icon 🔳.
15. Select *Front* for *Reference 1* and then the surface shown in Figure 1.18.
16. Choose "top" for Reference 2. Then select the surface shown in the figure.
17. Click on *Saved Views* in the *Orientation* dialog box (Figure 1.18).
18. In the *Name* field, type in Front. This will be the name of the view.
19. Then click on the *Save* button. Select *OK*.

20. Restore the model to the default view by selecting the *Reorient* icon (or *CTRL + D*).
21. Then, select .
22. Scroll through the list and find and select *Front*. Did the model reorient to the defined view?

1.5 DISPLAYING A MODEL

As discussed in Section 1.1, a model may be represented in several different ways by pressing the buttons shown in Figure 1.7 or Figure 1.12. These ways are:

1. A wireframe, where all the edges are shown as solid lines
2. Hidden lines, where hidden edges are shown in gray
3. No hidden, where hidden lines are not shown at all
4. Shaded, where the surfaces are shaded with a color.

A solid model can be created for an active model by using the *View, Shade* (the *Shade* icon), or by selecting *View, Model Setup,* and *Render Control.* The option *Shade* produces a shaded solid model, while *Render Control* produces a solid model that is photo quality of the object. In this case, the model is displayed against a background.

The construction of the solid model is produced by using a virtual light source to illuminate the wireframe. The solid model that is produced depends on the location of this light source and its intensity. The options for the light source may be changed by using the *View, Model Setup,* and *Lights* options. The background (the "room") may be modified by using *Model Setup* and *Room Editor.*

Because the *Render Control* option is graphics-intensive, it consumes a large amount of run time on the computer. It will not be used in this book. Instead, we will focus our attention on the *Shade* option.

An especially good use of shading is in the visualization of an assembly that is made of up of multiple parts. In such a case, it is advantageous to have each part a different color for greater visibility. Examine Figures 1.19(a) and 1.19(b). Notice how the use of different colors leads to a more visible model.

The appearance of a shaded model may be changed by selecting a different color scheme from the palette. If the desired color scheme does not exist in the palette, it may be created and added to the palette.

FIGURE 1.19 *The individual parts in the assembly are more visible in (b) because separate colors have been assigned to each part.*

(a)   (b)

As we will see in the next tutorial, the texture and shininess of the model may be set. Models tend to look nonrealistic if too shiny. To make the model look more realistic, reduce the shininess and intensity of the highlight. Furthermore, set the matte setting at the end of its scale and add a texture map.

1.5.1 TUTORIAL 1.4: A SOLID MODEL OF "BASEORIENT"

This tutorial illustrates:

- Shading a model
- Creating and adding a color to the color palette
- Shading the model with a new color

1. Select *View* and *Shade* (or insert ▣). The software will use the default settings to create a shaded model of the part.
2. Create a new color scheme. Select *View*, *Color*, and *Appearance*. Note the default color scheme in the palette (Figure 1.20).
3. Open the *Properties* box if it is not already open (see Figure 1.20).
4. Click on the "+" (the *Add* button) and enter a name for the color in the name field. In this case, say "New color."
5. Press the color button. The *Color Editor* will appear as in Figure 1.20.
6. Set the RGB values by using the slide controls or enter the desired values directly in the appropriate cell. In this case, use red = 69, green = 145, and blue = 179.
7. Press the Apply button. Note that if many parts are present, as in an assembly, you will need to choose the part to which the color should be applied.
8. Choose *Close*.
9. Shade the model by selecting ▣.

1.6 THE USE OF MAPKEYS

Pro/E allows for the use of hot keys (*Mapkeys*). This helps in navigating the menus. The default configuration file (Config.pro) supplied by PTC has numerous predefined *Mapkeys*. You might want to check this file for the appropriate *Mapkeys* by using the path *Tools* and *Options*.

The software also allows you to define new keys or modify old ones. This can be accomplished by using the *Tools* and *Mapkeys* commands. The software will load the *Mapkeys* dialog box shown in Figure 1.21. Notice that the currently defined *Mapkeys* are listed under the heading "Mapkeys in Session." By using the *Modify* button you can modify these Mapkeys.

The mapkey is created by recording a sequence of steps made by the user. The *Record Mapkey* box in Figure 1.21 contains options for recording the sequence, pausing the recording, or stopping it altogether.

It is not necessary to have an active model when creating a mapkey. In this case, however, the model is needed to see the result of using the mapkey. In the following tutorial, we will map the keys "IS" so that depressing these keys will reorient the model in isometric orientation.

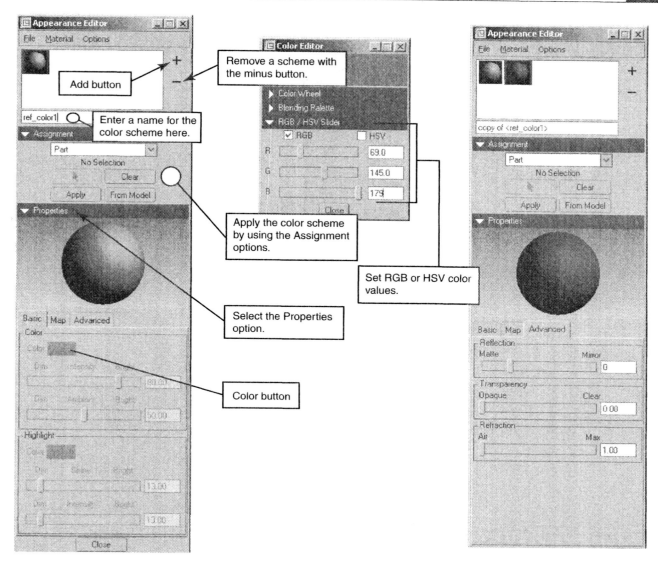

FIGURE 1.20 *The color scheme for shading the model may be changed by using the options in the* Appearance Editor *and* Color Editor. *The current available color schemes are shown in the palette.*

1.6.1 TUTORIAL 1.5: CREATING MAPKEYS INTERACTIVELY

This tutorial illustrates:

- Creating a mapkey
- Using a mapkey

1. Retrieve the part "BaseOrient." Use ☑.
2. Click on *Tools* and *Mapkeys*.
3. Select the *New* button.
4. Type "IS" into the *Key Sequence* field provided. Also, add the text "Change to Isometric" in the description field. Add a name if you want.

FIGURE 1.21 *Mapkeys are modified or created interactively by using the* Mapkey *and* Record Mapkey *dialog boxes, respectively.*

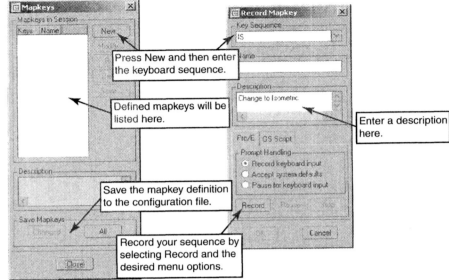

5. Then click on *Record*.

6. Using your mouse, select the sequence *Tools, Environment, Isometric,* and *OK*.

7. Then select *Stop* and *OK* from the *Record Mapkey* dialog box.

8. Scroll through the list of *Mapkeys* in the *Mapkeys* dialog box. You should see the letters "IS."

9. Select *Close*.

10. Now let us check if the new mapkey works. If the model was re-oriented during the process of creating the *Mapkey*, select *Tools, Environment, Trimetric,* and *OK*.

11. Then type in the letters "IS." The model should reorient to an isometric orientation.

1.7   PRINTING

In order to print (plot) in Pro/E you must have an active object. Therefore, if you want to print a certain model, you must first retrieve it.

Because most organizations use PostScript printers, we suggest using the *MS Printer Manager*. This choice will use your system printer drivers. The *Print* dialog box is shown in Figure 1.22.

1.7.1 TUTORIAL 1.6: PRINTING A COPY OF "BASEORIENT"

This tutorial illustrates:

• Printing

1. Retrieve "BaseOrient.ort" by selecting *File* and *Open* (or 📷) and then clicking on the file.

2. Select *File* and *Print* or click on the *Print* icon.

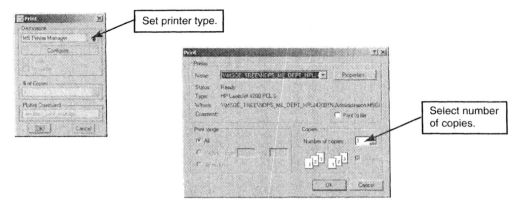

Set printer type.

Select number of copies.

FIGURE 1.22 *The* Print *box is used to set the properties for the print job.*

3. Set your printer type to *MS Printer Manager* or choose the appropriate printer available on your system.
4. Choose *OK*.
5. The software will load the *MS Print* box. Select the number of copies. If you wish to change the orientation or the paper size, use the *Properties* button.
6. Click on *OK*.

1.8 SUMMARY AND STEPS FOR USING VARIOUS INTERFACE OPTIONS

Our objectives in this chapter were related to the need to introduce the user to the Pro/E interface. We gained experience in customizing the interface by using the options contained within the *Tools* command.

The first of these options was the *Environment* option, which is used to activate or deactivate the display of various features including datum features. The steps for accessing the *Environment* option are:

1. Select *Tools* and *Environment*.
2. Turn the desired toggle on or off.
3. Choose *OK*.

We also spent some time investigating the methods and options for changing the colors of the display. The colors of the palette may be modified by using the *View* and *Color and Appearances* options. The built-in system color schemes may be set by using the path: *View*, *Display Settings*, *System Colors*, and *Scheme*.

In Pro/E, a part is saved in a part file with the extension ".prt." This part file may be loaded into memory by using the sequence:

1. Select *File* and *Open*, or 🖻.
2. Scroll through the list of available files and highlight the desired one.
3. Double-click on the file or select the *Open* button.

It is quite possible that you may have been assigned your own work directory. The user may connect to the work directory. This saves time in searching for a particular file. In addition, all saved files will be placed in the work directory. The user may connect to the work directory by doing the following:

1. Select *File* and *Work Directory*.
2. Scroll through the list and find the appropriate folder. Change the drive designation if necessary.
3. Choose *OK*.

Because Pro/E is a feature-based modeler, it is a simple matter to change the dimensions of a part. The change is made by doing the following:

1. Select the feature from the screen or from the model tree. Then press your right mouse button and choose *Edit*.
2. Click on the desired dimension.
3. Enter the new value of the dimension.
4. Select *Edit* and *Regenerate*.

A model may be reoriented by using the mouse button or by defining specific views of the model. These views may be saved. In order to define a view:

1. Select *View*, *Orientation*, and *Reorient* (or the *Reorient* icon).
2. Choose the reference type and then select the reference.
3. Select the second reference type and then the second reference.
4. Click on the *OK* button.

In order to save the view:

1. Click on *Saved Views* in the *Orientation* dialogue box (Figure 1.18).
2. In the *Name* field, type in the name of the view.
3. Then click on the *Save* button.
4. Select *OK*.

1.9 ADDITIONAL EXERCISES

1.1 List the primitives shown in Figure 1.3 that may be used to create the three-dimensional model of the part "BaseOrient." Sketch how the primitives would be arranged to form the model.
1.2 List and sketch the extrusions that would be used to construct the part "BaseOrient" if a constraint-based modeler is used.
1.3 In Tutorial 1.3, the part "BaseOrient" was reoriented using the *Front* and *Top* options. List three more combinations that may be used to achieve the same orientation.
1.4 Reorient the part "BaseOrient" so that the orientation in Figure 1.23 is obtained. Print out a copy of your work.

FIGURE 1.23

1.5 Reorient the part "BaseOrient" so that the orientation in Figure 1.24 is obtained. Print out a hard copy of your result.

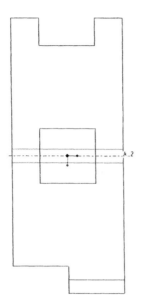

FIGURE 1.24

1.6 Reorient the part "BaseOrient" so that the orientation in Figure 1.25 is obtained. Print out a hard copy of your result. *Hint*: One of the surfaces is hidden; query select by pressing your right mouse button until the surface is highlighted and then the left mouse button.

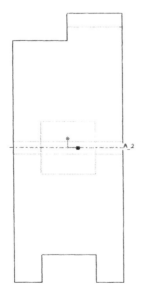

FIGURE 1.25

1.7 Reorient the part "BaseOrient" so that the orientation in Figure 1.26 is obtained. Print out a hard copy of your result. *Hint*: One of the surfaces is hidden; query select by pressing your right mouse button until the surface is highlighted and then the left mouse button.

FIGURE 1.26

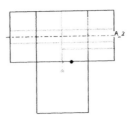

1.8 Modify the diameter of the hole in the part "BaseOrient" from 0.50" to 0.75". Print out a copy for credit.

1.9 Change the 4.00 dimension in Figure 1.15 to 4.50. Print out a copy of the result.

CHAPTER ■ 2

BASIC MODEL CONSTRUCTION WITH FEATURES

INTRODUCTION AND OBJECTIVES

Geometric modeling with a feature-based modeler like Pro/E requires the user to identify the cross section of the feature, and sketch and constrain the feature. The extension of the feature into three dimensions is done by the software, provided the user enters the appropriate information such as the depth, degree of rotation, and so on.

In Pro/E, the drawing of the cross section of a feature is performed in the Sketcher. The Pro/E user constructs a two-dimensional sketch by using some basic two-dimensional sketching tools. In Figure 2.1, the Sketcher is shown along with the icons that may be used to draw, modify, and constrain the sketch. Note the description of each icon. Many of the icons on the right-hand side of the screen, when selected, lead to additional drawing tools. These icons are shown in Figure 2.2.

The objectives of this chapter are to

1. Introduce the user to the Sketcher interface
2. Construct some simple features using the Sketcher and datum planes

2.1 SOLID AND THIN FEATURES

In Pro/E, a feature can be either *solid* or *thin* (consider Figure 2.3). Solid is the default. If a sketch is thin this means it has a constant thickness. The *thin* option allows the user to create a feature with a constant thickness faster because the feature may be constructed in one step. For example, in Figure 2.3, if the thin feature is created as a solid, an additional step is needed to remove the material from the inside of the feature.

An object created with the *Solid* option must be closed. That is, there must be an inside and an outside to the two-dimensional sketch. Thin

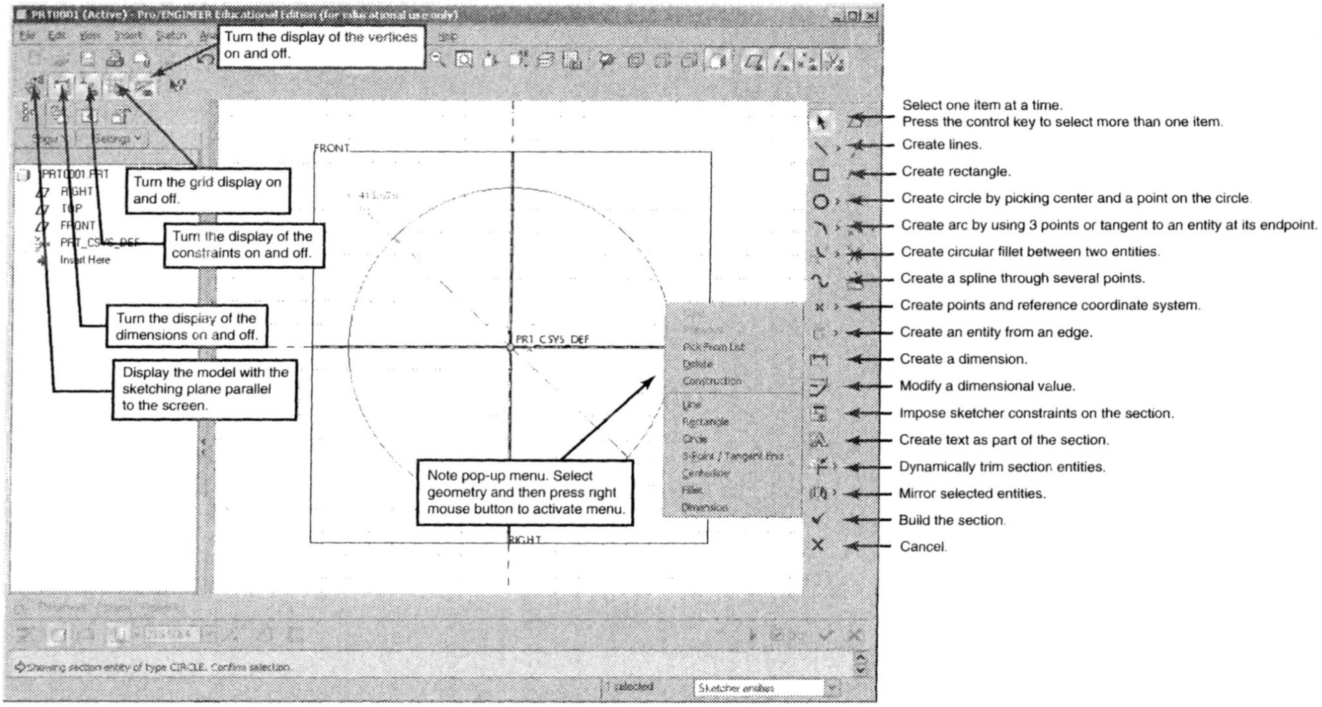

FIGURE 2.1 *The interface of the Sketcher is shown along with the corresponding icons and geometry toolbar. The pop-up menu is accessible by selecting an entity in the Sketcher and pressing the right mouse button.*

sections do not need to be closed. In the following sections, we will consider both solid and thin sections.

2.2 DATUM PLANES

In order to generate a three-dimensional model from the cross section, the software has been designed with three default planes, which may be used to sketch the section (Figure 2.4a). The datum planes are mutually orthogonal; that is, they are 90° to each other. Thus, if one datum is used to sketch the cross section, another may be used to reorient the model. This is illustrated in Figure 2.4. In this case, the Front datum plane is the sketching plane and datum plane Right is defined as the right-hand side of the model. This makes datum plane Front parallel to the screen. Using the datum plane Top as the top of the model will also give the same orientation.

2.3 SKETCHING

In the Sketcher, composite geometry is generated using simple entities. The tools shown in Figure 2.1 may be used to generate the geometry. The *Select* icon ▶ may be used to select an entity on the screen. After the entity

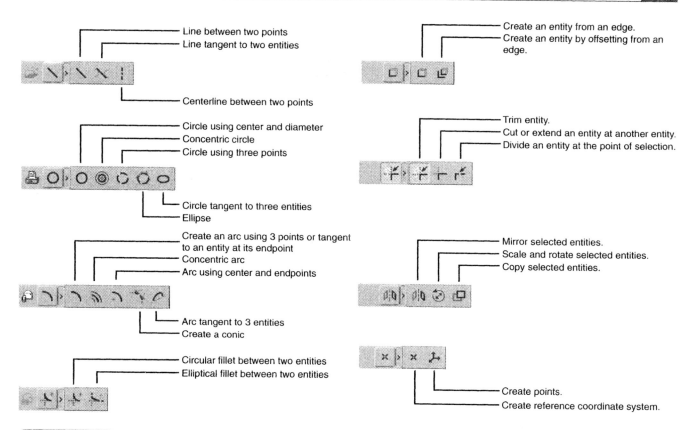

FIGURE 2.2 *In the Sketcher, tools are available via icons for generating basic geometric entities.*

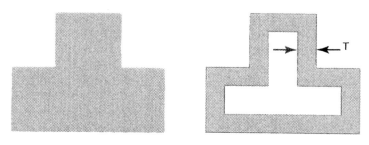

Thin section. The thickness is T.

FIGURE 2.3 *Both solid and thin sections may be constructed in the Sketcher. Thin sections have a constant thickness.*

has been selected, the pop-up menu shown in Figure 2.1 may be accessed by pressing the right mouse button. This menu allows the user to conveniently access tools for drawing additional simple entities or deleting the selected entity. Sketching of basic entities using the drawing tools is shown in Figure 2.5. More complicated geometric shapes may be constructed by using the basic entities of lines, circles, arcs, and so on and then trimming the

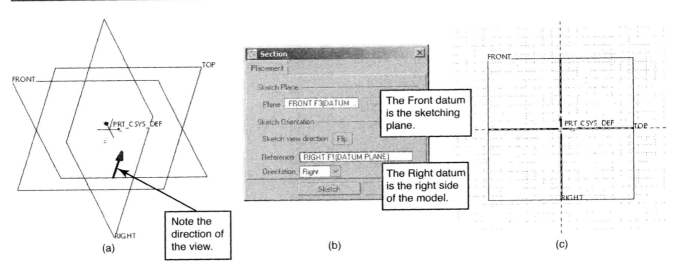

(a)

Note the direction of the view.

(b)

The Front datum is the sketching plane.

The Right datum is the right side of the model.

(c)

FIGURE 2.4 *The default template has datum planes with the names Top, Front, and Right. A model may be reoriented in the sketcher by using these datum planes or planar surfaces on the model.*

entities as needed. Notice in Figures 2.1 and 2.2 the *Trim* button ⚒. This button may be used to trim an entity at another entity or a point, extend an entity to another entity, or to divide an entity where it intersects another entity. The section for the Tension Plate shown in Figure 2.6 may be created by using circles and lines. The *Trim* button is used to break the entities where they intersect. This allows unneeded portions of the entities to be simply deleted.

2.4 REFERENCES

A two-dimensional drawing cannot be generated into a three-dimensional feature unless it is fully dimensioned. This requires two things. First, references must be defined from which the sketch may be dimensioned and second, dimensions must be added to the sketch using the defined references. When entering the Sketcher, the software will define the references based on the orientation of the model in the Sketcher. In general, these references will be two datum planes as shown in Figure 2.7. Of course, the user may change the references by pressing the button shown in the figure and then clicking on the desired reference. Note that *Use Edge/Offset* is the default reference type. As with the references themselves, the reference type may be changed as well.

2.5 DIMENSIONING A SKETCH

With references defined, the sketch may be dimensioned. Pro/E will dimension the sketch. However, the user may delete these dimensions and add

○ Circle tool

1. Click on screen to locate
   center of the circle.
2. Drag mouse to size circle.

＼ Line tool

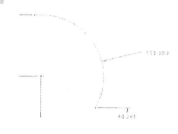

1. Click to locate first endpoint.
2. Drag mouse to size line.
3. Click to locate second endpoint.

▢ Rectangle tool

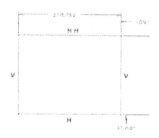

1. Click to locate upper right-hand corner
   of rectangle.
2. Drag mouse to size rectangle.
3. Click to locate lower left-hand corner
   of rectangle.

🖍 Arc tool

1. Click to locate first endpoint of the arc.
2. Move mouse to the location of the
   second endpoint.
3. Drag mouse to size arc and locate center.

FIGURE 2.5 *Drawing geometry in the Sketcher involves the use of the drawing tools and the mouse.*

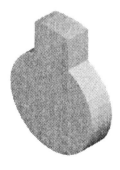

FIGURE 2.6 *Shaded and hidden models of the tension plate.*

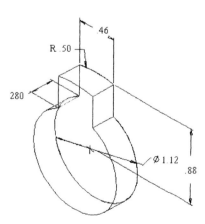

FIGURE 2.7 *The* References *box contains the references being used to dimension the section.*

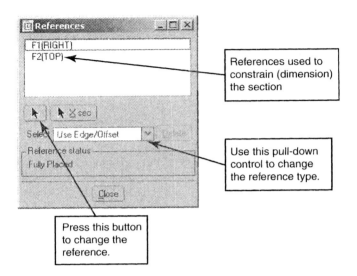

References used to constrain (dimension) the section

Use this pull-down control to change the reference type.

Press this button to change the reference.

new ones. Note that the *Dimension* icon 🖾 may be used to add a dimension to the sketch (Figure 2.2).

In order to actually create the dimension, the user needs to click on the entity to be dimensioned. This is done by using the mouse. A series of clicks with the mouse tells the software what entity the user wants to dimension and how it should be dimensioned. The middle mouse button is always used to place the dimension on the screen. Figure 2.8 illustrates how to dimension basic entities.

2.6   CHANGING THE UNITS OF A PART

Parameters describing the model, including the units, material properties, and the number of significant digits of the part, can be established by using the sequence *Edit* and *Set Up*. The Units Manager box (Figure 2.9) may be used to change the units of measurement.

2.7   THE DASHBOARD

Many menus and dialog boxes so familiar to users of older versions of Pro/E have been centralized in a single toolbar known as the Dashboard (Figure 2.10). The dashboard is used to set the options for making the sketch into a three-dimensional part. The *Sketch* icon 🖾 is used to initiate the Sketcher. Note that the section may be made into a thin section by using the *Thicken section* icon 🖾.

2.8   TUTORIAL 2.1: A VISE HANDLE

This tutorial illustrates:

- Constructing a simple part using datum planes and the Sketcher

1. The finished part is shown in Figure 2.11. This part has only one feature. What is the cross section for this feature?
2. Select *File* and *New* (or simply the *Create new object* icon 🗋).

 Dimensioning tool

A point

1. Click with left mouse button on desired point.
2. Click with left mouse button on reference.
3. Click with middle mouse button at the desired location. This places the dimensional value.

A line

1. Click with left mouse button on desired point.

Use the scheme for dimensioning points on each endpoint, or:
1. Select the line with the left mouse button.
2. Click with the middle mouse button at the desired location.

An arc

1. Click on the arc with the left mouse button.
2. Click with the middle mouse button at the desired location.

A circle

1. Click on two sides of the circle using the left mouse button.
2. Click at the desired location with the middle mouse button to locate the dimension.

An angle

1. Click on each line.
2. Click with the middle mouse button to locate the dimension. Click between the lines to place the included angle.

Diameter dimension for a revolved section

1. Click on the entity to be dimensioned.
2. Click on the centerline.
3. Click on the entity again.
4. Click with the middle mouse button to place the diametral, not radial, dimension (12.741 in figure).

FIGURE 2.8 *The Dimensioning tool may be used to dimension basic geometric entities.*

FIGURE 2.9 *The* Units Manager *dialog box may be used to change the units of a part dialog box.*

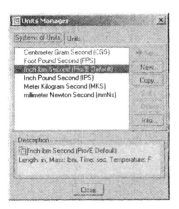

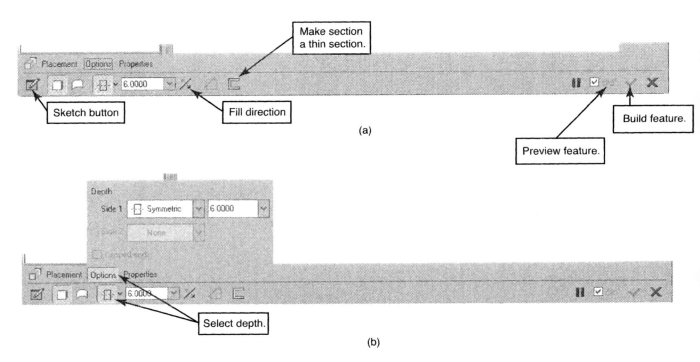

Make section a thin section.

Sketch button

Fill direction

Build feature.

Preview feature.

(a)

Depth

Side 1   Symmetric   6.0000

None

Placement   Options   Properties

Select depth.

(b)

FIGURE 2.10 *Tools in the dashboard are used to generate the cross section of a feature.*

3. For this model, indeed for the entire book, use the *Use default template* option. As shown in Figure 2.12, enter the name "Handle" in the *Name* field.

4. Select *OK*.

5. Note the default datum planes. Select *Insert* and *Extrude*.

6. The software will load the dashboard. Press the *Sketch* icon in the dashboard (see Figure 2.10).

7. With your mouse, click on the Front datum plane label. This will select the datum as the sketching plane.

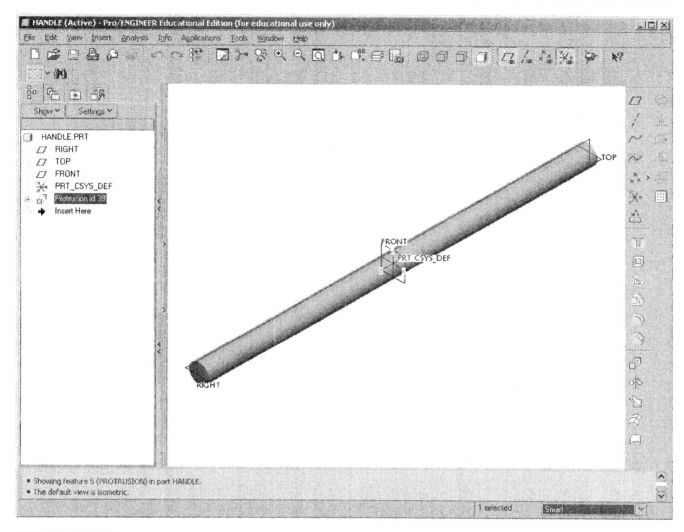

FIGURE 2.11 *The handle after extruding the section.*

8. If necessary, consult Figure 2.4 and use the pull-down for the *Orientation* field ⌄. Change the setting to "right" and then click on datum plane Right.

9. Make sure that the arrow on your screen points in the same direction as the one in Figure 2.4. If not, press the *Flip* button.

10. Select the *Sketch* button ⌷.

11. Note the *Reference* box (Figure 2.7). The status *Fully Placed* indicates that the section is fully constrained and that the software will dimension the section with respect to the Right and Top datum planes.

12. Select the *Circle* icon ◎ and draw a circle whose center is aligned with the edges of datum planes Top and Right as shown in Figure 2.1.

13. Select the diameter dimension. Press the *Modify* button ⍝ and change the value of the dimension to 0.30 as shown in Figure 2.13.

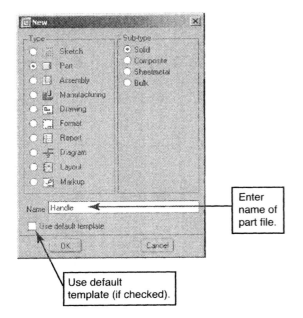

Enter name of part file.

Use default template (if checked).

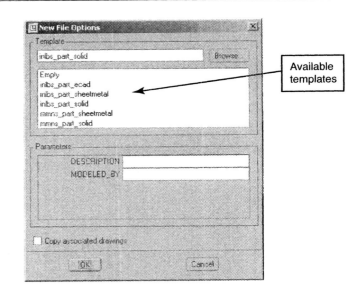

Available templates

FIGURE 2.12 *Use the* New *dialog box with the* Part *setting to create a part file.*

14. Press the *Check mark* button in the *Modify Dimensions* box . This will regenerate the section using the new dimension.

15. Build the section by pressing the *Check mark* in the Sketcher.

16. Press the pull-down below the *Options* button in the dashboard and choose *Symmetric* (see Figure 2.10).

17. Enter a depth of 6.00 (see Figure 2.10). This will extrude the section both sides from the sketching plane.

18. Press the *Check mark* button in the dashboard to build the model.

19. Congratulations! You have completed your first model using Pro/E. Save the part by selecting *File* and *Save* (or simply the Save icon).

2.9 TUTORIAL 2.2: A TENSION PLATE

This tutorial illustrates:

• Feature construction using several basic entities

1. Select *File* and *New* (or simply).

2. Make sure that the *Use default template* option is checked.

3. Enter the name "TensionPlate" in the *Name* field.

4. Select *OK*.

5. Select *Insert* and *Extrude* (or simply).

6. Choose the *Sketch* button in the dashboard.

7. With your mouse, choose datum plane Front as the sketching plane.

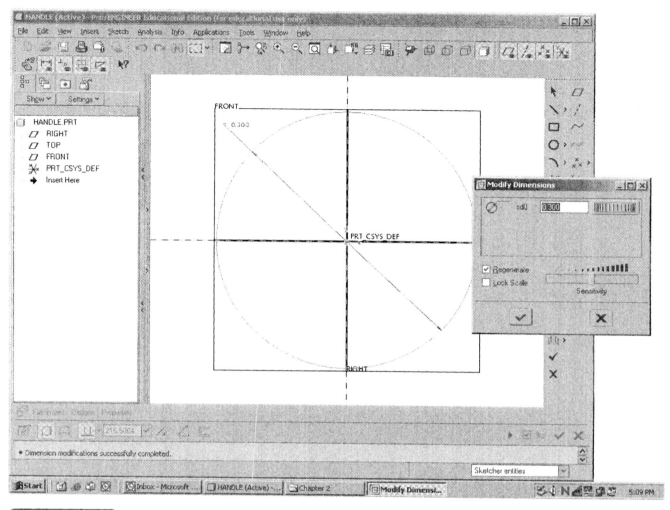

FIGURE 2.13 *The* Modify Dimensions *box may be used to change the value of one or more dimensions in a sketch.*

8. If necessary, use for the *Orientation* field and to "right." Then, click on datum plane Right.

9. Again, the arrow should point into the screen. Use *Flip* if it does not.

10. Select the *Sketch* button .

11. As shown in Figure 2.14, draw two circles with the center of each circle along the edge of datum plane Right.

12. Consulting Figure 2.15, draw two vertical lines an equal distance from datum plane Right.

13. Select the next to the *Trim* icon .

14. Now pick the *Divide* icon .

15. Select the intersections of the lines and circles as shown in Figure 2.16.

16. Now, press the *Select* button .

FIGURE 2.14 *Constructions needed to make the tension plate.*

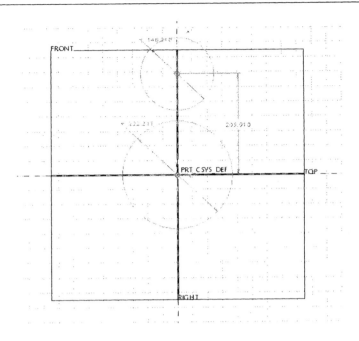

FIGURE 2.15 *Draw two vertical lines. Note the lines are equidistant from datum plane Right.*

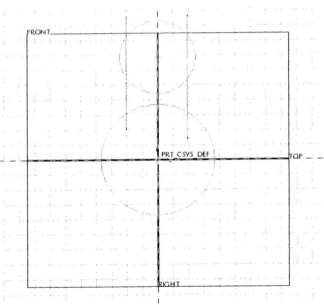

17. While pressing the CTRL key on your keyboard, use your mouse to pick the unwanted pieces of the lines and circles (use Figure 2.17 as a reference).

18. Press the *Delete* key on your keyboard.

19. Use the *Dimension* button ▣ to dimension as shown in Figure 2.18.

20. Build the section by pressing ✔.

21. Press ▾ below *Options* in the dashboard and choose symmetric (see Figure 2.10). Enter a depth of 0.280.

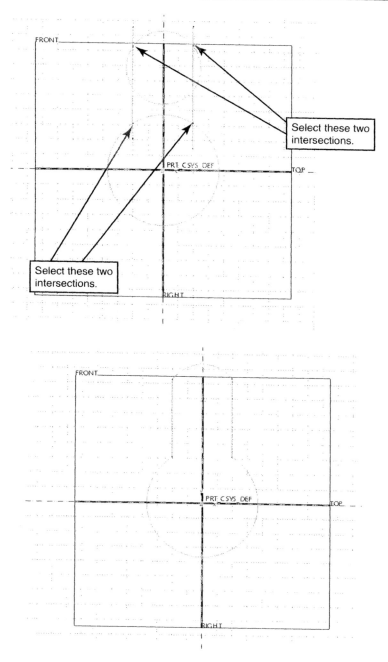

FIGURE 2.16 *Constructions with intersections shown. The intersections and endpoints of entities are shown by the software as solid black circles.*

Select these two intersections.

Select these two intersections.

FIGURE 2.17 *The geometry of the tension plate after breaking the entities and deleting the unneeded portions of the lines and circles.*

22. Press ☑ in the dashboard to build the model.

23. Save the part by pressing ▣.

2.10 TUTORIAL 2.3: HOLD-DOWN CLAMP ARM

This tutorial illustrates:

- Feature construction using several basic geometric entities
- Sketching on datum plane Right
- Changing the number of decimal points visible in the Sketcher

FIGURE 2.18 *Tension plate sketch with dimensions.*

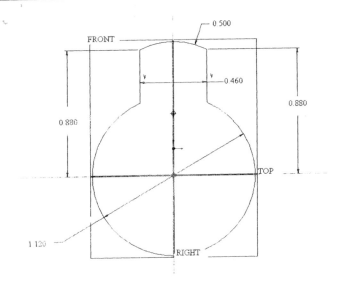

1. Select *File* and *New* (or simply).
2. Make sure that the *Use default template* option is checked.
3. Enter the name "HoldDownClamp_Arm" in the *Name* field.
4. Select *OK*.
5. Select *Insert* and *Extrude* (or simply).
6. Choose the *Sketch* button in the dashboard.
7. With your mouse, choose datum plane Front as the sketching plane.
8. If necessary, use for the *Orientation* field and to "right." Then, click on datum plane Right.
9. The arrow should point to the left. Use *Flip* if it does not.
10. Select the *Sketch* button .
11. This section has dimensions using four decimal places. Change the visibility of the number of decimal places by selecting *Sketch, Options,* and *Parameters.*
12. As in Figure 2.19, enter 4 in the *Num Digits* field. Select the check mark button.
13. Select the *Line* geometry tool and draw a line as shown in Figure 2.20(a).
14. Now pick the *Arc* tool and draw an arc as shown in Figure 2.20(b).
15. Using the and buttons, as necessary, complete the geometry of the section as shown in Figure 2.21.
16. Use and dimension the sketch so that it corresponds to the one shown in Figure 2.21. To dimension the angle, see Figure 2.8.
17. Use the *Modify* button . Click on each dimension. Enter the values shown in Figure 2.21.
18. Build the section by pressing .
19. Press below the *Options* button and choose . Enter a depth of 0.125.
20. Press in the dashboard to build the model.
21. Save the part by pressing . The extrusion is shown in Figure 2.22.

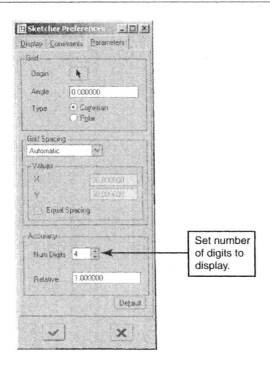

FIGURE 2.19 *Preferences for the Sketcher may be set by using the Sketcher Preferences dialog box.*

Set number of digits to display.

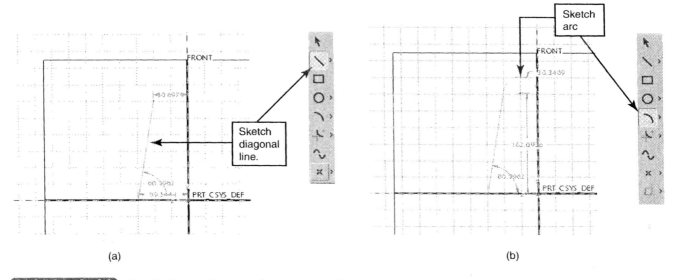

(a)   (b)

FIGURE 2.20 *Use the* Line *and* Arc *tools to generate the geometry.*

2.11 TUTORIAL 2.4: AN ANGLE BRACKET

This tutorial illustrates:

- Feature construction using *Thicken section* button

1. Select *File* and *New* (or simply ⬜).
2. Make sure that the *Use default template* option is checked.
3. Enter the name "AngleBracket" in the *Name* field.

FIGURE 2.21 *The geometry of the hold-down clamp arm is shown with dimensions.*

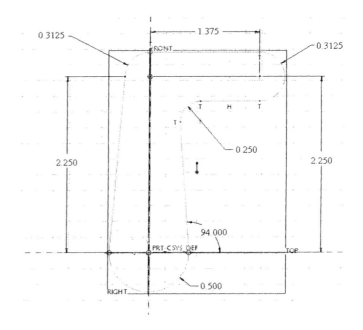

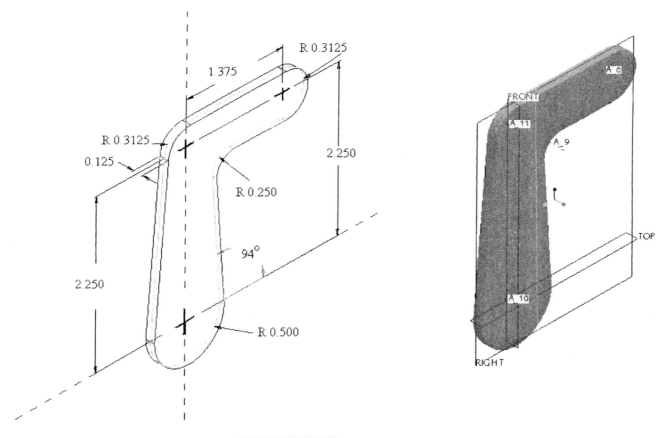

FIGURE 2.22 *The model of the hold-down clamp arm after generating the cross section into a three-dimensional feature.*

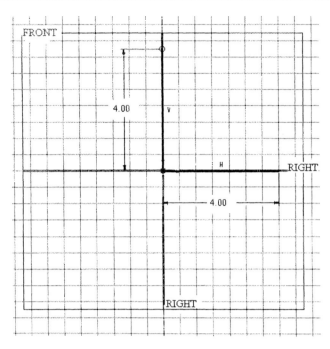

4. Select *OK*.

5. Select *Insert* and *Extrude* (or simply 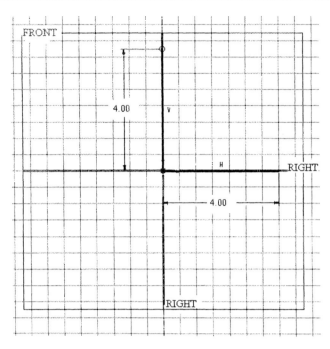).

6. Select the *Thicken section* button shown in Figure 2.10.

7. Enter a thickness of 0.500.

8. Choose in the dashboard.

9. With your mouse, choose datum plane Front as the sketching plane.

10. If necessary, use for the *Orientation* field and to "right." Then, click on datum plane Right.

11. Again, the arrow should point into the screen. Use *Flip* if it does not.

12. Select the *Sketch* button .

13. Sketch an "L" shape as shown in Figure 2.23 using .

14. Use and dimension the sketch as shown in the Figure 2.23.

15. Build the section by selecting .

16. Press the *Fill direction* button until the fill direction is inward as shown in Figure 2.24.

22. Press below the *Options* button in the dashboard and choose symmetric . Enter a depth of 4.00.

23. Press in the dashboard to build the model.

24. Save the part by pressing .

2.12 REDEFINING A FEATURE WITH EDIT DEFINITION

In Chapter 1, we saw how the value of a dimension may be modified to change a feature after it has been constructed. Now that we have gained some experience in constructing a feature, we may examine how the feature

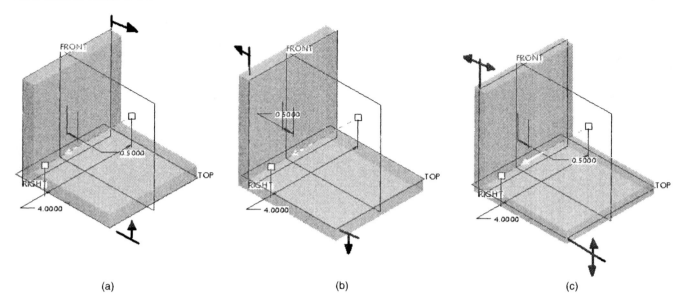

(a)   (b)   (c)

FIGURE 2.24 *The arrows indicate the various directions for adding thickness to the thin feature.*

may be redefined after it has been constructed. In previous versions of Pro/E this task was accomplished through the *Redefine* option. In Wildfire, the *Edit Definition* option is used instead. This feature may be accessed by selecting the feature in the model tree, clicking the right mouse button, and then selecting the option from the pop-up menu.

When the option is selected, the dashboard will reappear. The options that were available during the construction of the feature will be accessible once again and may be used to change any aspects of the feature. For example, the geometry of the section may be redrawn by choosing ▨ and redrawing the section as desired.

In addition, the *Edit Definition* option is very useful if you find that a feature was extruded in the wrong direction. In the dashboard, simply press ▨ until the correct direction is obtained and then rebuild the feature.

2.12.1 TUTORIAL 2.5: REDEFINING A FEATURE

This tutorial illustrates:

- Redefining a feature by using *Edit Definition*
- Deleting all the geometric entities in a sketch

1. Retrieve the model "BaseOrient." Select ▨ and double-click on the file.
2. In the model tree, select the rectangular protrusion (see Figure 2.25).
3. Press your right mouse button and choose *Edit Definition* from the pop-up menu.
4. In the dashboard, select ▨ .
5. We will not change the sketching or reference plane. But, note that you could do so at this point. Select ▨ .

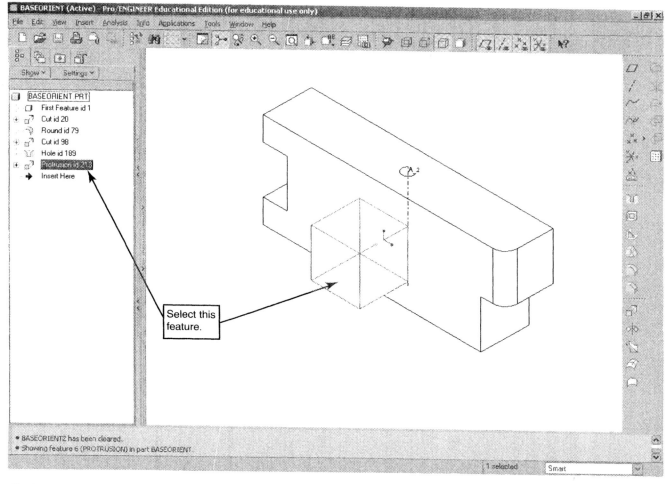

FIGURE 2.25 *The rectangular protrusion may be redefined.*

6. We will now delete all the entities in this section. Choose *Edit, Select,* and *All.* Then hit *Delete.*

7. Select the *Circle* tool ⭕ and sketch the geometry in Figure 2.26.

8. Select 🔲 and dimension the geometry as shown in Figure 2.26.

9. Choose ☑, click on each dimension, and enter the values shown in Figure 2.26.

10. Build the section by pressing ✔.

11. Press *OK* in the *Section* box.

12. We will leave the extrusion direction and depth the same as before, but note that both these aspects of the feature may be changed at this point.

13. Press ✔ in the dashboard to build the model.

14. Select *File* and *Save a Copy.* Enter the name: NewBaseOrient. Choose *OK.*

FIGURE 2.26 *Change the geometry of the rectangular protrusion to a circle.*

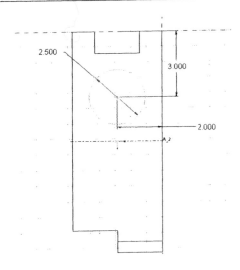

2.13 SUMMARY AND STEPS FOR USING THE SKETCHER

In Pro/E, the Sketcher is used to draw and dimension the cross section of a feature. After the section has been regenerated into a parametric two-dimensional model, it can be made three-dimensional by extrusion or revolution about an axis, or swept along a trajectory. Therefore, it is very important that the user of the software understands and is familiar with the Sketcher.

In this chapter, we concentrated on single solid and thin sections that were extruded to form a three-dimensional feature. Solid sections are closed and have an inside and an outside. Thin sections, on the other hand, are open sections. Thin sections are used in creating features that have a constant thickness.

Sections are drawn in the sketcher by using elementary entities such as lines, arcs, and circles. More advanced entities such as conic and spline sections are also available. In general, the procedure for using the Sketcher is:

1. Select *Insert*.
2. Choose the type of feature, for example, *Extrude* (or the *Extrude* icon ⬛).
3. Select the sketching plane.
4. Select the reference plane.
5. Draw the section.
6. Dimension the section to the proper values.
7. Build the section by pressing ✔ in the Sketcher.
8. Define the depth of the section using *Options* in the dashboard.
9. Build the model by pressing ✔ in the dashboard.

The units for a part may be set by using a template when the part file is created. However, the units may be changed at any time. In order to change the units for an existing part use this sequence:

1. Choose *Edit* and *Set Up*.
2. In the PART SETUP menu, select *Units*.
3. Select from *Systems of Units* or create your own system.
4. Use the *Set* button.
5. Select *Close*.

After a feature is built it may be redefined by selecting the feature in the model tree, pressing the right mouse button, and selecting *Edit Definition* from the pop-up menu. The dashboard will reappear. Make any necessary changes and then rebuild the feature by selecting ☑.

2.14 ADDITIONAL EXERCISES

2.1 For the block models shown in Figure 2.27, determine the number of features, the cross section of each feature, the suitable sketching plane for the feature, and the suitable reference plane.

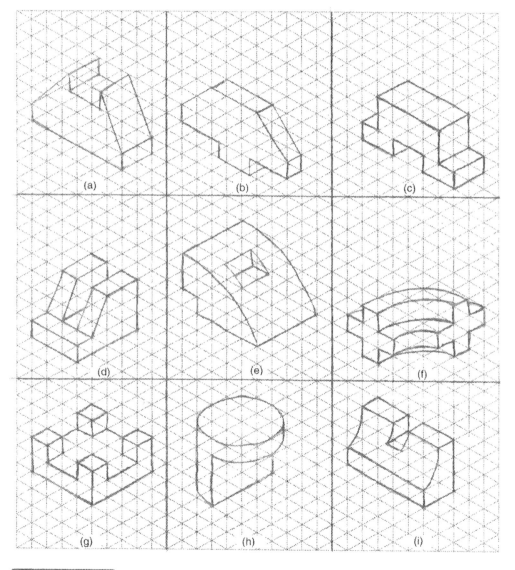

FIGURE 2.27

2.2 As assigned, construct the models shown in Figure 2.28. Use the grid to determine the dimensions of the parts.

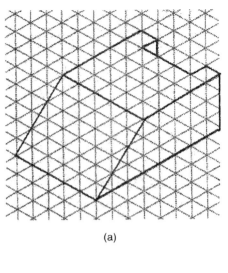

(a)

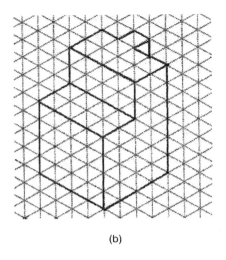

(b)

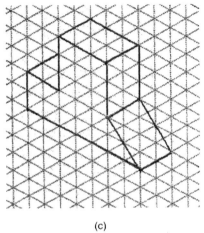

(c)

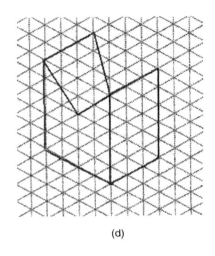

(d)

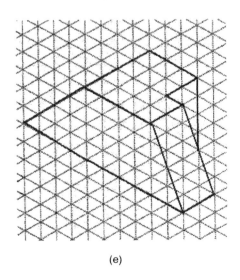

(e)

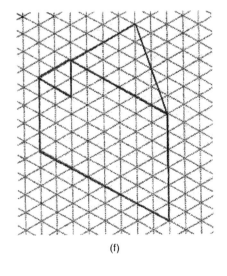

(f)

FIGURE 2.28

2.3 Construct the washer shown in Figure 2.29.

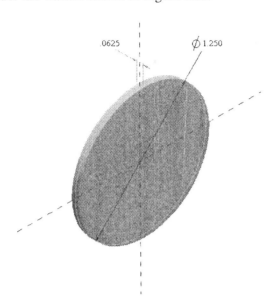

FIGURE 2.29

.0625   Ø 1.250

2.4 Use the *Thin* option to construct a model of the retaining ring illustrated Figure 2.30.

THIN EXTRUSION
GREATER TO USE ON CONSTANT WIDTH

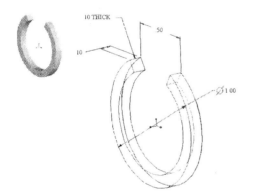

FIGURE 2.30

10 THICK
50
10
Ø 1 00

DIA?
OPEN AREA?

2.5 Construct the model of the woodruff key using Figure 2.31.

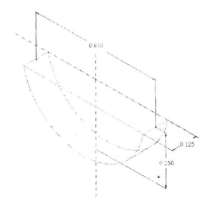

FIGURE 2.31

0.610
0.125
0.250

2.6 Using Figure 2.32, construct the model of the retainer. The holes will be added in Chapter 4.

FIGURE 2.32

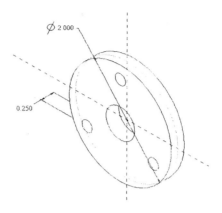

2.7 Construct the model of the angle block. Use Figure 2.33.

FIGURE 2.33

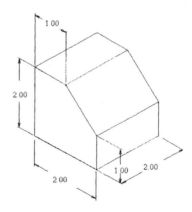

HW

2.8 Using Figure 2.34, create a model of the hold down.

FIGURE 2.34

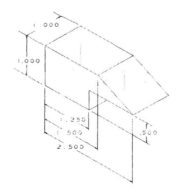

2.9 Create the lever arm using the geometry shown in Figure 2.35.

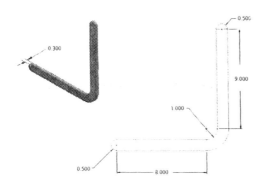

FIGURE 2.35

2.10 Construct the model of the cam shown in Figure 2.36.

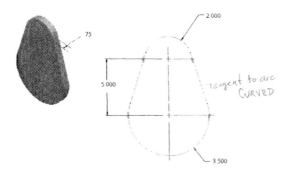

FIGURE 2.36

2.11 Using the *Extrude* option, create the tension test specimen, whose geometry is given in Figure 2.37.

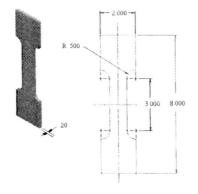

FIGURE 2.37

2.12 Create a model of the gib head key illustrated in Figure 2.38.

FIGURE 2.38

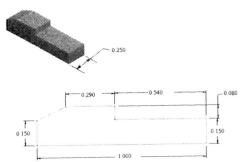

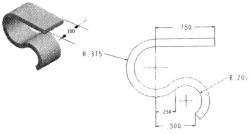

2.13 Use the *Thin* option to construct a model of the steel pin illustrated in Figure 2.39.

FIGURE 2.39

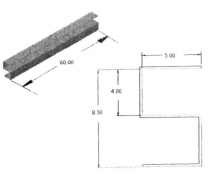

2.14 The "S-Stiffener" shown in Figure 2.40 may be created using a single *Thin* protrusion. Construct a model of the stiffener.

FIGURE 2.40

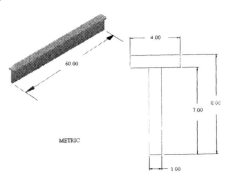

2.15 The "T-Beam" has metric units. Create a model of the beam using the *Thin* option and Figure 2.41.

FIGURE 2.41

2.16 Construct a model of the "I Beam" using the geometry shown in
      Figure 2.42. Note that the units of this part are metric.

FIGURE 2.42

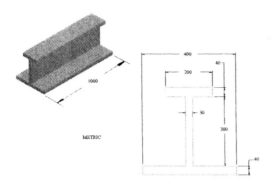

CHAPTER ▮3▮

MODELS WITH MULTIPLE
FEATURES

INTRODUCTION AND OBJECTIVES

In Chapter 2, we used datum planes and the Sketcher to create some simple base features. The objects are called base features, because they do not contain any holes or cutouts. The objects were simple in the sense that a protrusion of a two-dimensional sketch into the third dimension was sufficient to create the basic shape of the object. However, most practical applications require multiple protrusions. In some cases, the default datum planes are not sufficient to create the model; additional datum planes called user-defined datum planes must be used.

The objectives of this chapter are to

1. Use datum planes to construct models with multiple protrusions
2. Construct user-defined datum planes and be able to use these datum planes in the construction of one or more features

▮3.1▮ DATUM PLANES AND MULTIPLE FEATURES

As discussed in Section 2.4, the software will automatically select two datum planes that are perpendicular to the sketch so that the sketch may be constrained. Furthermore, additional features may be dimensioned to existing features.

In the context of geometric dimensioning, if the datum planes used to construct the feature coincide with the geometric datum planes, then constraining the sketch to these datum planes makes sense. The user does have the option to change the dimensioning scheme by introducing new dimensions. Often this is motivated whenever a preexisting feature is used to constrain a new feature.

3.2 Symbols Used in the Regeneration Process

Regeneration is the most critical phase in the sketching process. Depending on your sketch and dimensions, you may or may not get what you want. The software makes some assumptions in building the section called constraints. These constraints can be changed or new ones added by selecting the *Constraints* icon ▩.

It must be remembered that when you regenerate the sketch, Pro/E is attempting to interpret your rough sketch to produce the desired feature. If the feature fails to regenerate, it is because Pro/E is unable to interpret the sketch, most likely because the sketch is improperly constrained. For example, the value of two different dimensions may conflict.

After a successful regeneration, Pro/E will display symbols that denote the various constraints used in interpreting the sketch. A listing of these constraints is given in Table 3.1. These constraints are based on several rules:

1. If you draw two arcs or circles with approximately the same radius, the arcs or circles are assumed to be of the same size.
2. If you draw a centerline, entities will be assumed symmetric about that centerline.
3. Lines approximately horizontal or vertical are assumed horizontal or vertical.
4. Entities will be assumed to be tangent if sketched approximately tangent.
5. Lines of roughly the same length will be assumed to be of the same length.
6. Point entities near other entities such as lines, arcs, and circles will be assumed to lie on that entity.
7. Endpoints of arcs and lines may be assumed to have the same coordinates.

Notice the "V" symbol in Figure 3.1. This indicates that the lines are vertical (within an accepted tolerance). These symbols are based on the software's interpretation of the sketch drawn by the user.

Table 3.1 SKETCHER CONSTRAINT SYMBOLS.

Symbol	Constraint
"V"	Entity assumed vertical
"H"	Entity assumed horizontal
"L" with index	entities with same index are equal length
"⊥"	Entities with perpendicular symbol are perpendicular
"//"	entities with parallel symbol are parallel
Thick dashes	points have equal coordinates
"T"	entities are tangent
→ ←	symmetric
"R" with index	entities are of equal radii
─○─	point entity

FIGURE 3.1 *The software uses symbols to convey its interpretation of the sketch. The symbol "V" in the figure indicates that the software is assuming that the line is vertical.*

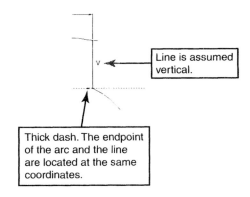

Line is assumed vertical.

Thick dash. The endpoint of the arc and the line are located at the same coordinates.

3.3 THE USE OF RELATIONS

The *Relations* option may be used to dimension a feature that depends on another feature. It may also be used to dimension a feature with a given equation. Relations are used whenever the design intent warrants their use.

For example, suppose that parameter sd0 depends on sd1, such that sd0 is always one half of sd1. We can ensure that sd0 is always one half of sd1 by setting a relation between the two parameters in the form of a mathematical expression such as: sd0 = sd1/2. The *Relations* box shown in Figure 3.2 allows the user to define a relation.

The *Relations* box operates like a mini–word processor. Expressions for the relations are entered in the box as shown in Figure 3.2. The *Insert Function* button may be used to insert common expressions such as cosine (cos). In Figure 3.2, note the button used to select dimensions from the screen. This option allows the user to formulate equations without having to remember or retype the parameter name.

FIGURE 3.2 *The* Relations *box is used to define relationships between the various parameters.*

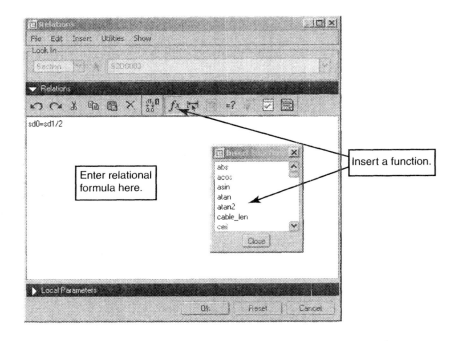

Enter relational formula here.

Insert a function.

3.4 Tutorial 3.1: A Bushing

This tutorial illustrates:

- Constructing a part with multiple extrusions
1. Select *File* and *New* (or simply 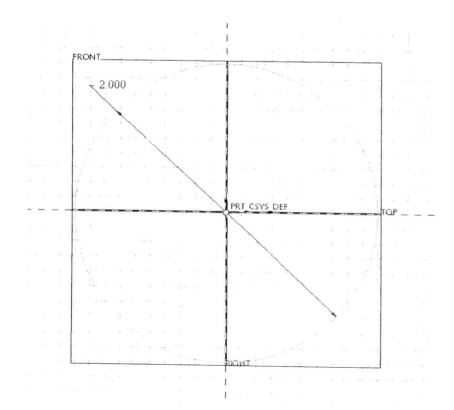).
2. Make sure that the *Use default template* option is checked.
3. Enter the name "BushingT3" in the *Name* field.
4. Select *OK*.
5. Select *Insert* and *Extrude* (or simply).
6. Choose the *Sketch* button in the dashboard.
7. With your mouse, choose datum plane Front as the sketching plane.
8. If necessary, use for the *Orientation* field and to "right." Then, click on datum plane Right.
9. Again, the arrow should point into the screen. Use *Flip* if it does not.
10. Select the *Sketch* button .
11. In the *Sketcher*, sketch the circle shown in Figure 3.3.
12. Select the diameter dimension. Press the *Modify* button and change the value of the dimension to 2.00 as shown in Figure 3.3.
13. Press to build the section.
14. Press the pull-down below the *Options* button in the dashboard and choose *Blind* .

FIGURE 3.3 *The geometry for the first extrusion is a simple circle.*

FIGURE 3.4 *In the figure, note the direction of the first and second extrusions.*

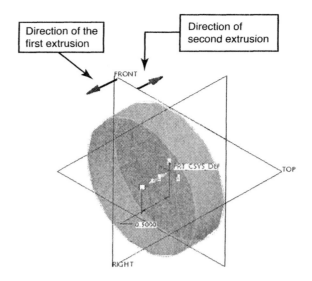

15. Enter a depth of 0.50. This will extrude the section one side from the sketching plane.

16. Now, compare your model to the one shown in Figure 3.4. Press ⚡ if your extrusion does not correspond to the one in the figure.

17. Press ✔ in the dashboard to build the model.

18. Construct the second protrusion by selecting 🗗.

19. Choose the *Sketch* button ✎ in the dashboard.

20. With your mouse, choose datum plane Front as the sketching plane.

21. If necessary, use ⌄ for the *Orientation* field and to "right." Then, click on datum plane Right.

22. Select the *Sketch* button ▬▬.

23. Again, the arrow should point into the screen. Use *Flip* if it does not.

24. In the sketcher, use the *Circle* icon ⭕ and sketch a second circle as shown in Figure 3.5.

25. Select the diameter dimension. Press the *Modify* button ✑ and change the value of the dimension to 1.50 as shown in Figure 3.5.

26. Press the pull-down below the *Options* button in the dashboard and choose *Blind* ⫰.

27. Enter a depth of 4.00.

28. Compare your model to the one shown in Figure 3.4. Press ⚡ if your extrusion does not correspond to the one in the figure.

29. Press ✔ in the dashboard to build the model.

30. Save the part by pressing 🖫. The model is shown in Figure 3.6.

3.5 TUTORIAL 3.2: BASE FEATURES FOR A PROPELLER

This tutorial illustrates:

- Constructing a part with multiple extrusions
- Constructing a metric part

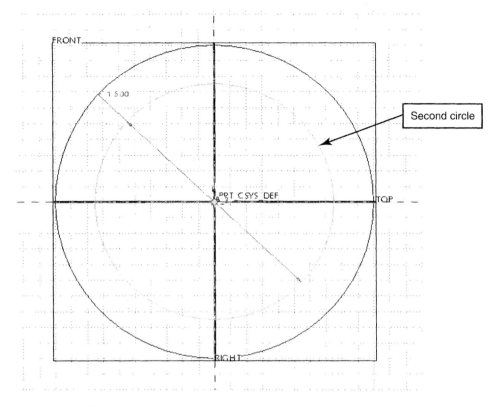

FIGURE 3.5 *A second circle is used as the sketch for the second extrusion.*

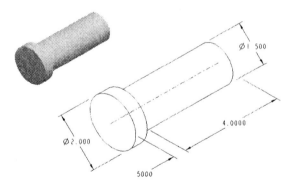

FIGURE 3.6 *The final form of the bushing is shown along with the dimensions of the model.*

1. Consider the model shown in Figure 3.7. Then, select [image].
2. Make sure that the *Use default template* option is unchecked.
3. Enter the name "Propeller" in the *Name* field.
4. Select *OK*.
5. This is a metric part so select "mmns_part_solid" from the list. Then choose *OK*.
6. Select [image]. Choose [image] in the dashboard.
7. With your mouse, choose datum plane Front as the sketching plane and datum plane Right as the "right" orientation.

FIGURE 3.7 *The completed shaded model of the propeller.*

FIGURE 3.8 *The geometry for each of the first and second protrusions is a circle, but they are extruded in different directions.*

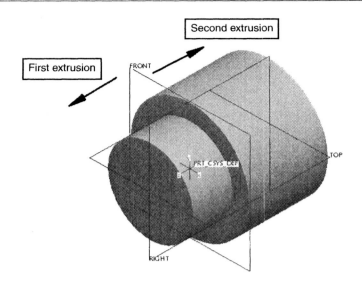

8. Again, the arrow should point into the screen. Use *Flip* if it does not.

9. Select [▭] and sketch a circle whose center is aligned with the datum planes.

10. Double-click on the diameter dimension and change the value to 65.

11. Select [▥] and enter a depth of 30. Use [▨] to extrude the section in the direction shown in Figure 3.8.

12. Build the section by pressing [✓].

13. Press [✓] in the dashboard to build the model.

14. Repeat steps 6 through 13. The diameter of the second protrusion is 90. Extrude in the direction shown in Figure 3.8 to a depth of 60.

15. Save the part by pressing [▣].

3.5 TUTORIAL 3.3: A SHIFTER FORK

This tutorial illustrates:

- Constructing a part with multiple extrusions

1. Select *File* and *New* (or simply [▣]).

2. Make sure that the *Use default template* option is checked.

3. Enter the name "ShifterFork" in the *Name* field.

4. Select *OK*.

5. Select *Insert* and *Extrude* (or simply [▣]).

6. Select the *Thicken section* button [▣].

7. Enter a thickness of 0.500 in the cell next to the icon.

8. Choose the *Sketch* button [▨] in the dashboard.

9. With your mouse, choose datum plane Front as the sketching plane.

10. If necessary, use [▼] to change to the orientation in the *Orientation* field to "right." Then, click on datum plane Right.

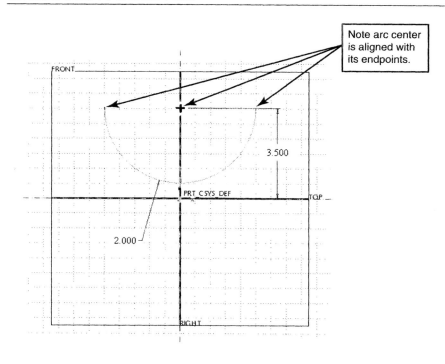

Note arc center is aligned with its endpoints.

FIGURE 3.9 *Sketch a semicircle for the first section. Note the alignment of the semicircle center with datum plane Right.*

11. Again, the arrow should point into the screen. Use *Flip* if it does not.

12. Select the *Sketch* button ▭.

13. Sketch the semicircle in the sketcher by using the *Arc* button ◠ as shown in Figure 3.9.

14. If necessary use the *Dimension* tool ▭ and dimension the arc as shown in Figure 3.9.

15. Use the *Modify* tool ▱ click on each dimension and modify the dimensions to the ones shown in Figure 3.9.

16. Press ✔ to build the section.

17. Check the direction of the fill. Change the orientation of the model by selecting ▱ and then "Front." Compare your model with that in Figure 3.10.

18. Press ▱ (next to ▯) until your model corresponds to the one in Figure 3.10.

19. Select ▱.

20. Select *Options* in the dashboard and then *Blind* ▱. Enter a depth of 0.750 in the cell.

21. Check the direction of the extrusion. Compare your model with that in Figure 3.10.

22. Press ▱ (next to ▱) until your model corresponds to the one in Figure 3.10.

23. Build the part by pressing ✔.

24. Save the part by pressing ▱. The model will be completed in the following tutorials.

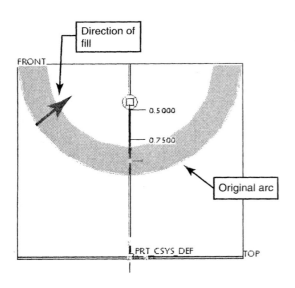

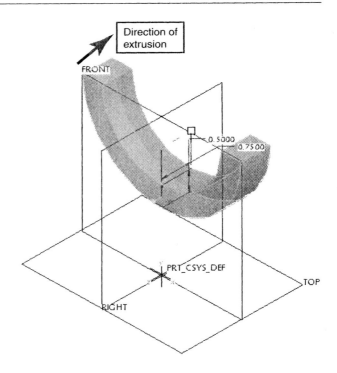

FIGURE 3.10 *Fill the thin section as shown. Note the direction of the extrusion.*

3.6 TUTORIAL 3.4: A SECOND EXTRUSION FOR THE SHIFTER FORK

This tutorial illustrates:

- Adding a second extrusion to a model
- Constructing a two sided blind extrusion

1. If the model is not already in memory, select ⊞ and double-click on the file.
2. Select ⊡.
3. Choose the *Sketch* button ⊡ in the dashboard.
4. With your mouse, choose datum plane Front as the sketching plane.
5. If necessary, use ⊡ to change to the orientation in the *Orientation* field to "right." Then, click on datum plane Right.
6. Again, the arrow should point into the screen. Use *Flip* if it does not.
7. Select the *Sketch* button ▭.
8. Sketch the circle in the sketcher by using the *Circle* button ◯ as shown in Figure 3.11.
9. Double-click on the diameter dimension and type in 2.000.
10. Hit *return* on your keyboard.
11. Build the section by pressing ✓.
12. Now, select *Options* as shown in Figure 2.10.

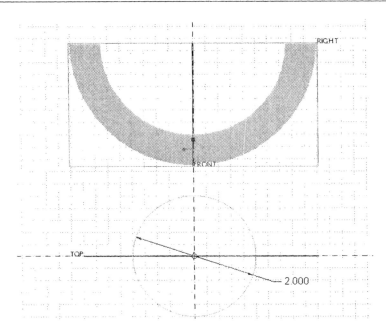

FIGURE 3.11 *The geometry for the second feature is a circle. Align the center of the circle as shown.*

13. For "Side 1," choose *Blind* . Enter a depth of 5.650 in the cell.

14. For "Side 2," choose *Blind* . Enter a depth of 0.350 in the cell.

15. Note the direction of the extrusion in Figure 3.12. If your model does not correspond to the one in the figure, press .

16. Build the part by pressing .

17. The model after the extrusion is shown in Figure 3.13. Save the model by selecting .

37 TUTORIAL 3.5: A THIRD EXTRUSION FOR THE SHIFTER FORK

This tutorial illustrates:

* Adding a third extrusion to a model
* Using a concentric arc

1. If the model is not already in memory, select and double-click on the file.

2. Select .

3. Choose the *Sketch* button in the dashboard.

4. With your mouse, choose datum plane Front as the sketching plane.

5. If necessary, use to change to the orientation in the *Orientation* field to "right." Then, click on datum plane Right.

6. Again, the arrow should point into the screen. Use *Flip* if it does not.

7. Select the *Sketch* button .

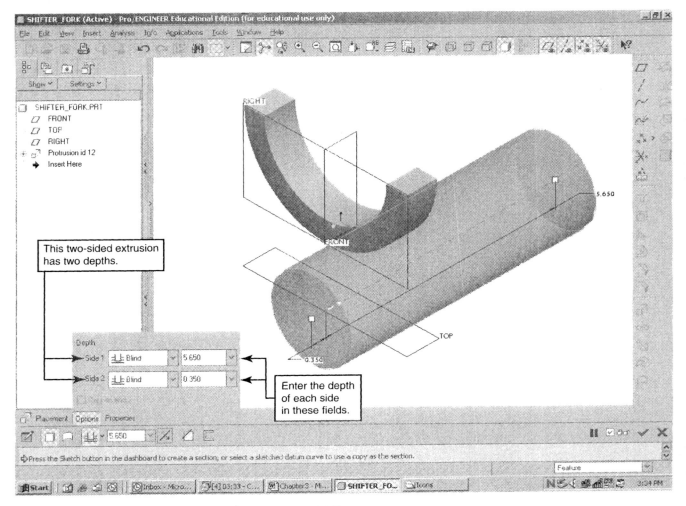

FIGURE 3.12 *The second protrusion is constructed using a nonsymmetric blind. This requires the user to enter the depth in both directions of the sketching plane.*

8. Examine Figure 3.14. Use the *Line* tool and sketch two lines that are symmetric to datum plane as shown in the figure.

9. Select the *Dimension* tool and dimension as shown in Figure 3.14.

10. Choose the *Pointer* and double-click on the dimension. Enter a new value of 1.800 and then hit the *return* key on your keyboard.

11. Select next to the *Arc* icon and change to the *Concentric arc* icon.

12. Now, click on either of the two circles shown in Figure 3.15.

13. Drag the endpoint of the arc until it intersects the endpoint of the desired line. Anchor the arc to the line endpoint by left mouse clicking at the intersection.

14. Drag the arc endpoint to the remaining line endpoint. Again, click to anchor to the endpoint.

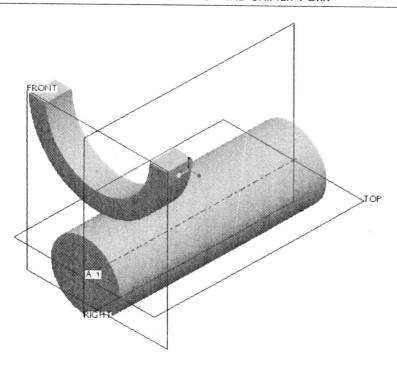

FIGURE 3.13 *After the second protrusion, the model takes on the shape shown in the figure.*

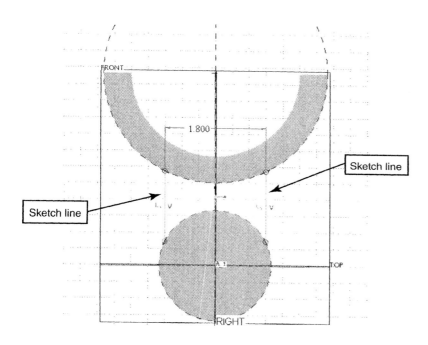

FIGURE 3.14 *Begin sketching the third protrusion by drawing two vertical lines. Note that the lines are symmetric with respect to the edge of datum plane Right.*

15. Now, repeat the previous process. Select the *Concentric arc* icon.

16. Click on the circle shown in Figure 3.16.

17. Drag the endpoints of the arc until they intersect the endpoints of the two lines. Click the left mouse button to anchor the arc to the endpoints.

FIGURE 3.15 *Add an arc coincident with the section of the first extrusion.*

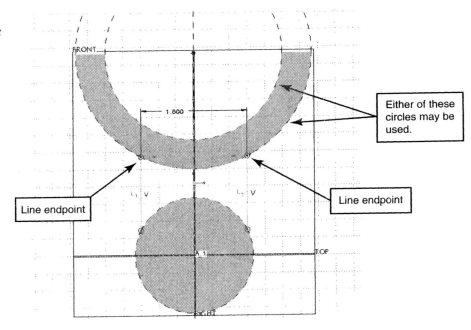

18. Press ✔ to build the section.
19. Select 🔲.
20. Select *Options* in the dashboard and then *Blind* 🔺. Enter a depth of 0.500 in the cell.
21. Check the direction of the extrusion. Compare your model with that in Figure 3.17.
22. Press 🔺 (next to 🔺) until your model corresponds to the one in Figure 3.17.
23. Build the part by pressing ✔.
24. Save the part by pressing 🔲. The model will be completed in future tutorials.

FIGURE 3.16 *Add an arc coincident with the section of the second extrusion.*

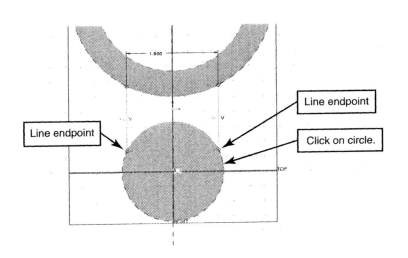

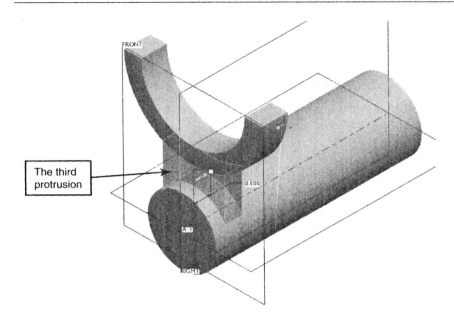

FIGURE 3.17 *The model shown with the third protrusion.*

The third protrusion

3.8 TUTORIAL 3.6: A FOURTH EXTRUSION FOR THE SHIFTER FORK

This tutorial illustrates:

- Adding a fourth extrusion to a model
- Sketching on datum plane Right

1. If the model is not already in memory, select ☞ and double-click on the file.

2. This part has a curvilinear protrusion, and the option "show_axes_for_extr_arcs" must be set to "yes" in the configuration file. Check if this option is set by choosing *Tools* and *Options*.

3. If the option is not set to "yes," follow the next six steps. Otherwise, cancel and go to step 10.

4. When the *Options* box is loaded, select the pull-down for the "Showing" field and change to "Current Session" as shown in Figure 3.18.

5. As in Figure 3.18, uncheck "Show only options loaded from file."

6. Scroll down the list until you find "show_axes_for_extr_arcs." Click on the option.

7. In the *Value* field enter "yes" or use the pull-down next to the field and select "yes."

8. Press the *Add /Change* button ▦.

9. Press *OK*.

10. Select ▦.

11. Choose the *Sketch* button ▦ in the dashboard.

12. With your mouse, choose datum plane Right as the sketching plane.

13. If necessary, use ▦ to change to the orientation in the *Orientation* field to "Top." Then, click on datum plane Top.

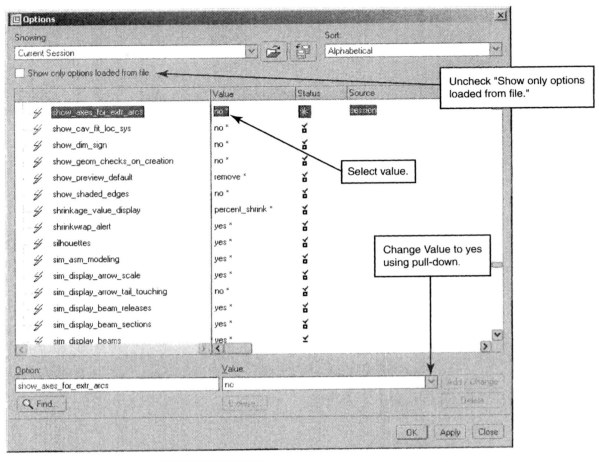

FIGURE 3.18 *Use the* Options *box to change the setting for "show_axes_for_extr_arcs" from "no" to "yes."*

14. Again, the arrow should point into the screen. Use *Flip* if it does not.

15. Select the *Sketch* button .

16. Use the *Line* tool and sketch the three lines shown in Figure 3.19.

17. Choose the *Arc* tool and sketch the arc shown in Figure 3.19.

18. Now, choose the *Dimension* tool and dimension the sketch as shown in Figure 3.19.

19. Press to build the section.

20. Select .

21. Select *Options* in the dashboard and then *Symmetric* . Enter a depth of 2.000 in the cell.

22. Build the part by pressing .

23. Save the part by pressing . Additional features will be added to the model in later chapters.

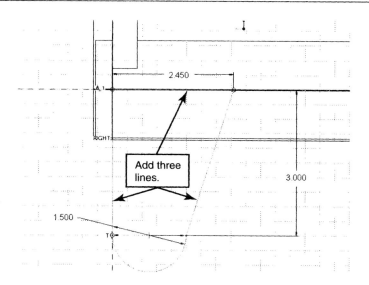

FIGURE 3.19 *Generate the geometry of the fourth extrusion by using three lines and an arc.*

3.9 DATUM PLANES ADDED BY THE USER

Consider the model of the rod support shown in Figure 3.20(a). The support contains an inclined plane (surface A). For most parts, inclined or oblique planes are difficult to construct unless special sketching planes are created. These planes can be made on the fly during creation of the feature or by constructing the plane ahead of time using the *Insert, Model Datum,* and *Plane* options or by simply pressing the *Datum* tool icon ▱ .

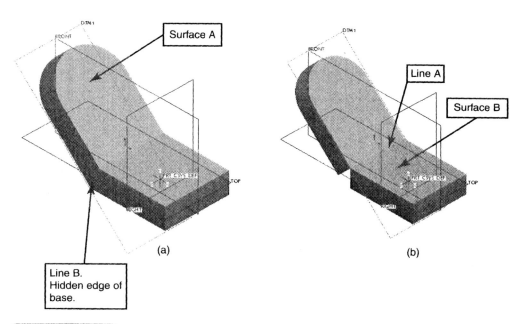

FIGURE 3.20 *The model contains an inclined plane. In order to create this inclined plane, the user must create a user-defined datum. In (b), the inclined surface created by using a sketching datum plane through line A. Note the protrusions do not blend into one another.*

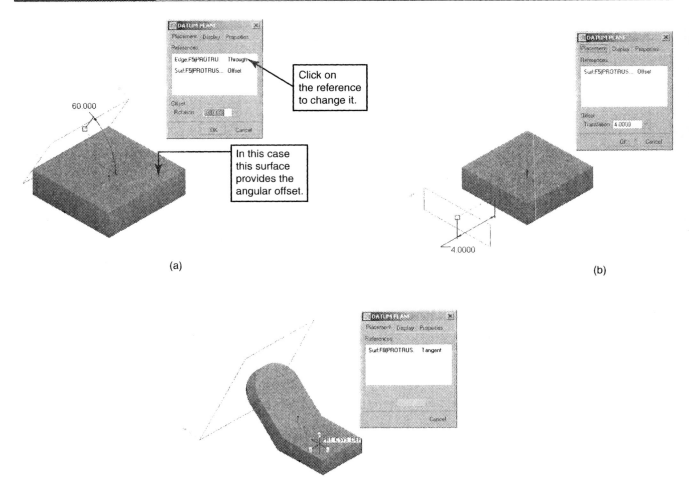

FIGURE 3.21 *A user-defined datum plane is constructed by constraining the datum plane to existing features in the model.*

User-defined datum planes are created by using a simple procedure involving: selecting ▱, picking the reference with the mouse, choosing the reference from the drop-down list, and adding additional references using the CTRL key as the references are selected with the mouse. This procedure will be illustrated in the next tutorial.

In generating a datum plane, the user must be aware of the constraints used to properly place a datum. The datum plane must be properly constrained by choosing appropriate references. These references are:

1. *Through* (Figure 3.21(a)). The new datum goes through a selected reference.

2. *Normal.* The new datum is placed normal to the selected reference.

3. *Parallel.* The new datum is created parallel to a selected reference.

4. *Offset* (Figure 3.21(b)). The new datum is created at an offset linear distance or angle from a selected reference.

5. *Tangent* (Figure 3.21(c)). The new datum is placed tangent to the selected reference.

Some options will fully constrain the plane without any additional added constraints. Pro/E calls these stand-alone constraints. The stand-alone constraints are:

1. *Through/Plane*. A datum is constructed coinciding with a plane.

2. *Offset/Plane*. A datum plane is constructed parallel to an existing plane at an increment provided by the user.

3. *Offset/Coord Sys*. The datum plane is created perpendicular to one of the axes of the coordinate system and offset from the origin.

For some feature the placement of the user-defined datum should be carefully considered. For example, in Figure 3.20(b), creating a datum plane through Line A and at 45° to the surface B would lead to the shown extrusion. Because the sketch is extruded perpendicular to the sketching plane, the protrusion does not pass fully through the base, as it should.

The protrusion may be properly created by placing the sketching plane so that it passes through Line B instead of Line A as in Figure 3.20(a). The perpendicular direction of this new plane would pass through the base and so would the protrusion.

3.9.1 TUTORIAL 3.7: A ROD SUPPORT

This tutorial illustrates:

- Creating a user-defined datum plane
- Sketching on a user-defined datum plane

1. Select *File* and *New* (or simply ▣).

2. Make sure that the *Use default template* option is checked.

3. Enter the name "RodSupport" in the *Name* field.

4. Select *OK*.

5. This part has a curvilinear protrusion, and the option "show_axes_for_extr_arcs" must be set to "yes" in the configuration file. Check if this option is set by choosing *Tools* and *Options*.

6. If the option is not set to "yes," follow the next six steps.

7. When the *Options* box is loaded, select the pull-down for the "Showing" field and change to "Current Session" as shown in Figure 3.18.

8. As in Figure 3.18, uncheck "Show only options loaded from file."

9. Scroll down the list until you find "show_axes_for_extr_arcs." Click on the option.

10. In the *Value* field enter "yes" or use the pull-down next to the field and select "yes."

11. Press the *Add/Change* button ▦ .

12. Press *OK*.

13. Now create the base feature. Select the *Extrude* tool ▣).

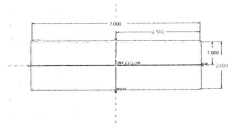

FIGURE 3.22 *The geometry of the base feature.*

14. Choose the *Sketch* button ▦ in the dashboard.
15. With your mouse, choose datum plane Front as the sketching plane.
16. If necessary, use ▾ for the *Orientation* field and to "right." Then, click on datum plane Right.
17. Again, the arrow should point into the screen. Use *Flip* if it does not.
18. Select the *Sketch* button ▭.
19. In the Sketcher, use the *Rectangle* tool ▢ and sketch the geometry shown in Figure 3.22.
20. Select the Dimension tool ⊡ and dimension your section so that it corresponds to the one shown in Figure 3.22.
21. Press the *Modify* button ⇗ and change the value of each dimension to the values shown in Figure 3.22.
22. Build the section by pressing the check mark in the Sketcher ✓.
23. Press the pull-down below the *Options* button in the dashboard and choose *Symmetric* ⊟.
24. Enter a depth of 6.00 and hit the *return* key.
25. Press ✓ in the dashboard to build the model.
26. Now, add a user-defined sketching plane. Select ▱.
27. Use Figure 3.23. Click on Line B. In the *References* collector of the Datum Plane box, make sure the reference is set to "Through." If it is not, click on the reference and use the pull-down to change to "Through."

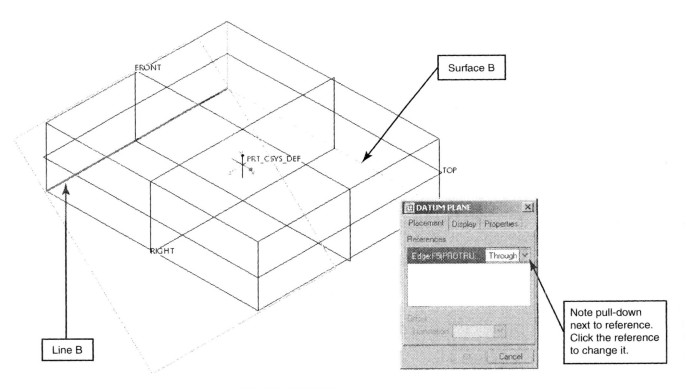

Surface B

Note pull-down next to reference. Click the reference to change it.

Line B

FIGURE 3.23 *Use Line B and the* Through *option.*

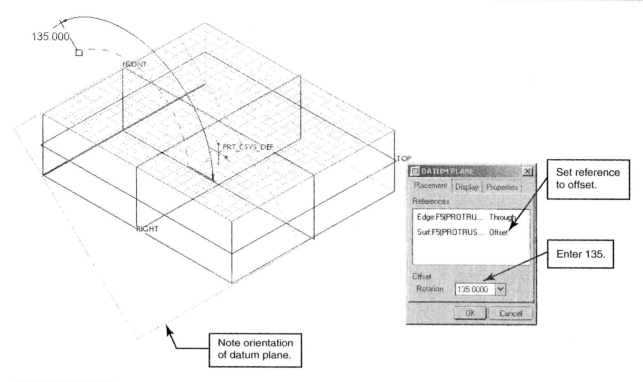

135.000

FRONT

PRT_CSYS_DEF

TOP

RIGHT

Set reference
to offset.

DATUM PLANE

Placement | Display | Properties

References

Edge:F5(PROTRU... Through
Surf:F5(PROTRUS... Offset

Offset
Rotation 135.0000

OK   Cancel

Enter 135.

Note orientation
of datum plane.

FIGURE 3.24 *The second reference is an angular offset using Surface B.*

28. While pressing the CTRL key select Surface B, shown in Figure 3.23.

29. In the *References* collector, make sure the reference is set to "Offset." If it is not, click on the reference and use the pull-down to change to "Offset."

30. Enter 135 for the rotation angle as shown in Figure 3.24.

31. Press *OK*.

32. The base feature with the datum plane (DTM1) is shown in Figure 3.25. We now add an additional datum for dimensioning purposes. Once again, press 🔲.

33. Pick Line B and change, if necessary, the reference to "Through."

34. Now, while pressing the CTRL key, click on datum plane DTM1.

35. In Figure 3.26, the model has been reoriented to show the relative orientations of the datum planes. Note that the second reference is "Normal." Make sure that your reference is set to "Normal" as well.

36. A second datum plane called DTM2 has been created as shown in Figure 3.26.

37. Select 🔲.

38. Choose the *Sketch* button 🔲 in the dashboard.

39. With your mouse, choose datum plane DTM1 as the sketching plane.

40. If necessary, use 🔲 for the *Orientation* field and to "left." Then, click on Surface C shown in Figure 3.27.

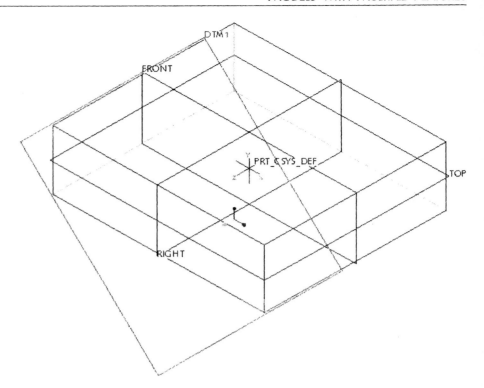

FIGURE 3.25 *Note the orientation of the new datum. The name of the new datum is "DTM1."*

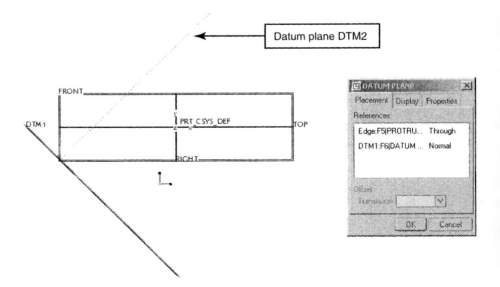

FIGURE 3.26 *The relative orientation of datum plane DTM2 and DTM1 is easily seen by rotating the model.*

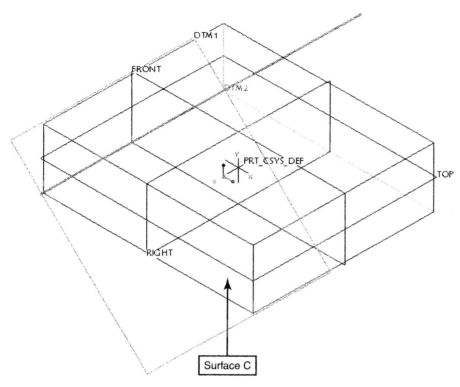

FIGURE 3.27 *Use Surface C as the "Left" to orient the model in the Sketcher.*

41. Again, the arrow should point into the screen. Use *Flip* if it does not.

42. Select the *Sketch* button [▨].

43. Because of the orientation of the sketching plane, you will need to define two edges that may be used to constrain the model. As shown in Figure 3.28, press the *References* button, and then select datum plane Front and then datum plane DTM2.

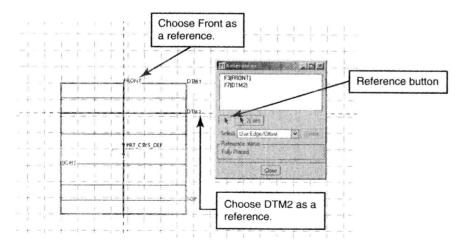

FIGURE 3.28 *Datum planes Front and DTM2 may be used to dimension the new section.*

FIGURE 3.29 *Use three lines and an arc to create the geometry of the inclined feature.*

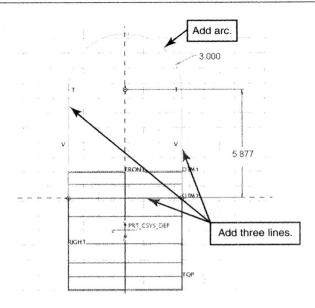

44. In the *References* box, the message *Fully Placed* should be visible. Press the Close button in the References box.

45. Use ▧ and sketch three lines as shown in Figure 3.29.

46. Use ▧ and sketch an arc as shown in Figure 3.29.

47. Select ▧ and dimension your section so that it corresponds to the one shown in Figure 3.29.

48. Press the *Modify* button ▧ and change the value of each dimension to the values shown in Figure 3.29.

49. Build the section by pressing the check mark in the *Sketcher* ▧.

50. Press the pull-down below the *Options* button in the dashboard and choose *Blind* ▧.

51. Enter a depth of 2.00 and hit the *return* key.

52. Press ▧ in the dashboard to build the model.

53. Save the part by pressing ▧. Additional features will be added to the model in later chapters.

3.10 SUMMARY AND STEPS FOR USER-DEFINED DATUM PLANES

For models with multiple features, the datum planes may be used to sketch the required section for feature. Often, the default datum planes are not sufficient for constructing the desired feature. In such cases, the user may define additional datum planes. Such datum planes may be added to the model at any time, provided the user is able to locate the datum. This task is accomplished by using the *Insert, Model Datum,* and *Plane* options or by simply pressing the *Datum* tool icon ▧.

The general steps for adding a user-defined datum to a model are as follows:

1. Select *Insert, Model Datum,* and *Plane* or ▧.

2. Pick the required feature from the model.

3. Change the reference type, if necessary.

4. If needed, select additional references by pressing the CTRL key as you pick the reference with your mouse.

5. Choose *OK* to create the datum.

3.11 ADDITIONAL EXERCISES

3.1 Construct the model of the Heat Sink Plate shown in Figure 3.30. Use user-defined datums.

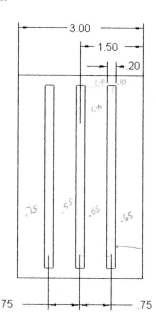
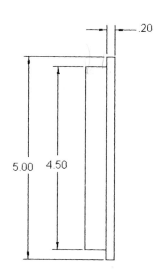

FIGURE 3.30

3.2 Create the bushing shown in Figure 3.31. Do not create the hole.

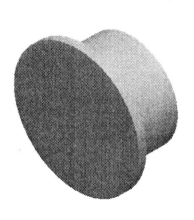
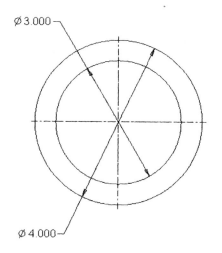
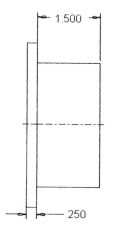

FIGURE 3.31

3.3 Construct the axle shown Figure 3.32.

FIGURE 3.32

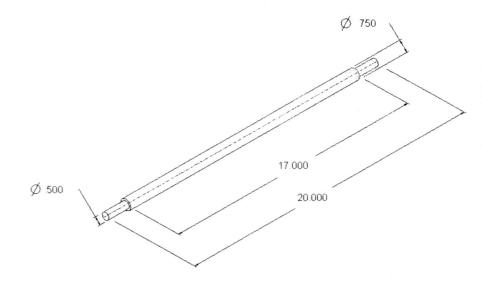

3.4 The volume control dial illustrated in Figure 3.33 may be created using the datum planes. Construct the model.

FIGURE 3.33

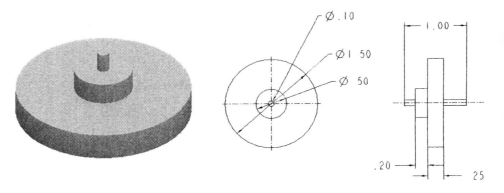

3.5 Construct the base feature for a cement trowel as shown in Figure 3.34.

FIGURE 3.34

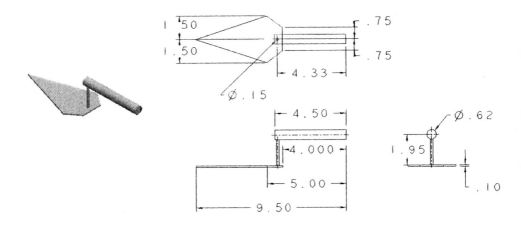

3.6 Create the base features for a control handle as given in Figure 3.35.

FIGURE 3.35

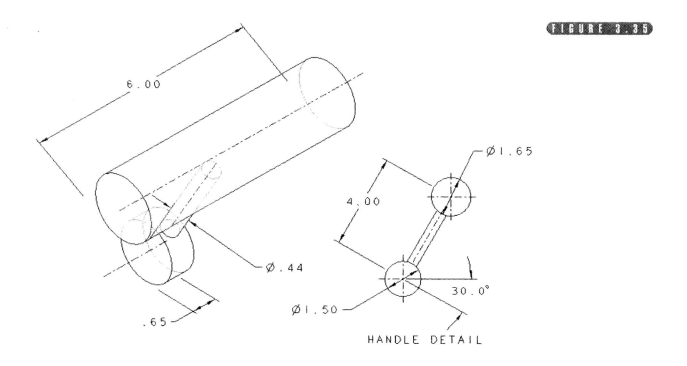

HANDLE DETAIL

3.7 The base features for angled rod support may be created by using the default datum planes as well as any additional user-defined datum. Create the model, using Figure 3.36 as a guide.

FIGURE 3.36

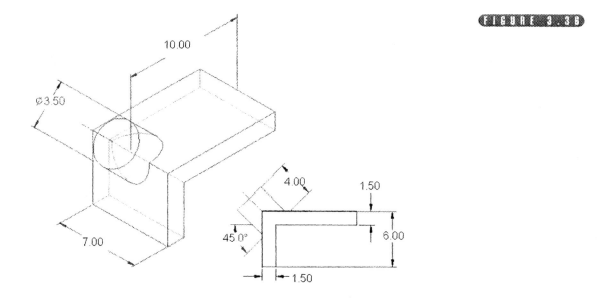

3.8 Construct the features for the side guide as shown in Figure 3.37.

FIGURE 3.37

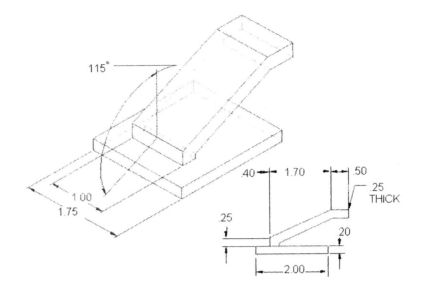

3.9 Use the geometry shown in Figure 3.38 to create the model of the offset bracket. Note an additional user-defined datum is required.

FIGURE 3.38

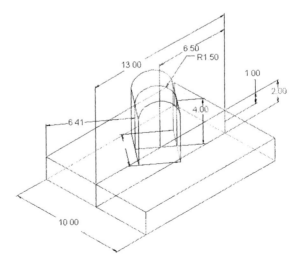

3.10 Construct the model of the side support reproduced in Figure 3.39. *Hint*: Create a datum plane through the axis A_2 and the lower edge of the rectangular base. Then project *Blind* 3.00.

FIGURE 3.39

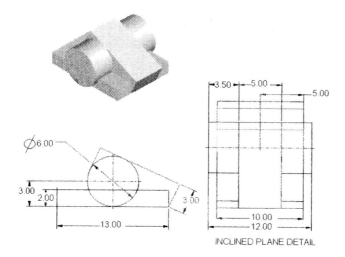

INCLINED PLANE DETAIL

3.11 Create the model of the double side support as shown in Figure 3.40.

FIGURE 3.40

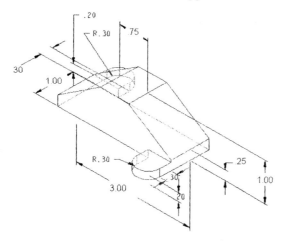

3.12 The slotted support in Figure 3.41 is a part with metric units. Construct the model.

FIGURE 3.41

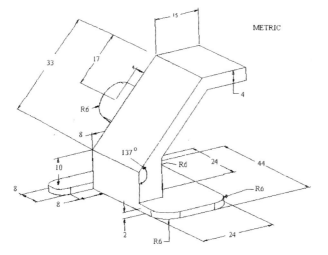

METRIC

3.13 The clamp base in Figure 3.42 may be created using a user-defined datum. Note that the units of the part are metric.

FIGURE 3.42

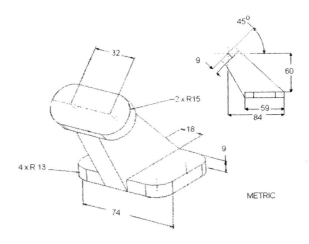

METRIC

3.14 Construct the model of the support. Use Figure 3.43.

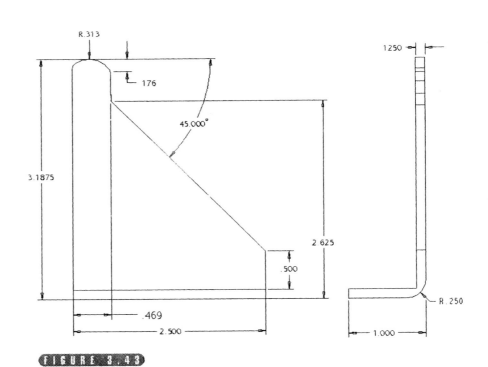

FIGURE 3.43

3.15 Construct the mirror image of the support from Exercise 3.14. Use Figure 3.44 as a guide and the geometry from Figure 3.43.

FIGURE 3.44

3.16 Construct the three pins shown in Figure 3.45.

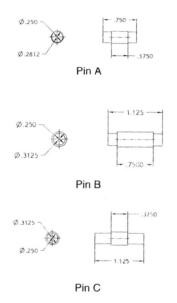

FIGURE 3.45

Pin A

Pin B

Pin C

CHAPTER ■ 4

ADDING HOLES TO BASE FEATURES

INTRODUCTION AND OBJECTIVES

So far, all the models that we have constructed are composed of protrusions. These base parts do not contain any holes. In this chapter, we will discuss how to create a hole. With Pro/E, holes can be placed by locating their centers with two linear dimensions, along a known axis, at a point, or on a radial direction.

In creating a hole, the user must define three distinct aspects of the hole. The first is the geometry of the hole—that is, whether or not the hole has a constant diameter. The second is the location of the hole. The third aspect is the depth. These details, in creating a hole, will be considered in more detail in the sections that follow.

The objectives of this chapter are to

1. Understand the different methods for locating a hole on a base feature
2. Know the difference between a *Straight* and *Sketched* hole
3. Specify the depth of a hole

▶ 4.1 ◀ ADDING HOLES IN PRO/E

A hole may be added as a feature by using *Insert* and *Hole* or the tool icon ▦ . Pro/E allows the user to define a *Simple, Sketched, or Standard Hole* (Figure 4.1) using an interface similar to the one used to construct basic features. A sketched hole is user-defined with a varying diameter, whereas straight holes are holes with a constant diameter. *Sketched* holes are created by using the Sketcher. Revolving the sketch around the axis of the hole creates the sketched hole. *Sketched* holes are always blind and one-sided.

The *Standard* hole option is used to create a standard size hole. Standard threads may be placed with this option when creating the hole. Because it

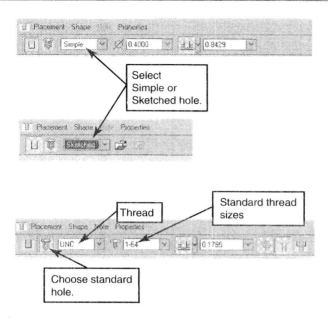

FIGURE 4.1 *The options in the dashboard may be used to create a hole feature.*

is advantageous in design to use standard size holes and threads, this option will usually be used in this book.

Pro/E provides four options for placement of holes using the tools shown in Figure 4.1. The first of these methods is *Linear* placement—that is, the center of the circle is defined by linear dimensions from two boundaries or edges. In the second method (the *Radial* option), the hole is placed along a radius. This is useful for placing holes on a bolt circle. The third way to place a hole is by using the *Coaxial* option. This option is convenient for holes along the center axis of a cylinder or shaft. Finally, the hole can be placed on a defined point (the *On Point* option).

As shown in Figure 4.1, the depth of the hole must be defined in order for the feature to be created. The depth of the hole may be prescribed in several ways. These are:

1. *Blind.* Hole A is generated to a specific numerical depth (sd0 in Figure 4.2) provided by the user.

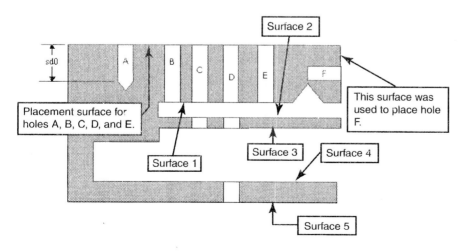

FIGURE 4.2 *A cross section of an object with holes placed to varying depths.*

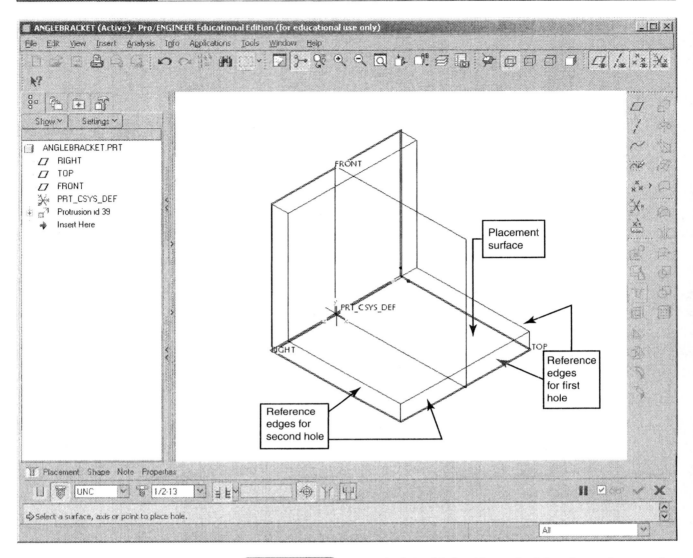

FIGURE 4.3 *A standard-sized hole with standard threads may be created by choosing the option from the dashboard.*

2. *Symmetric.* The hole is drilled on both sides of the placement plane.

3. *To Next.* The hole E is projected so that it passes and stops at the next surface (Surface 1 in Figure 4.2).

4. *Through All.* For this option, Hole D is projected through all features and surfaces.

5. *Through Until.* The hole (C in Figure 4.3) is created through all surfaces until Surface 3 is reached.

6. *To Selected.* The hole is created until a surface (Hole B), curve, or point/vertex (Hole F) is reached.

4.2 TUTORIAL 4.1: ADDING HOLES TO THE ANGLE BRACKET

This tutorial illustrates:

- Adding a standard hole to a part
- Adding a straight hole to part

1. If the model "AngleBracket" is not already in memory, select [image] and double-click on the file.
2. Select [image].
3. Select the *Standard hole* tool in the dashboard [image].
4. Select UNC thread (see Figure 4.3).
5. Pick ½-13 screw size as shown in Figure 4.3.
6. Choose *Through all* for the depth. Again, consider Figure 4.3.
7. Make sure that the countersink and counter bore tools are unselected as shown in Figure 4.3.
8. Now, click on the placement surface, that is, the surface shown in Figure 4.3.
9. The center of the hole is to be defined relative to the two edges shown in Figure 4.3.
10. Consider Figure 4.4 and note the reference handles. One way to locate the hole is by simply dragging the handles.
11. Drag the first reference handle until it coincides with the edge shown in Figure 4.5.
12. Double-click on the dimensional value of the handle and enter 1.5. Hit *return*.

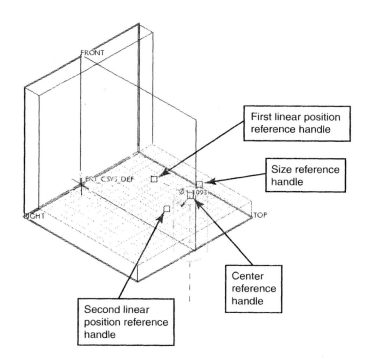

FIGURE 4.4 *Place the hole as shown. Note the reference handles used to define the size and location of the hole.*

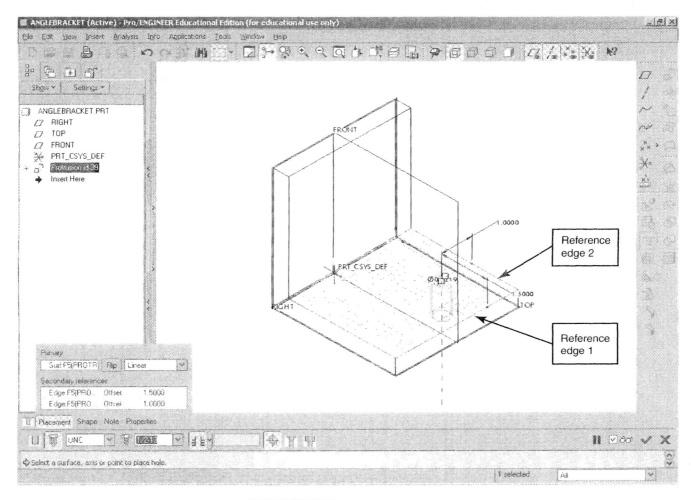

FIGURE 4.5 *Drag the reference handles. This will place the hole in its proper location.*

13. Then drag the second handle until it coincides with the second edge shown in Figure 4.5.

14. Double-click on the dimensional value of the handle and enter 1.0. Hit *return*.

15. Press the *Placement* button in the dashboard. Note the references and the corresponding values of 1.5 and 1.0.

16. Build the feature by pressing ✔ .

17. A second hole is a *Straight* hole and is not threaded. Place this hole by selecting ▣ .

18. In the dashboard, select *Straight* and *Simple*.

19. Enter a diameter of 0.5 and select the *Through all* option as shown in Figure 4.5.

20. Click on the placement surface shown in Figure 4.6.

21. Drag the handles as shown in Figure 4.6 and 4.7.

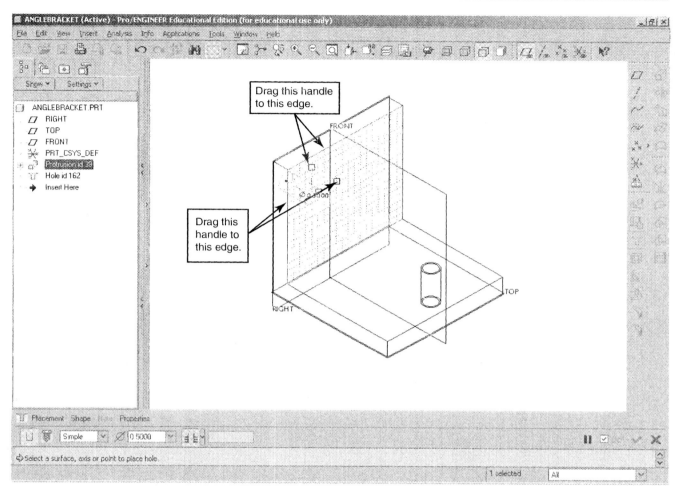

FIGURE 4.6 *Place the second hole.*

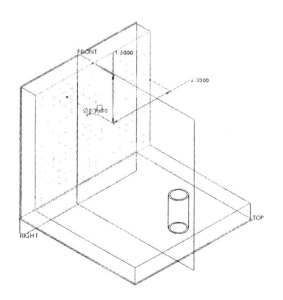

FIGURE 4.7 *Drag the reference handle for the second hole to locate it.*

22. Double-click on each dimensional value and enter 2.0 and 1.5 as appropriate.
23. Build the feature by pressing 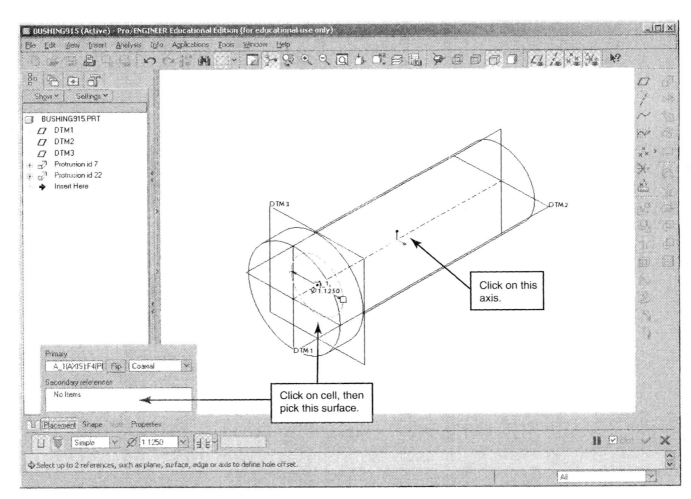.

4.3 TUTORIAL 4.2: A COAXIAL HOLE FOR THE BUSHING

This tutorial illustrates:

- Adding a coaxial hole to a part

1. If the model "BushingT3" is not already in memory, select and double-click on the file.
2. Select .
3. As shown in Figure 4.8, select *Straight, Simple,* and *Through all* options.
4. Enter a diameter of 1.125.
5. With your mouse click on the axis A_1.

FIGURE 4.8 *Add the coaxial hole to the bushing from Chapter 3.*

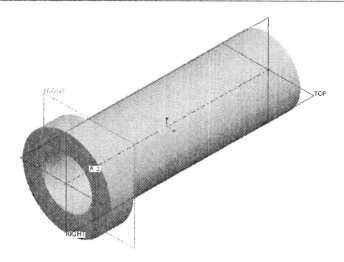

FIGURE 4.9 *The bushing from Chapter 3 after adding the 1.125-diameter coaxial hole.*

6. Press the *Placement* option in the dashboard. Click in the Secondary references field as shown in Figure 4.8.

7. Click on the surface shown in Figure 4.8.

8. Build the feature by pressing ☑.

9. The bushing with the coaxial hole is shown in Figure 4.9.

10. Save the part by pressing ▢.

4.4 TUTORIAL 4.3: A COAXIAL HOLE FOR THE PROPELLER

This tutorial illustrates:

- Adding a coaxial hole to a part

1. If the model "Propeller" is not already in memory, select ☞ and double-click on the file.

2. Select ▥.

3. Select *Straight, Simple,* and *Through all* options.

4. Enter a diameter of 15.

5. With your mouse, click on the axis A_1.

6. Press the *Placement* option in the dashboard. Click in the Secondary references field.

7. Click on the surface shown in Figure 4.10.

8. Build the feature by pressing ☑.

9. Save the part by pressing ▢.

4.5 TUTORIAL 4.4: A PLATE WITH COUNTERBORED HOLES

This tutorial illustrates:

- Constructing a sketched hole

1. Select ▢.

2. Make sure that the *Use default template* option is checked.

3. Enter the name "PlateWithCounterboreHoles" in the *Name* field.

FIGURE 4.10 *Add a coaxial hole to the propeller.*

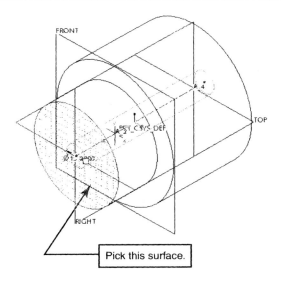

Pick this surface.

4. Select *OK*.
5. Select [icon].
6. Choose the *Sketch* button [icon] in the dashboard.
7. With your mouse, choose datum plane Front as the sketching plane.
8. If necessary, use [icon] for the *Orientation* field and to "right." Then, click on datum plane Right.
9. Again, the arrow should point into the screen. Use *Flip* if it does not.
10. Select the *Sketch* button [icon].
11. Use the *Rectangle* tool [icon] and sketch the geometry shown in Figure 4.11. Build the section.
12. Now, choose the *Dimension* tool [icon] and dimension the sketch as shown in Figure 4.11.
13. Press the pull-down below the *Options* button in the dashboard and choose *symmetric* [icon].
14. Enter a depth of 5.00.
15. The plate is shown in Figure 4.12. A nonstandard counterbore hole will be added next.

FIGURE 4.11 *In order to construct the base feature, sketch and dimension the rectangle shown.*

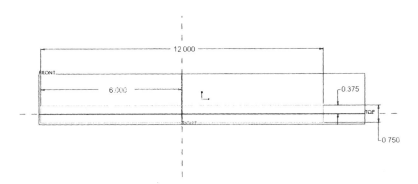

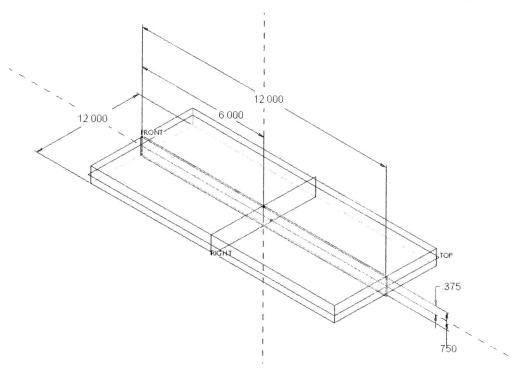

FIGURE 4.12 *The geometry of the base feature is shown.*

16. Place the hole by selecting .

17. In the dashboard, select *Straight* and *Sketched*.

18. Press the *Sketch* icon to launch the Sketcher.

19. Press ▶ next to ◥ and change the line type to *Centerline* .

20. Draw a vertical centerline.

21. Now, change the line type back to *Line* type ◥.

22. Using ◥, sketch the geometry shown in Figure 4.13.

23. Now, choose the *Dimension* tool and dimension the sketch as shown in Figure 4.13. Note that the dimensions are diametral, not radial. Consult Figure 2.8 on how to create diametral dimensions for revolved sections.

24. Press the *Modify* button and change the value of each dimension to the ones shown in Figure 4.13.

25. Build the section by pressing the check mark in the Sketcher .

26. Click on the placement surface shown in Figure 4.14.

27. Consulting Figure 4.14, drag the reference handles to the edges shown.

28. Your model should now correspond to the one shown in Figure 4.15. Double-click on each dimensional value and enter the numbers shown in Figure 4.15.

29. Press in the dashboard to build the model.

30. Save the part by pressing .

FIGURE 4.13 *Sketch the geometry shown to create the Sketched hole.*

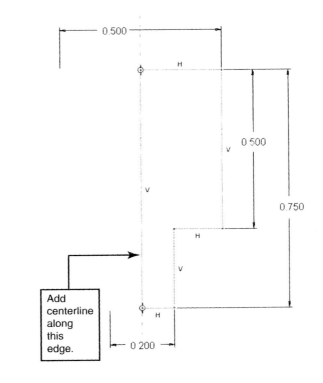

Add centerline along this edge.

FIGURE 4.14 *Place the hole near the corner of the plate as shown.*

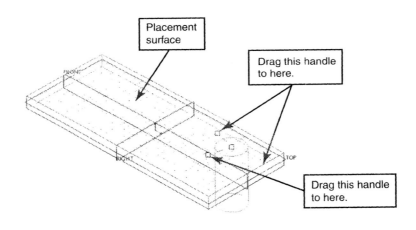

Placement surface

Drag this handle to here.

Drag this handle to here.

FIGURE 4.15 *Corner of the plate with dimensions locating the hole.*

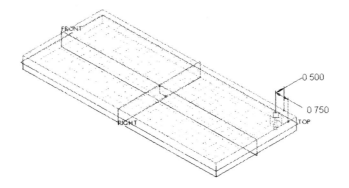

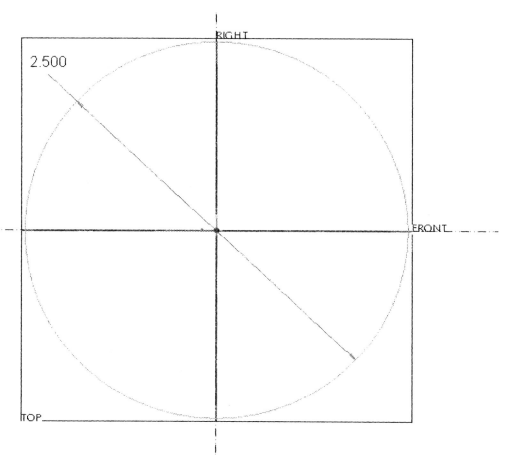

FIGURE 4.16 *Draw a circle to create the base feature for the pipe flange.*

4.6   TUTORIAL 4.5: A PIPE FLANGE WITH A BOLT CIRCLE

This tutorial illustrates:

- Constructing a radial hole
- Constructing a coaxial hole

1. Select 🔲.
2. Make sure that the *Use default template* option is checked.
3. Enter the name "PipeFlange" in the *Name* field.
4. Select *OK*.
5. Select 🔲.
6. Choose the *Sketch* button 🔲 in the dashboard.
7. With your mouse, choose datum plane Top as the sketching plane.
8. If necessary, use 🔲 for the *Orientation* field and to "top." Then, click on datum plane Front.
9. Again, the arrow should point into the screen. Use *Flip* if it does not.
10. Select the *Sketch* button 🔲.
11. In the Sketcher, sketch the circle shown in Figure 4.16.

12. Select the diameter dimension. Press the *Modify* button ⧉ and change the value of the dimension to 2.50 as shown in Figure 4.16.

13. Press ✔ to build the section.

14. Press the pull-down below the *Options* button in the dashboard and choose *Blind* ⧉.

15. Enter a depth of 0.25. This will extrude the section one side from the sketching plane.

16. Extrude the feature down.

17. Build the section by pressing the check mark in the *Sketcher* ✔.

18. Add a section feature by selecting ⧉.

19. Choose the *Sketch* button ⧉ in the dashboard.

20. With your mouse, choose datum plane Top as the sketching plane.

21. If necessary, use ⧉ for the *Orientation* field and to "top." Then, click on datum plane Front.

22. Again, the arrow should point into the screen. Use *Flip* if it does not.

23. Select the *Sketch* button ⧉.

24. In the Sketcher, sketch the circle shown in Figure 4.17.

25. Double-click on the diameter dimension and change the value of the dimension to 1.00 as shown in Figure 4.17.

26. Press ✔ to build the section.

27. Press the pull-down below the *Options* button in the dashboard and choose *Blind* ⧉.

28. Enter a depth of 1.50. Extrude the feature up as shown in Figure 4.18.

29. First, add the coaxial hole to the second extrusion. Select ⧉.

30. Select *Straight, Simple,* and *Through all* options.

31. Enter a diameter of 0.750.

32. With your mouse, click on the axis A_2.

33. Press the *Placement* option in the dashboard. Click in the Secondary references field and then select the surface shown in Figure 4.19.

34. Click on the surface shown in Figure 4.19.

35. Build the feature by pressing ✔.

36. Now, let us add the hole on the bolt circle. Once again, select ⧉.

37. Select *Straight, Simple,* and *Through all* options.

38. Enter a diameter of 0.500.

39. Now click on the placement surface shown in Figure 4.20.

40. Press the *Placement* option and change to *Diameter* as shown in Figure 4.20.

41. Click in the Secondary references field and then on datum Front. Note the angular dimension in this case; leave the value at 0.000°.

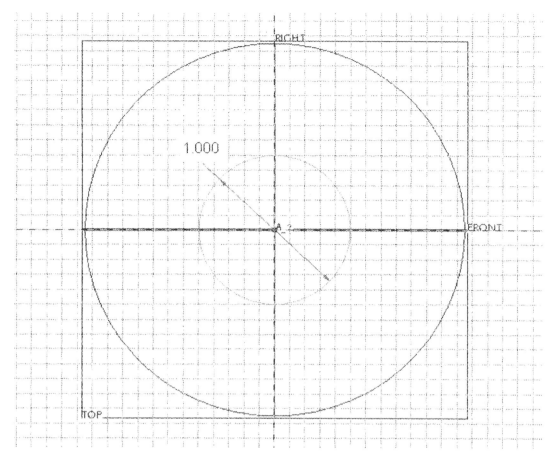

FIGURE 4.17 *Draw a circle for the second protrusion in the pipe flange.*

42. In Figure 4.21, note the diameter position handle. Drag this handle until it intersects with axis A_2.

43. A diameter dimension will be activated when the handle intersects the axis. Change the value of this dimension to that of the bolt circle—that is, 1.600 as shown in Figure 4.22.

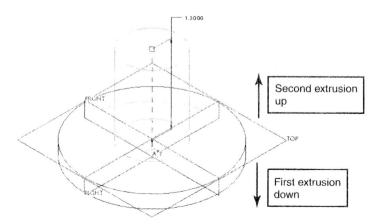

FIGURE 4.18 *The pipe flange is shown with the two protrusions.*

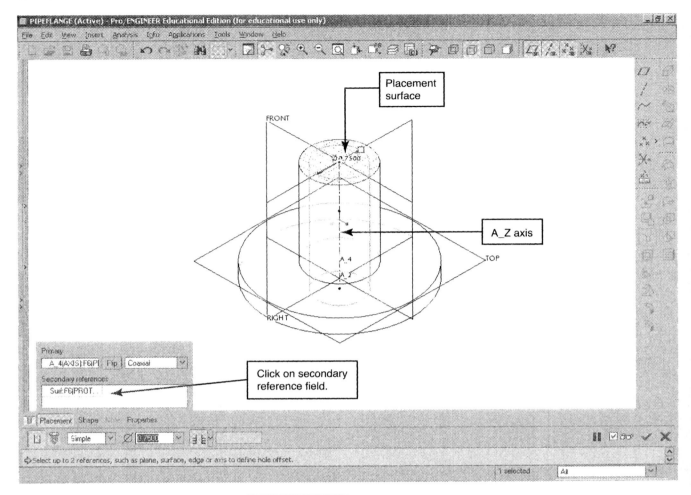

Add a coaxial hole to the second protrusion.

44. Press 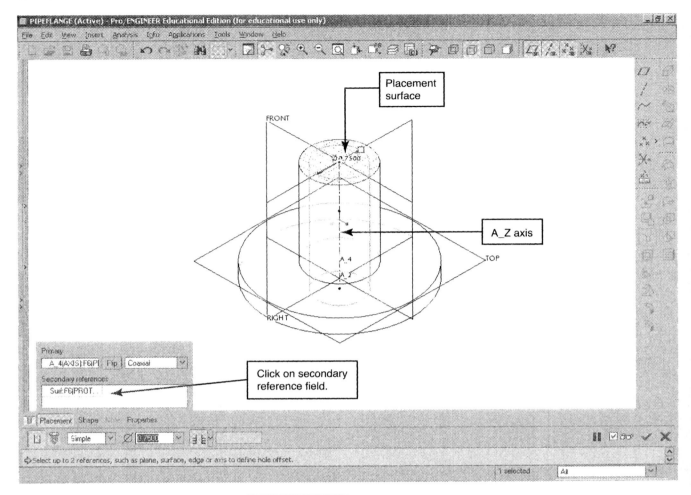 in the dashboard to build the model.

45. Save the part by pressing . The model with both holes is shown in Figure 4.23.

4.7 TUTORIAL 4.6: A HOLE ON THE INCLINED PLANE OF THE ROD SUPPORT

This tutorial illustrates:

- Constructing a radial hole
- Constructing a coaxial hole

1. If the model "RodSupport" is not already in memory, select and double-click on the file.

2. Select .

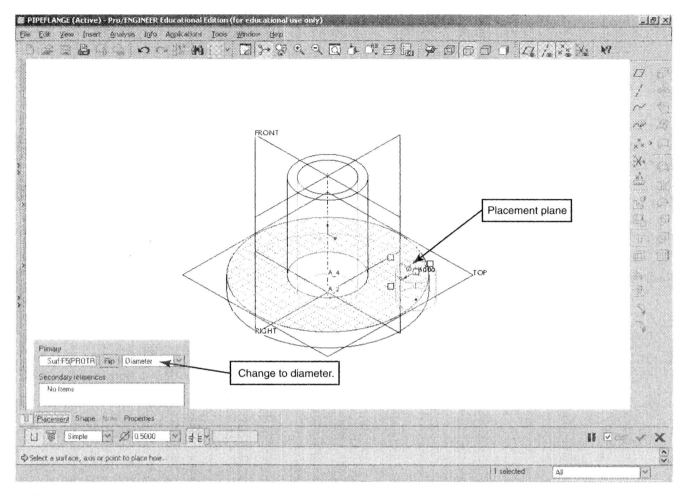

FIGURE 4.20 *Add a diametral hole to the first protrusion of the pipe flange.*

3. Because the model was created with the "show_ axes_for_extr_arcs" option set to "yes," a hole may be easily added to the inclined surface of the model by using the *Coaxial* option. As shown in Figure 4.24, select *Straight, Simple,* and *Blind* options.

4. Enter a depth of 2.00 and a diameter of 3.00.

5. With your mouse, click on the axis A_1.

6. Press the *Placement* option in the dashboard. Click in the Secondary references field as shown in Figure 4.24.

7. Click on the surface shown in Figure 4.25.

8. Build the feature by pressing ✔.

9. Add the two additional holes shown in Figure 4.25. Press ▦.

10. For each hole, the procedure is the same, except that the linear dimensions are different. Select *Straight, Simple,* and *Blind.*

11. Enter a diameter of 1.500 and a depth of 2.000.

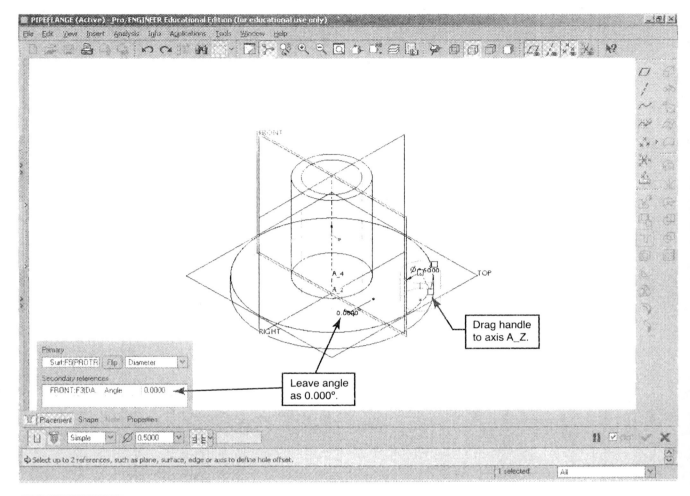

FIGURE 4.21 *Drag the handle to the axis.*

12. Select the surface in Figure 4.25 as the placement surface.

13. Drag the reference handles to intersect the edges shown in Figure 4.25.

14. Double-click and enter the corresponding dimensional values shown in Figure 4.25.

4.8 SUMMARY AND STEPS FOR ADDING HOLES

In this chapter, we examined the options available in Pro/E for placing a hole on a model. Holes in Pro/E may be constructed as a *Standard* hole, which is a hole of standard size. The Standard hole may have a constant or varying diameter along its length. For nonstandard size holes, Pro/E has two additional options: the *Straight* hole and the *Sketched* hole. For the *Straight*

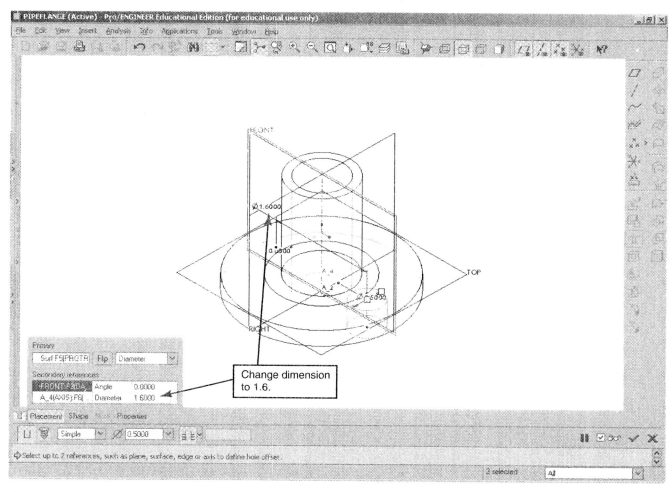

Change dimension to 1.6.

FIGURE 4.22 *Change the diametral dimension to 1.600.*

PATTERN!
THE HOLE

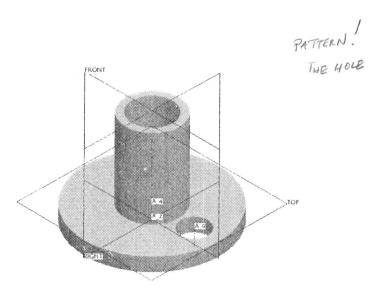

FIGURE 4.23 *A pipe flange with the diametral hole.*

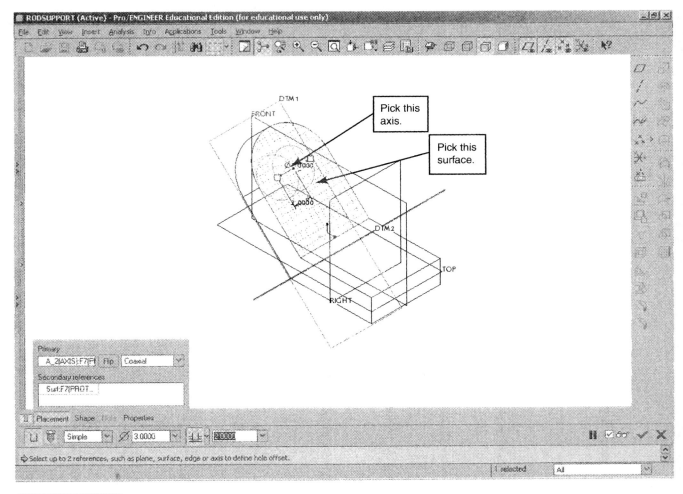

FIGURE 4.24 *The rod support with the coaxial hole.*

hole, the diameter is constant along its length, whereas in a Sketched hole the diameter may vary.

In Pro/E, holes are placed using one of four possible options. These options are *Linear, Radial* (or Diameter), and *Coaxial.* The *Linear* option is used to place a hole when the linear location of the center of the hole is known relative to two surfaces or edges. The *Radial* or *Diameter* option may be used to create a hole along a radial or diametral direction such as in a bolt circle. If the hole extends along a known axis, then the *Coaxial* option may be used.

In general, a hole may be placed on a model by using the following steps:

1. Select ⬚.
2. Choose either a *Standard* or *Straight* hole. If the type of hole is *Sketched* type, draw the geometry of the hole using the sketcher. Do not forget to place a centerline. Then dimension to the appropriate values.

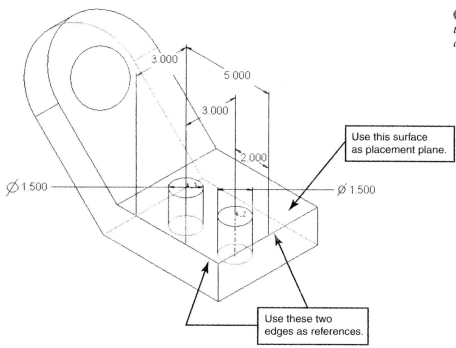

FIGURE 4.25 *This figure shows the model of the rod support after adding all three holes.*

3. Select the placement surface.
4. Drag the reference handles to the appropriate references.
5. Double-click on the reference dimensions to modify their value.
6. Build the feature.

4.9 ADDITIONAL EXERCISES

The following problems require the placement of various types of holes. In some cases, we have chosen problems from previous chapters. In such cases, their source is indicated in parentheses.

4.1 Add the hole to the washer shown in Figure 4.26 (Exercise 2.3).

FIGURE 4.26

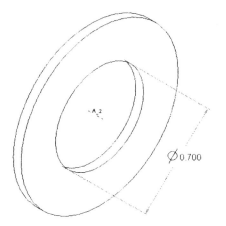

4.2 Using Figure 4.27, complete the model of the hold down (Exercise 2.8).

FIGURE 4.27

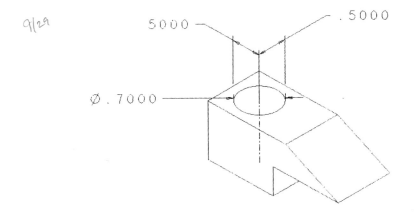

4.3 Add the hole to the bushing from Exercise 3.2. Use Figure 4.28.

FIGURE 4.28

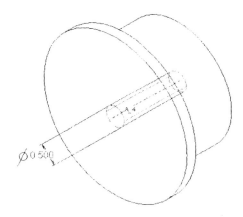

4.4 Create the model of the bumper using Figure 4.29.

FIGURE 4.29

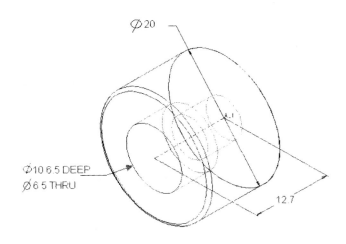

4.5 Complete the model of the angled rod support from Exercise 3.7. Use Figure 4.30 for reference.

FIGURE 4.30

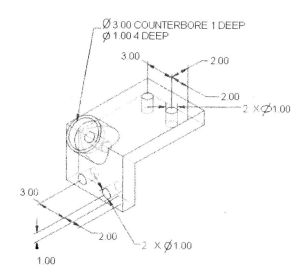

Ø 3.00 COUNTERBORE 1 DEEP
Ø 1.00 4 DEEP

3.00

2.00

2.00

2 X Ø1.00

3.00

2.00

2 X Ø1.00

1.00

4.6 Add the holes shown in Figure 4.31 to the model of the offset bracket (Exercise 3.9).

FIGURE 4.31

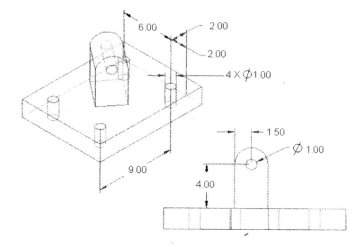

6.00

2.00

2.00

4 X Ø1.00

1.50

Ø 1.00

9.00

4.00

4.7 For the model of the side support constructed in Exercise 3.10 add the holes as shown in Figure 4.32.

FIGURE 4.32

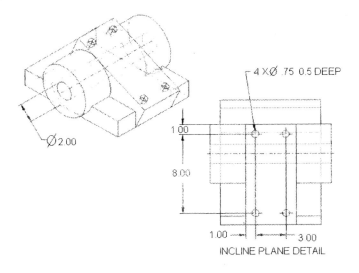

4 X Ø .75 0.5 DEEP

1.00

8.00

1.00   3.00

Ø2.00

INCLINE PLANE DETAIL

4.8 Complete the model of the control handle from Exercise 3.6. Use the geometry shown in Figure 4.33.

FIGURE 4.33

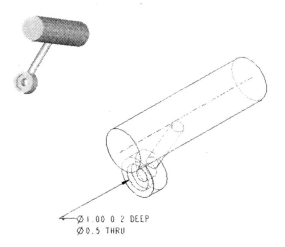

Ø1.00 0.2 DEEP
Ø0.5 THRU

4.9 Add the counterbored hole to the model of the angle block from Exercise 2.7. Use the geometry shown in Figure 4.34.   *Due Mon 9/29*

FIGURE 4.34

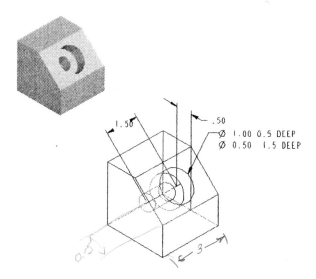

1.50   .50
Ø 1.00 0.5 DEEP
Ø 0.50 1.5 DEEP
0.5
3

4.10 Add the four holes to the model of the side guide from Exercise 3.8. Use Figure 4.35 as a reference.

FIGURE 4.35

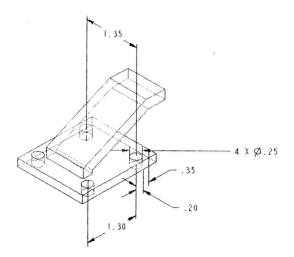

1.35
4 X Ø.25
.35
.20
1.30

4.11 For the model of the double side support from Exercise 3.11, add the two holes. The geometry of the holes is given in Figure 4.36.

FIGURE 4.36

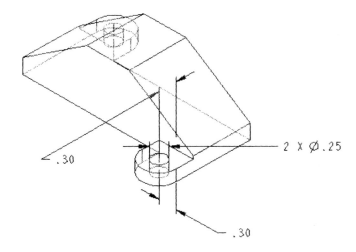

4.12 Add the holes to the model of the slotted support from Exercise 3.12. Use the geometry shown in Figure 4.37.

FIGURE 4.37

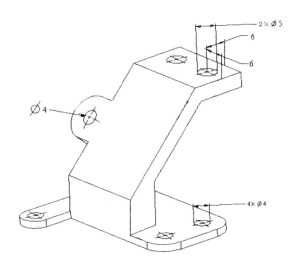

4.13 For the model of the clamp base from Exercise 3.13, add the holes. Use Figure 4.38 for the geometry of the holes.

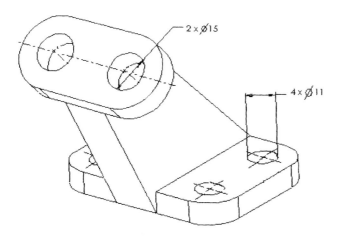

FIGURE 4.38

4.14 Add the holes to the model of the retainer using Figure 4.39 as a guide. The retainer was originally constructed in Exercise 2.6.

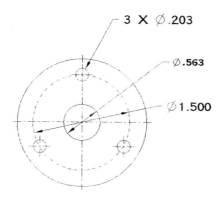

FIGURE 4.39

4.15 Add the holes to the support from Exercise 3.14. Use Figure 4.40.
4.16 Add the holes to the support from Exercise 3.15. Use Figure 4.40.

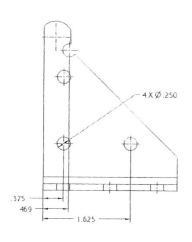
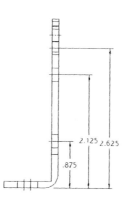

FIGURE 4.40

CHAPTER 5

Options that Remove Material: Cut, Neck, and Shell

Introduction and Objectives

In this chapter, we consider options that remove material from a base feature. The first option that we examine is *Cut*. Simply put, a cut is created using a bulk material removal process such as by a punching or sawing. Keyways, key seats, and kerfs are special types of internal cuts called slots.

Pro/E has another feature for the removal of material, the *Neck* option. A neck is a slot that is revolved around an axis of revolution. Another option, called *Shell,* removes a surface and then removes material below the surface.

The objective of this chapter is to

1. Become acquainted with the material removal options *Cut, Neck,* and *Shell*

5.1   Adding a Cut

A *Cut* may be placed on a base feature using the same procedure as was used to generate a solid protrusion, except that the *Cut* icon 🔲 is selected in the dashboard. This feature may be either Solid or Thin, in a manner analogous to a protrusion. However, in a *Thin* cut, the thickness of the feature is the thickness of the material removed from the model.

5.1.1 Tutorial 5.1: using Cut to remove material from a U-bracket

This tutorial illustrates:

• Removing material with a cut

1. Consider Figure 5.1. Now, select *File* and *New* (or simply 🔲).
2. Make sure that the *Use default template* option is checked.

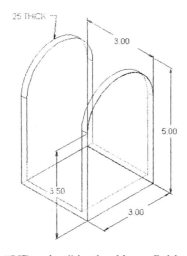

FIGURE 5.1 *The U bracket.*

25 THICK

3.00

5.00

3.50

3.00

3. Enter the name "UBracket" in the *Name* field.

4. Select *OK*.

5. Select *Insert* and *Extrude* (or simply).

6. Select the *Thicken section* button .

7. Enter a thickness of 0.25.

8. Choose the *Sketch* button in the dashboard.

9. With your mouse, choose datum plane Front as the sketching plane.

10. If necessary, use for the *Orientation* field and to "right." Then, click on datum plane Right.

11. Again, the arrow should point into the screen. Use *Flip* if it does not.

12. Select the *Sketch* button .

13. Sketch a "U" shape as shown in Figure 5.2 using the line tool .

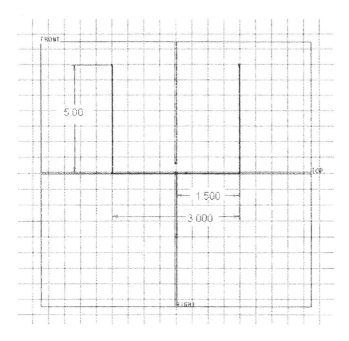

FIGURE 5.2 *The sketch used in generating the base feature for the U Bracket.*

FRONT

5.00

1.500

3.000

RIGHT

FIGURE 5.3 *Note the fill direction for the thin feature.*

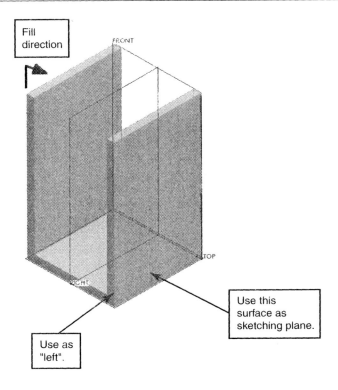

Fill direction

FRONT

Use this surface as sketching plane.

Use as "left".

14. Use the *Dimension* button and dimension the sketch as shown in the Figure 5.2.
15. Build the section by selecting ✔.
16. Press the *Fill direction* button until the fill direction is inward as shown in Figure 5.3.
17. Press the pull-down below the *Options* button in the dashboard and choose *Symmetric* ⬦.
18. Enter a depth of 3.00.
19. Build the section by pressing ✔.
20. Select.
21. Set the depth to *Through all*.
22. Select.
23. Choose the *Sketch* button in the dashboard.
24. With your mouse, choose the surface shown in Figure 5.3 as the sketching plane.
25. Use for the *Orientation* field and change to "left." Then, click on the surface shown in Figure 5.3.
26. Again, the arrow should point into the screen. Use *Flip* if it does not.
27. Select the *Sketch* button.
28. Use and sketch the three lines shown in Figure 5.4.
29. Use and sketch the arc shown in Figure 5.4.
30. Now, choose the *Dimension* tool and dimension the sketch as shown in Figure 5.4.

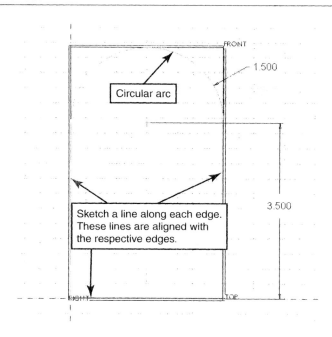

FIGURE 5.4 *Sketch the geometry shown for the cut.*

31. Press the *Modify* button ⬚, select each dimension, and change the values to those shown in Figure 5.4.

32. Build the section by pressing ✔.

33. Now, the software will either remove material from the outside or the inside of the sketch. This side may be changed by using ⬚ (next to ⬚) as necessary.

34. Furthermore, the direction of the extruded cut should be through the model. You may need to change this direction. Use the ⬚ to the left of ⬚ for this purpose.

35. Press ⬚ in the dashboard to build the model.

36. Save the part by pressing ⬚ .

5.1.2 TUTORIAL 5.2: A WASHER WITH A SLOT

This tutorial illustrates:

- Constructing a slot as a cut

1. Consider Figure 5.5, the part to be constructed. Now, select ⬚ .

2. Make sure that the *Use default template* option is checked.

3. Enter the name "Washer_T52" in the *Name* field.

4. Select *OK*.

5. Select ⬚ .

6. Choose the *Sketch* button ⬚ in the dashboard.

7. With your mouse, choose datum plane Front as the sketching plane.

8. If necessary, use ⬚ for the *Orientation* field and to "right." Then, click on datum plane Right.

9. Again, the arrow should point into the screen. Use *Flip* if it does not.

FIGURE 5.5 *The completed model of the washer.*

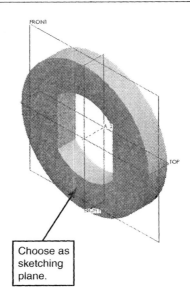

Choose as sketching plane.

10. Select the *Sketch* button ⬚ .
11. Sketch the first circle shown in Figure 5.6.
12. Select the diameter dimension and change the diameter to 0.438.
13. Build the section by pressing ✔ .
14. Choose *Symmetric* ⬚ .
15. Enter a depth of 0.063.
16. Press ✔ in the dashboard to build the model.
17. Now, select ⬚ once again.
18. Select ◁ .
19. Choose a depth of *Through all* ⬚ .

FIGURE 5.6 *Draw a circle.*

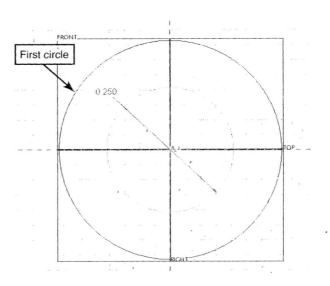

First circle

0.250

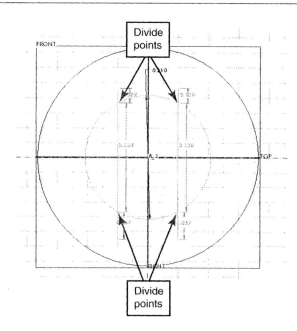

20. Select the *Sketch* button 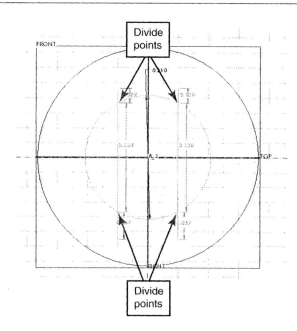.
21. Choose the surface shown in Figure 5.5 as the sketching plane.
22. If necessary, use ⬚ for the *Orientation* field and to "right." Then, click on datum plane Right.
23. Use ⬚ and sketch the second circle shown in Figure 5.6.
24. Double-click on the diameter dimension and change the value to 0.25.
25. Use ⬚ and sketch the two lines as shown in Figure 5.7.
26. Select the ⬚ next to the *Trim* icon ⬚.
27. Now pick the *Divide* icon ⬚.
28. Select the intersections of the lines and circles as shown in Figure 5.7.
29. Now, press the *Select* button ⬚.
30. While pressing the control key on your keyboard, use your mouse to pick the unwanted pieces of the lines and circles (use Figure 5.8 as a reference).
31. If necessary, use the *Dimension* tool ⬚ and dimension the sketch as shown in Figure 5.8
32. Press the *Modify* button ⬚, select each dimension, and change the values to those shown in Figure 5.8.
33. Build the section by pressing ⬚.
34. Now, remove material from the inside of the sketch. This side may be changed by using ⬚ (next to ⬚) as necessary.
35. Furthermore, change the direction of the extruded cut, if necessary, by using ⬚ to the left of ⬚.
36. Press ⬚ in the dashboard to build the model.
37. Save the part by pressing ⬚.

FIGURE 5.8 *Dimension the cut to the values given in this figure.*

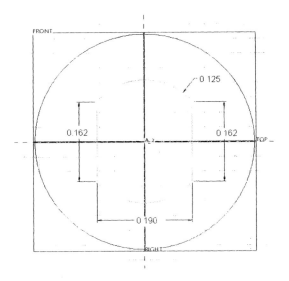

5.1.3 TUTORIAL 5.3: A CUT FOR THE PROPELLER

This tutorial illustrates:

- Constructing a thin cut

1. If the model "Propeller" is not already in memory, select and double-click on the file.
2. Add an annular cavity to the surface shown in Figure 5.9 by creating a cut. Select .

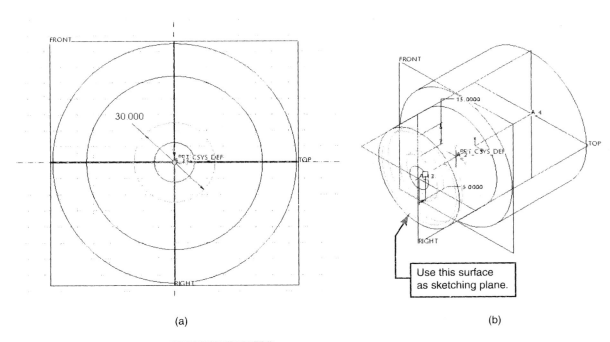

(a)   (b)

FIGURE 5.9 *The cavity in the propeller hub is generated by drawing a circle as in (a). In (b), note the geometry of the cavity.*

3. Select ⬛. Enter a depth of 5.

4. Select ⬛. Select ⬛. Enter a thickness of 15.

5. Choose the *Sketch* button ⬛ in the dashboard.

6. With your mouse, choose the surface shown in Figure 5.9(b) as the sketching plane. Select datum plane Right as "right." The arrow should point into the screen. Use *Flip* if it does not.

7. Select the *Sketch* button ⬛.

8. Sketch a circle as shown in Figure 5.9(a).

9. Double-click on the dimension and change the value to 30.

10. Build the section by pressing ⬛.

11. Press ⬛ in the dashboard to build the model.

12. Save the part by pressing ⬛.

5.2 ADDING A NECK

A *Neck* is a revolved slot; the angle of rotation is defined by the user. Rotation of the slot is performed around a centerline that the user must add to the sketch when creating the geometry of the slot. All neck features are sketched with *Open* sections.

Make sure that the *Neck* option is available on your version of Pro/E. By using the path *Tools* and *Options* you may check whether or not the *Neck* as well as other Advanced features are active. These options will be active if the option allow_anatomic_features is set to "yes." If it is not set to "yes" simply scroll down the list until the option is found. Then, select it. Enter "yes" in the value field and then press ⬛.

5.2.1 TUTORIAL 5.4: AN ADJUSTMENT SCREW

This tutorial illustrates:

- Constructing a neck
- Use of an advanced feature option

1. Consider Figure 5.10, the part to be constructed. Now, select ⬛.

2. Make sure that the *Use default template* option is checked.

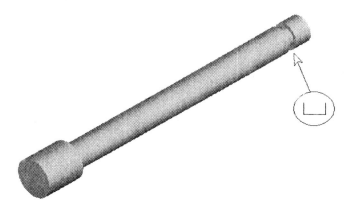

FIGURE 5.10 *The figure shows a blown-up view of the slot geometry.*

3. Enter the name "AdjustmentScrew" in the *Name* field.

4. Select *OK*.

5. Select ▣.

6. Choose the *Sketch* button ▣ in the dashboard.

7. With your mouse, choose datum plane Front as the sketching plane.

8. If necessary, use ▾ for the *Orientation* field and to "right." Then, click on datum plane Right.

9. Again, the arrow should point into the screen. Use *Flip* if it does not.

10. Select the *Sketch* button ▭.

11. Use ◯ and sketch a circle whose center is aligned with datum planes Right and Top.

12. Double-click on the diameter dimension and change the value to 0.750.

13. Build the section by pressing ✔.

14. Press the pull-down below the *Options* button in the dashboard and choose *Blind* ⬇.

15. Enter a depth of 0.8. Note the direction of the extrusion in Figure 5.10. If necessary, change the direction by using ▨.

16. Press ✔ in the dashboard to build the model.

17. Select ▣.

18. Choose the *Sketch* button ▣ in the dashboard.

19. With your mouse, choose the surface shown in Figure 5.11 as the sketching plane.

FIGURE 5.11 *The base features for the adjustment screw part.*

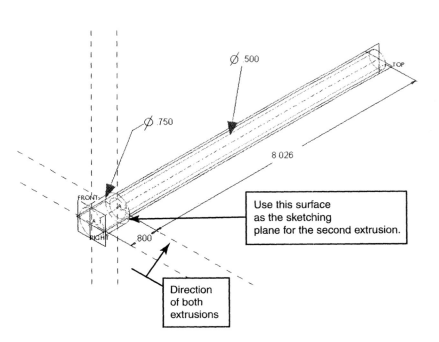

20. If necessary, use  for the *Orientation* field and to "right." Then, click on datum plane Right.

21. Again, the arrow should point into the screen. Use *Flip* if it does not.

22. Select the *Sketch* button .

23. Use and sketch a circle whose center is aligned with datum planes Right and Top.

24. Double-click on the diameter dimension and change the value to 0.500.

25. Build the section by pressing .

26. Press the pull-down below the *Options* button in the dashboard and choose *Blind* .

27. Enter a depth of 8.026. Note the direction of the extrusion in Figure 5.11. If necessary, change the direction by using .

28. Press in the dashboard to build the model.

29. Choose *Insert, Advanced,* and *Neck*.

30. Choose a rotation of 360 from the *Angle* menu (Figure 5.12). Select Done.

31. We will use datum plane Right as the sketching plane. So select datum plane Right and then *OK*.

32. Now, choose *Top* from the menu and then click on datum plane Top.

33. Press next to and change the line type to *Centerline* .

34. Draw a vertical centerline.

35. Now, change the line type back to *Line* type .

36. Using the line tool draw the three lines in Figure 5.13. Align the endpoints of the lines, as shown, to the silhouette of the edge.

37. Now, choose the *Dimension* tool and dimension the sketch as shown in Figure 5.13.

38. Press the *Modify* button and change the value of each dimension to the value shown in Figure 5.13.

39. Build the section by pressing the check mark in the Sketcher .

40. Press in the dashboard to build the model.

41. Save the part by pressing .

FIGURE 5.12 *Revolve the cut by using the values from the menu.*

FIGURE 5.13 *Draw the open section of the neck by using three lines.*

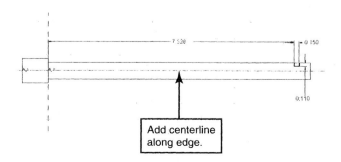

Add centerline along edge.

5.3   USING THE SHELL OPTION

The *Shell* option is another built-in Pro/E menu command for removing material from a base feature. This option removes a given surface and the material below it to create a hollow shell of specified thickness. It is important to remember that the *Shell* option removes material from all the features that exist on the model. Thus, if you do not want a certain feature to be shelled, you need to add that feature after shelling the rest of the model. If the feature cannot be added after shelling, the order in which it appears in the model tree must be changed. We will examine how to the change the order of which a feature is created in Chapter 13.

To see the effect of the order in which a shell is created, consider the emergency light housing shown in Figure 5.14. The design intent of the model is such that upon assembly, a washer and hexagonal nut will be placed within a group of four bosses and secure the housing to a mounting bracket.

We may easily generate each boss by first placing four holes in the recessed panel. This will add four cylindrical features to the model, each of which will become a boss when the shell process is performed (Figure 5.14(b)).

If, on the other hand, the holes are added after the shell process is performed, the holes do not exist in the model. Hence, no thickness is added to the cylindrical surface of the holes and therefore no boss is generated. What we end up with is a shelled base feature with four holes in it rather than four bosses (Figure 5.14(c)).

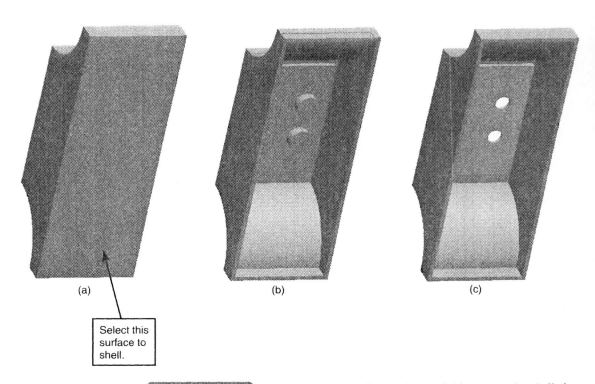

(a)   (b)   (c)

Select this surface to shell.

FIGURE 5.14 *The base is shown. In (a), the model has yet to be shelled. In (b), shelling removes material from the model and bosses are generated but only if the holes are added to the model before shelling. In (c), the holes were added after shelling and bosses are not created.*

5.3.1 TUTORIAL 5.5: AN EMERGENCY LIGHT COVER

This tutorial illustrates:

- Constructing a cut
- Generating a boss
- Using the shell process

1. Consider Figure 5.15, the part to be constructed. Now, select 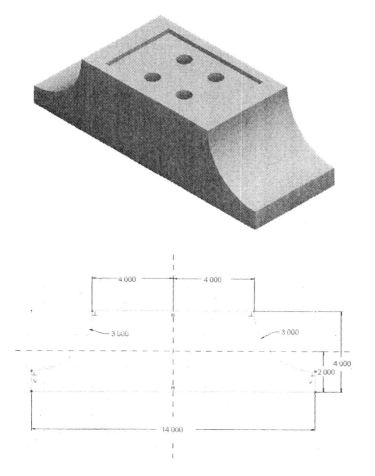.
2. Make sure that the *Use default template* option is checked.
3. Enter the name "EmergencyLightBase" in the *Name* field.
4. Select *OK*.
5. Select.
6. Choose the *Sketch* button in the dashboard.
7. With your mouse, choose datum plane Front as the sketching plane.
8. If necessary, use for the *Orientation* field and to "right." Then, click on datum plane Right.
9. Again, the arrow should point into the screen. Use *Flip* if it does not.
10. Select the *Sketch* button.
11. Use and and generate the geometry shown in Figure 5.16.

FIGURE 5.15 *This shaded model of the emergency light base shows the recessed surface containing the bosses*

FIGURE 5.16 *The geometry needed to construct the emergency light base.*

12. Now, choose the *Dimension* tool 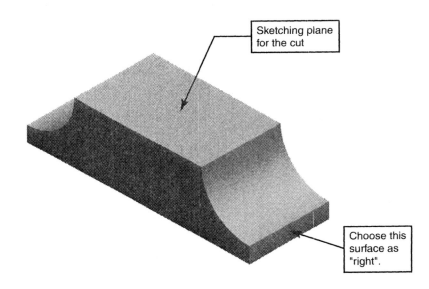 and dimension the sketch as shown in Figure 5.16.
13. Press the *Modify* button and change the value of each dimension to the value shown in Figure 5.16.
14. Build the section by pressing the check mark in the *Sketcher*.
15. Now, construct the recessed surface using a cut. Select once again.
16. Select.
17. Choose a depth of *Blind*. Enter a depth of 0.25.
18. Select the *Sketch* button.
19. Choose the surface shown in Figure 5.17 as the sketching plane.
20. Use for the *Orientation* field and to "right." Then, click on the surface shown in Figure 5.17.
21. Use the *Rectangle* tool and generate the geometry shown in Figure 5.17.
22. Now, choose the *Dimension* tool and dimension the sketch as shown in Figure 5.18.

FIGURE 5.17 *The model of the emergency light base after the first protrusion.*

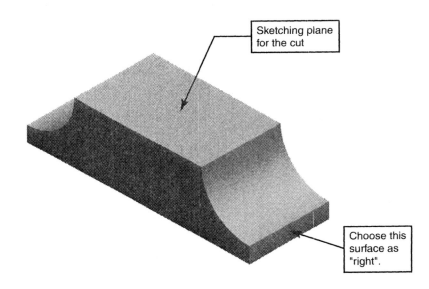

Sketching plane for the cut

Choose this surface as "right".

FIGURE 5.18 *The geometry of the cut is a rectangle.*

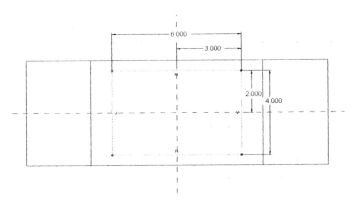

6.000
3.000
2.000
4.000

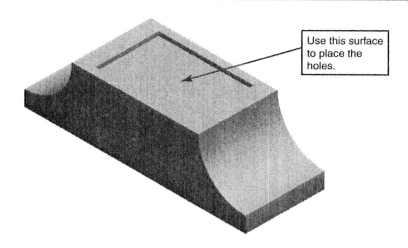

Use this surface to place the holes.

FIGURE 5.19 *Using the* Cut *option to remove material yields a rectangular recessed surface.*

23. Press the *Modify* button ⊠ and change the value of each dimension to the value shown in Figure 5.18.

24. Build the section by pressing ☑.

25. Add the four holes. You will need to repeat this process four times. Select ☑.

26. Select *Straight, Simple,* and *Blind.* Enter a diameter of 0.75 and a depth of 0.5.

27. Use the recessed surface shown in Figure 5.19 as the placement surface.

28. Drag the reference handles to the appropriate edges as desired. Use the dimensions in Figure 5.20 as a reference.

29. Build the hole feature by pressing the check mark in the *Sketcher* ✔.

30. Shell the model. Select ⊟.

31. Enter a thickness of 0.2.

32. Click on the surface shown in Figure 5.14(a).

33. Build the shell feature by pressing the check mark in the *Sketcher* ☑.

34. Finally, add a coaxial hole through each boss. Select ☑.

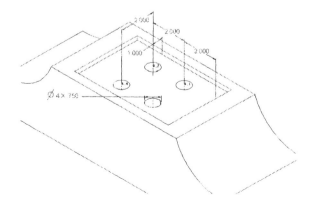

FIGURE 5.20 *The model is shown with the holes placed.*

FIGURE 5.21 *The placement plane for each hole is the top of the boss.*

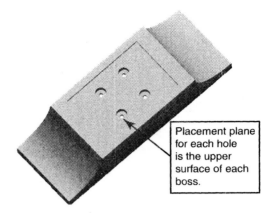

Placement plane
for each hole
is the upper
surface of each
boss.

35. Select *Straight, Simple,* and *Through All.* Enter a diameter of 0.25.

36. Click on the axis running through the desired boss.

37. Select *Placement* and click in the Secondary references cell. Use the upper surface of each boss (Figure 5.21) as the placement surface.

38. Build the feature by pressing ✔.

39. Repeat the process for the remaining three holes.

40. Save the part by pressing 🖫.

5.4 SUMMARY

In this chapter, we discussed the options *Cut, Neck*, and *Shell.* These options are used to remove material from a model.

A *Cut* is material removed from a given side of a base feature. A *Neck* is a revolved slot. The geometry of the neck is sketched using an *open* section. Then, the section is revolved about an axis of revolution.

The *Shell* option is used to remove material below a given surface. Because the *Shell* option removes material from all the features that exist on the model, this option may be used to generate a boss. In order to generate a boss, a hole must be placed on the model before it is shelled.

5.4.1 STEPS FOR USING THE CUT OR SLOT OPTIONS

The following sequence of steps may be used to create a *Cut*:

1. Select 🗗 or any feature tool.

2. Choose ◿.

3. Choose the direction of the cut using ⬚—that is, inside or the outside the sketch.

4. Enter the depth of the cut and the corresponding direction using ⬚.

5. In the Sketcher, sketch the section.

6. Dimension to the proper dimensional values. Build the section using ✔.

7. Build the feature using ✔.

5.4.2 STEPS FOR ADDING A NECK TO A BASE FEATURE

In general, a *Neck* may be created on a base feature by using the following steps:

1. Select *Insert, Advanced,* and *Neck*.
2. Choose the angle of rotation from the *ANGLE* menu.
3. Select *Done*.
4. Pick the sketching plane and an reference plane.
5. In the sketcher, sketch the section using an open section.
6. Dimension and modify to the appropriate values.
7. Build the section using ✔.
8. Build the feature using ✔.

5.4.3 STEPS FOR USING THE SHELL OPTION

The general sequence of steps for using the *Shell* option are:

1. Select ⊡.
2. Pick the shell surface.
3. Enter the thickness of the shell.
4. Build the feature using ✔.

5.5 ADDITIONAL EXERCISES

The problems chosen for this section require the use of the material-removal options discussed in this chapter. The base features will require the use of options from previous chapters in this book.

Construct the models as assigned. Save your work unless instructed not to do so.

5.1 Construct the guide plate. Use Figure 5.22 for reference.

FIGURE 5.22

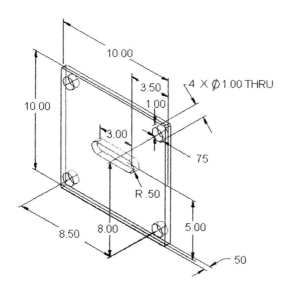

5.2 Add the cut to the side guide (see also Exercises 3.8 and 4.10). Use Figure 5.23 for the geometry of the cut.

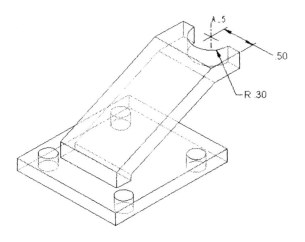

5.3 Construct the slotted hold-down. The geometry of this model is shown in Figure 5.24.

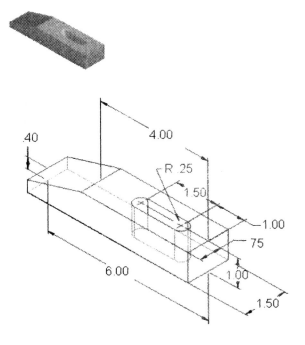

5.4 Create the model of the slide base illustrated in Figure 5.25.

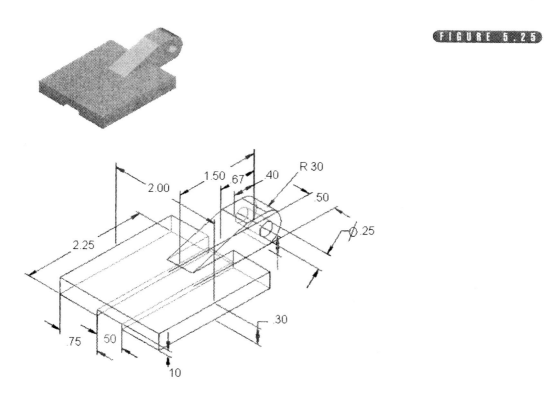

FIGURE 5.25

5.5 Use the *Neck* option to construct the groove on the V-pin shown in Figure 5.26.

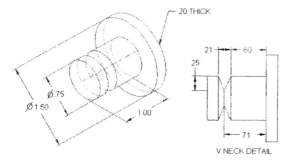

FIGURE 5.26

5.6 Construct the model of the coupling. Use Figure 5.27 for reference.

FIGURE 5.27

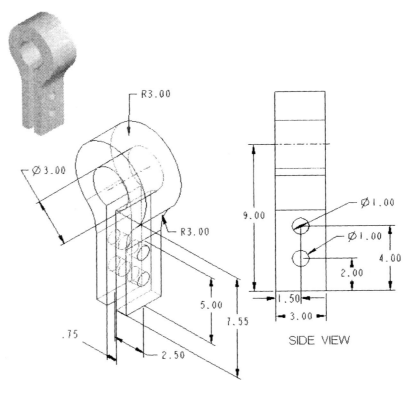

5.7 Create the model of the pipe clamp top component. The geometry is given in Figure 5.28.

FIGURE 5.28

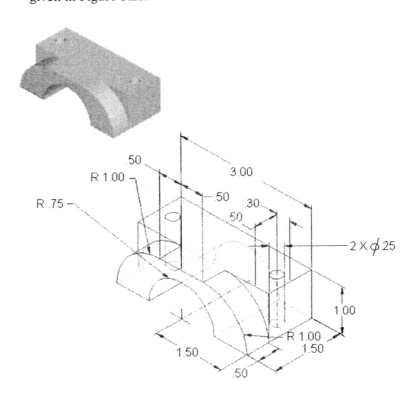

5.8 Construct the model of the pipe clamp bottom component. Use the geometry given in Figure 5.29.

FIGURE 5.29

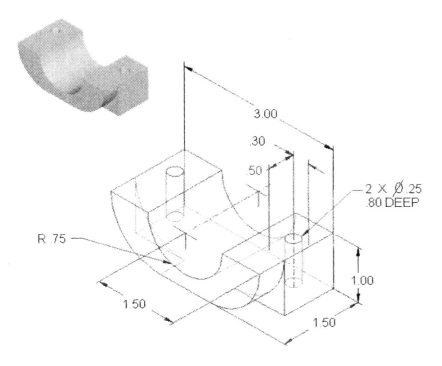

5.9 Use the *Protrusion* and *Shell* options to create the model of the robot arm housing. The geometry of the model is given in Figure 5.30.

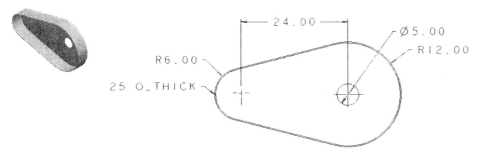

FIGURE 5.30

5.10 Use the *Shell* option to construct the bosses on the model of the electric motor cover. The geometry of the model is shown in Figure 5.31.

A shaded model of motor cover   The model rotated to show boss detail

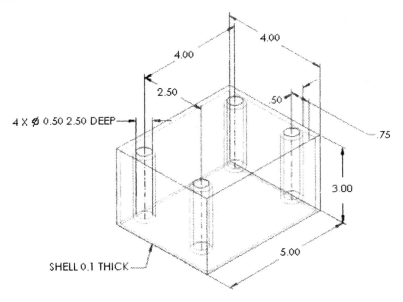

FIGURE 5.31

5.11 Add the slot shown in Figure 5.32 to the slotted support (see also Exercise 4.12).

FIGURE 5.32

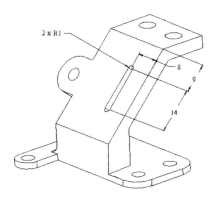

5.12 Use *Cut* to generate a short taper in the support from Exercise 4.15.
      Use Figure 5.33 (compare Figure 5.33 with Figure 4.40).

5.13 Generate a short taper to the support from Exercise 4.16 using
      Figure 5.33 (compare Figure 5.33 with Figure 4.40).

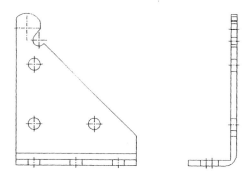

FIGURE 5.33

CHAPTER ■ 6

OPTIONS THAT ADD MATERIAL: FLANGE, RIB, AND SHAFT

INTRODUCTION AND OBJECTIVES

This chapter deals with options that may be used to add a feature to a base feature. The options discussed in this chapter are *Flange, Rib,* and *Shaft.*

A flange is a revolved protrusion. Flanges are often used in joints as load-bearing surfaces. Ribs are thin sections that act as structural elements. Shafts are revolved sections that are often used to align parts or as threaded studs.

The objectives of this chapter are to:

1. Attain an ability to construct a flange.
2. Place a shaft with varying cross sections on a base feature using the *Shaft* option.
3. Utilize the *Rib* option to place a thin sections between two base features.

6.1   FLANGES

A *Flange* is the opposite of a neck. In creating a flange, material is added by revolving a sketched section around a centerline. Consider again the adjustment screw in Figure 5.10. The larger cylinder could have been constructed by creating a flange around the smaller diameter cylinder. Of course, you would have had to construct the smaller-cylinder longer. That is, the length of the cylinder would be 8.026 + 0.800 = 8.826. Then the flange could be created along one end.

The procedure for creating a flange is the same as that for creating a neck. Flanges, like all necks, are sketched with an *open* section. A centerline, which becomes the axis of revolution, must be defined.

Make sure that the *Flange* option is available on your version of Pro/E. By using the path *Tools* and *Options* you may check whether or not the

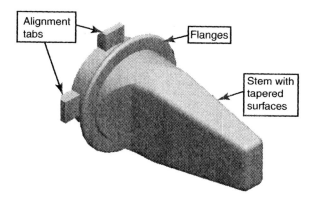

Alignment tabs

Flanges

Stem with tapered surfaces

FIGURE 6.1 *The lamp holder to be constructed in tutorial 6.1.*

Flange as well as other Advanced features are active. These options will be active if the option allow_anatomic_features is set to "yes." If it is not set to "yes" simply scroll down the list until the option is found. Then, select it. Enter "yes" in the value field and then press ▨.

As an example in illustrating the construction of a flange, consider the lamp holder shown in Figure 6.1. The holder is used in an emergency light. The holder fits into the base of a parabolic reflector and is locked into place by twisting the holder to engage the alignment tabs. The flange keeps the holder in place.

The construction of the flange begins with the creation of the base feature—in this case, a right circular cylinder. A series of cuts is used to generate the taped surfaces.

6.1.1 TUTORIAL 6.1: A LAMP HOLDER

This tutorial illustrates:

- Generating a tapered surface by using a series of cuts
- Construction of an extrusion on a curvilinear surface
- Creation of a flange

1. Select ▨.
2. Make sure that the *Use default template* option is checked.
3. Enter the name "EmergencyLightHolder" in the *Name* field.
4. Select *OK*.
5. Select ▨.
6. Choose the *Sketch* button ▨ in the dashboard.
7. With your mouse, choose datum plane Right as the sketching plane.
8. If necessary, use ▨ for the *Orientation* field and to "top." Then, click on datum plane Top.
9. Again, the arrow should point into the screen. Use *Flip* if it does not.
10. Select the *Sketch* button ▨.
11. Sketch a circle whose center is aligned with datum planes Top and Front. Modify the diameter dimension to 0.718.
12. Build the section by pressing ▨.

FIGURE 6.2 *The base feature for the lamp holder is a right circular cylinder.*

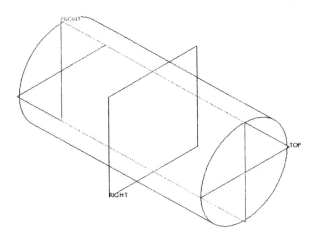

13. Extrude the section using 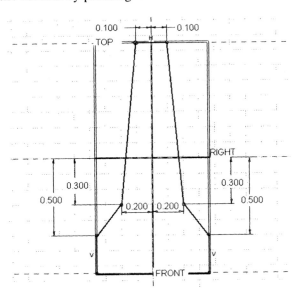. Enter a depth of 1.500.
14. Press ☑ in the dashboard to build the model. The model at this point is shown in Figure 6.2.
15. Select ▣.
16. Select ▢.
17. Choose *Symmetric* ⊟ for the cut. Enter 0.718 for the depth.
18. Select the *Sketch* button ▭.
19. Select datum plane Top as the sketching plane. Use datum plane Right as the "top" orientation. The arrow pointing outward from the model.
20. Draw the geometry shown in Figure 6.3 using the ◣.
21. Use ⊟ and dimension your sketch so it corresponds to the one shown in Figure 6.3.
22. Use the *Modify* tool ◪ and modify the dimensional values to the values given in Figure 6.3.
23. Build the section by pressing ☑.

FIGURE 6.3 *The sketch required for constructing the first cut.*

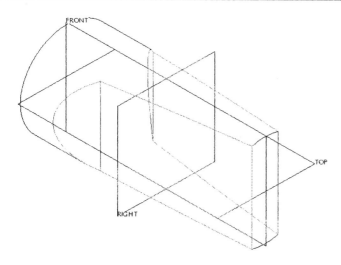

FIGURE 6.4 *The model after the first cut has a taper along the stem.*

24. Use the *Preview* tool 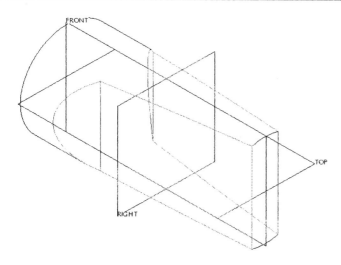 and check the direction of the cut. The cut should be outside of the sketch. Use ⚡ to change the direction if necessary.

25. Press ✔ in the dashboard to build the model. The model at this point is shown in Figure 6.4.

26. Once again, select 🔲 and then select ◿.

27. Choose *Symmetric* 🔳 for the cut. Enter 0.718 for the depth.

28. Select the *Sketch* button 🔲.

29. Select datum plane Front as the sketching plane. Use datum plane Top as the "top" orientation. The arrow should point outward from the model.

30. Draw the geometry shown in Figure 6.5 using the ◣.

31. Use 🔳 and dimension your sketch so it corresponds to the one shown in Figure 6.5.

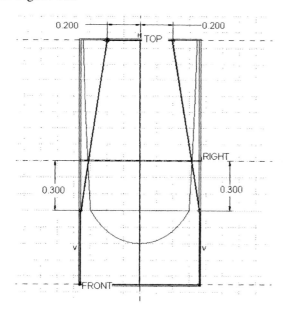

FIGURE 6.5 *The geometry for the second cut is given in this figure.*

FIGURE 6.6 *After using the geometry to cut the model a second time, the model takes on this shape.*

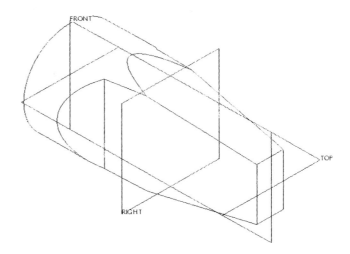

32. Use the *Modify* tool ⬚. Change the dimensional values to the values given in Figure 6.5.
33. Build the section by pressing ✔.
34. Use the Preview tool ⬚ and check the direction of the cut. The cut should be outside of the sketch. Use ⬚ to change the direction if necessary.
35. Press ⬚ in the dashboard to build the model. The model at this point is shown in Figure 6.6.
36. Now, construct the first of the two tabs. Select ⬚.
37. Choose *Symmetric* ⬚. Enter 0.185 for the depth.
38. Select datum plane Front as the sketching plane. Use datum plane Top as the "top" orientation. The arrow should point outward from the model.
39. Draw the geometry shown in Figure 6.7 using the ⬚. Note that the geometry extends 0.010 into the base feature. This is so the tab will blend into the curvilinear surface when it is extruded.

FIGURE 6.7 *The geometry of the first alignment tab with the appropriate dimensions.*

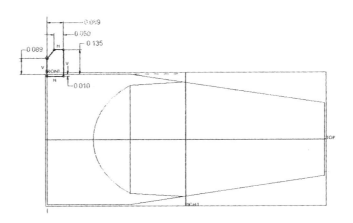

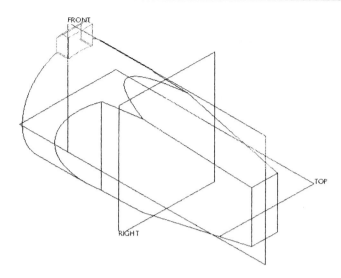

FIGURE 6.8 *The model after the first alignment tab has been added.*

40. Use ▣ to dimension the sketch. Use Figure 6.7 as a guide.

41. Use ▣. Change the dimensional values to the values given in Figure 6.7.

42. Build the section by pressing ✓.

43. Press ✓ to build the model. The model at this point is shown in Figure 6.8. The second tab will be added in Chapter 8 using the *Copy* option.

44. Now, add the Flange. Choose *Insert, Advanced,* and *Flange.*

45. From the *Options* menu, select 360° and One Side. Then, click on *Done.*

46. Select datum plane Front as the sketching plane. Choose Okay. Select "Top" from the menu and then click on datum plane Top.

47. Press ▣ next to ▣ and change to ▣. Add a centerline as shown in Figure 6.9.

48. Sketch the geometry shown in Figure 6.9 using the *Line* and *Arc* tools. Note that this is an open section.

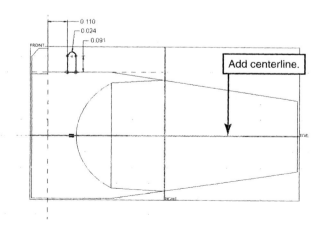

FIGURE 6.9 *The geometry of the flange consists of two vertical lines and an arc between them. Notice the location of the centerline.*

FIGURE 6.10 *After regenerating, this shaded model of the connecting arm can be created. The rib is the slender blended rectangular feature between the two cylinders.*

Rib

49. Use  and dimension to the values given in Figure 6.9.
50. Use and modify the values to the ones given in Figure 6.9.
51. Build the section by pressing .
52. Save the part by pressing .

6.2 RIBS

A rib or web is a thin protrusion attached to a base feature. A classic example of a part with a rib is the connecting arm shown in Figure 6.10. The base feature for the arm is constructed by making two cylinders of unequal diameter.

6.2.1 TUTORIAL 6.2: A CONNECTING ARM

This tutorial illustrates:

- Constructing a rib
- Cutting a rib
- Generating two features in one step

1. Select *File* and *New* (or simply).
2. Make sure that the *Use default template* option is checked.
3. Enter the name "ConnectingArm" in the *Name* field.
4. Select *OK*.
5. Select *Insert* and *Extrude* (or simply).
6. Choose the *Sketch* button in the dashboard.
7. With your mouse, choose datum plane Front as the sketching plane.
8. If necessary, use for the *Orientation* field and to "right." Then, click on datum plane Right.
9. Again, the arrow should point into the screen. Use *Flip* if it does not.

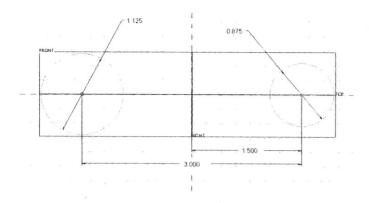

FIGURE 6.11 *Use this sketch of the two circles in the Sketcher.*

10. Select the *Sketch* button .
11. In the Sketcher, sketch the geometry shown in Figure 6.11. Draw both circles using 🔘.
12. Dimension using 📐 as shown in Figure 6.11.
13. Use 📝 and modify the dimensional values. Use Figure 6.11 as a guide.
14. Build the section by pressing ✔.
15. Choose *Symmetric* 🔲. Enter a depth of 0.625.
16. Build the model by pressing ✔.
17. Add the rib. Select *Insert* and *Rib* or press 🔲 in the toolbar.
18. Enter a thickness of 0.438. Press 📝.
19. Choose the datum plane Front as the sketching plane. Then select datum plane Top as the "top."
20. Use ❌ and sketch the geometry in Figure 6.12. Note that this is an open section.
21. Build the section by pressing ✔.
22. Fill the rib inward. Change the direction, if necessary, by using 🔲.

FIGURE 6.12 *Sketch showing the proper positioning of the lines to create the rib.*

FIGURE 6.13 *Connecting arm with added rib.*

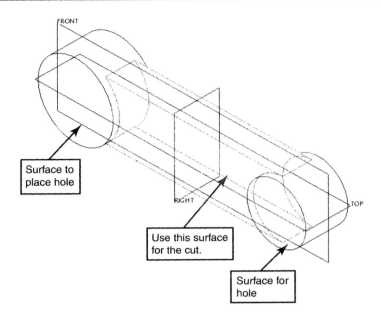

23. Press ✔ in the dashboard to build the model.

24. Add the coaxial holes to each cylindrical protrusion. Start with the cylinder on the left-hand side of the model. Select ☒. Then, select *Straight, Simple,* and ☷. Enter a diameter of 0.560.

25. Click on the axis of the desired cylinder. Choose *Placement* and click in the *Secondary references* field, then select the corresponding surface shown in Figure 6.13.

26. Press ✔ in the dashboard to build the model.

27. Repeat steps 24 through 26 for the right-hand-side cylinder. Use a diameter of 0.440 and the placement surface shown in Figure 6.13.

28. Select ☐ to add the 0.077 × 0.126 keyway in Figure 6.14. Then, select ◁.

29. Choose ⊥ for the cut. Enter 0.100 for the depth.

30. Select ☐ and datum plane Front as the sketching plane. Use datum plane Right as the "right" orientation.

31. Draw the geometry shown in Figure 6.14 using ☒.

FIGURE 6.14 *The 0.077 × 0.126 keyway may be added to the connecting arm with the geometry shown.*

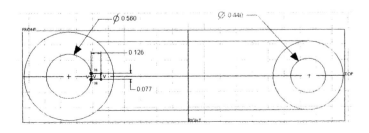

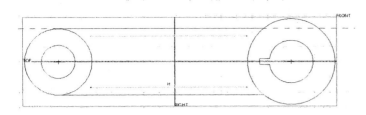

FIGURE 6.15 *The connecting arm in the sketcher with the horizontal lines located.*

32. Consulting Figure 6.14, use and dimension your sketch. Use to modify the values.

33. Select .

34. The cut should be inside of the sketch. Use to change the direction if necessary.

35. Press to build the model.

36. Complete the rib by adding a cut. Once again, select and .

37. Select and the surface shown in Figure 6.13 as the sketching plane. Use datum plane Top as the "top" orientation.

38. Use and sketch the two lines shown in Figure 6.15.

39. Use and click on the circle shown in Figure 6.16. Then, drag the arc from one line endpoint to the other.

40. Repeat step 68 but use the circle shown in Figure 6.17.

41. Use and dimension as shown in Figure 6.18.

42. Modify the dimensional values to those shown in Figure 6.18 by using .

43. Build the section by pressing .

44. Change the direction of the cut, if necessary, by using .

45. Press to build the model.

46. Save the part by pressing .

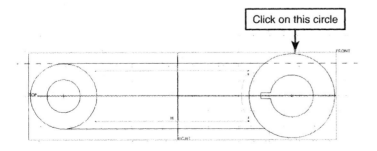

Click on this circle

FIGURE 6.16 *Add a concentric arc to the sketch.*

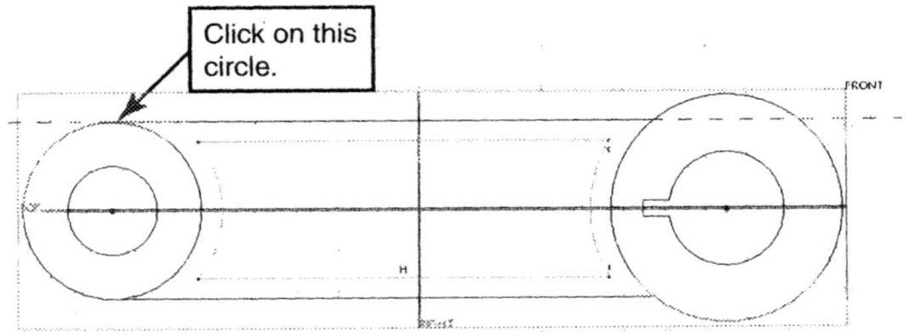

FIGURE 6.17 *Add a second concentric arc.*

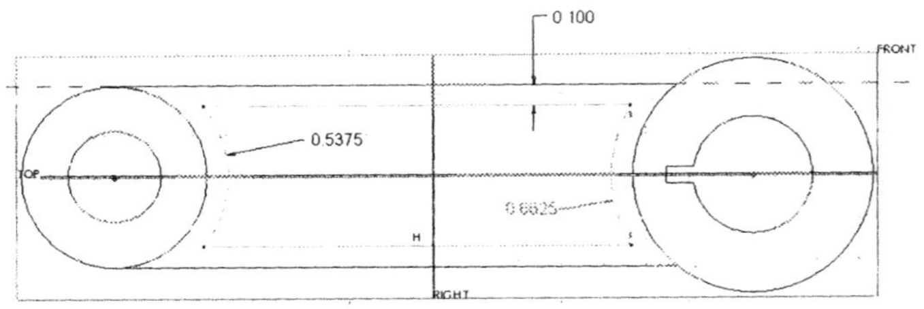

FIGURE 6.18 *Use these dimensions to properly generate the cut.*

6.3 SHAFTS

A shaft is a surface of revolution that extends from a placement surface. Shafts are created by sketching a cross section. This section must be closed. Then, the section is revolved about a centerline, which the Pro/E user must place in the sketch.

The shaft is located on the base feature using a method analogous to the placement of holes. That is, the shaft is located using the *Linear, Coaxial, On Point,* and *Radial* options discussed in Chapter 4.

Make sure that the *Shaft* option is available on your version of Pro/E. By using the path *Tools* and *Options* you may check whether or not the *Shaft* as well as other Advanced features are active. These options will be active if the option allow_anatomic_features is set to "yes." If it is not set to "yes" simply scroll down the list until the option is found. Then, select it. Enter "yes" in the value field and then press .

6.3.1 TUTORIAL 6.3: A SHAFT FOR THE U-BRACKET

This tutorial illustrates:

- Constructing a rib
- Cutting a rib
- Generating two features in one step

1. Retrieve the part "Ubracket" by using *File* and *Open* or .
2. Select *Insert, Advanced,* and *Shaft.*
3. Choose *Linear* and *Done* as your placement options.
4. In the Sketcher, draw the geometry shown in Figure 6.19. Do not forget to add a centerline . The section must be closed.

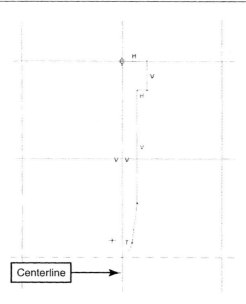

FIGURE 6.19 *Create a two-dimensional sketch of the shaft section using this geometry.*

Centerline

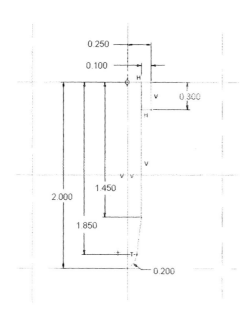

FIGURE 6.20 *The dimensions required for creating the shaft.*

5. Dimension using the dimensions given in Figure 6.20. Use ⬚. Modify the dimensional values using ⬚.

6. With your mouse, select the bottom of the base as the placement plane as shown in Figure 6.21(a).

7. The shaft is to be centered on the bottom of the part. Pick one of the two edges shown in Figure 6.21(b) and then enter 1.5.

8. Select the other edge and enter 1.5. Select *OK*. Your bracket should appear as shown in Figure 6.21(a).

9. Save the part using ⬚.

FIGURE 6.21 *After adding the shaft, a shaded model of the U-bracket with the shaft may be created.*

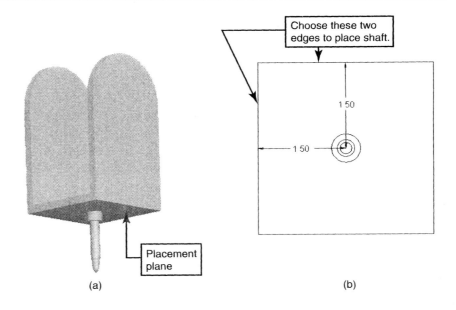

(a)

(b)

6.4 SUMMARY

In this chapter, we considered three options used to add material to a part. These options are *Flange, Rib,* and *Shaft.* A *Flange* is created by adding a revolved section around a centerline. Both the cross section and the centerline must be provided by the user. A *Rib* is a thin protrusion attached to one or more features. Ribs are constructed by sketching the section of the rib on a sketching plane and extruding the section to a given depth. A *Shaft* is a closed revolved section that extends from a surface. Shafts are located by the method used to locate holes. The user must sketch the profile of the shaft, as well as locate an axis of revolution.

6.4.1 STEPS FOR ADDING A FLANGE

A flange must be added to a base feature. In general, the sequence of steps for adding a flange is as follows:

1. Choose *Insert, Advanced,* and *Flange.*
2. Select the angle of revolution.
3. Choose the sketching plane and the orientation plane. Pro/E will load the model in the sketcher using you selections.
4. Sketch the geometry of the section. The section must be *open.*
5. Place a centerline. This centerline will be used as the axis of revolution.
6. Dimension the section.
7. Build the feature. The software will revolve the section.

6.4.2 STEPS FOR ADDING A RIB

After the base feature has been constructed, a rib may be added to the model. The general steps for adding the rib are:

1. Select *Insert* and *Rib* or .
2. Enter the thickness of the rib.

3. Select the sketching plane and the orientation plane.

4. After Pro/E has reoriented and displayed the model in the Sketcher, sketch the section.

5. Dimension and modify the section as needed.

6. Build the section. Then, select the fill direction.

7. Build the feature.

6.4.3 STEPS FOR ADDING A SHAFT

Like ribs and flanges, a shaft is added to a base feature. In general, the following sequence of steps may be followed in order to add a shaft to a model:

1. Select *Insert, Advanced,* and *Shaft.*

2. Pick the placement plane with your mouse.

3. Choose the placement option using the options from the *PLACEMENT* menu.

4. Pro/E will open a new window. Sketch the section of the shaft. Also, be sure to place a centerline.

5. Dimension and modify the section.

6. Build the section.

7. Place the shaft by entering or selecting the appropriate features as required by your selection in step 3.

8. Build the feature.

6.5 ADDITIONAL EXERCISES

The exercises in this section can be constructed using the options discussed in this chapter. Create models of the various parts. For some parts, additional options from previous tutorials are necessary to complete the part. Save your work unless instructed not to do so.

6.1 Construct the two-sided bushing, using the *Flange* option. Use Figure 6.22 for the geometry of the model.

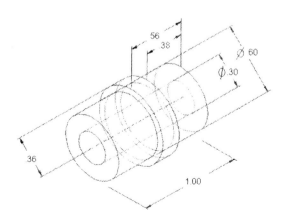

FIGURE 6.22

6.2 Use the *Flange* option and the geometry in Figure 6.23 to construct the model of the lock plunger.

FIGURE 6.23

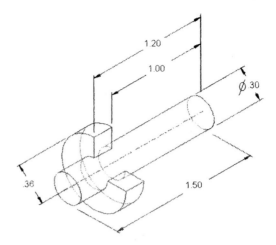

6.3 Construct the model of the lifter as shown in Figure 6.24.

FIGURE 6.24

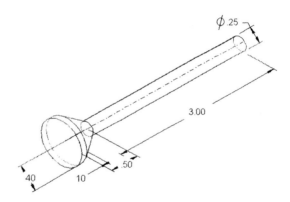

6.4 Create the model of the pin. Use the geometry in Figure 6.25.

FIGURE 6.25

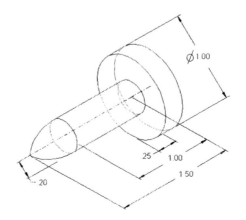

6.5 Use the geometry shown in Figure 6.26 to create the model of the cone head rivet.

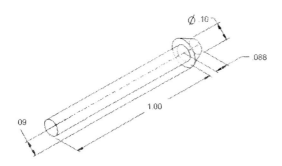

FIGURE 6.26

6.6 Using the geometry shown in Figure 6.27 create the model of the flat top countersunk rivet.

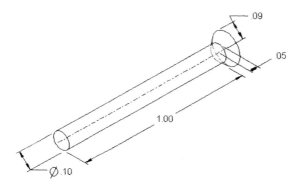

FIGURE 6.27

6.7 Construct the model of the valve poppet. Use the model shown in Figure 6.28.

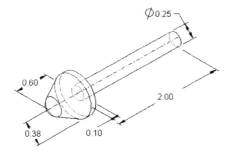

FIGURE 6.28

6.8 Use Figure 6.29 to create the model of the detent pin.

FIGURE 6.29

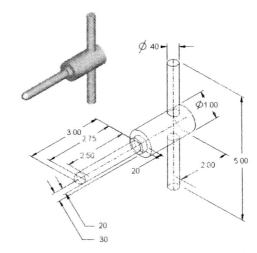

6.9 Construct the model of the coupler. Use Figure 6.30 as a reference.

FIGURE 6.30

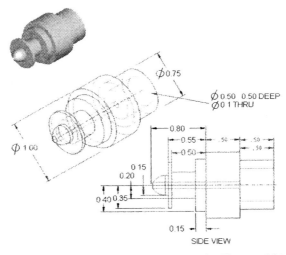

6.10 Create the model of the lift lever shown in Figure 6.31.

FIGURE 6.31

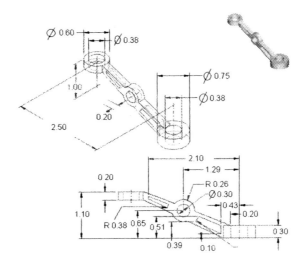

6.11 Construct the model of the angle bracket with side reinforcements.
      Use the geometry shown in Figure 6.32.

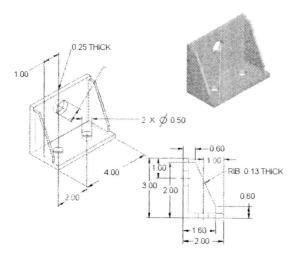

FIGURE 6.32

6.12 Use the geometry in Figure 6.33 to create the model of the single
      bearing bracket.

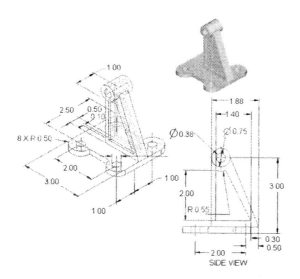

FIGURE 6.33

6.13 Using Figure 6.34, construct the model of the double bearing bracket.

FIGURE 6.34

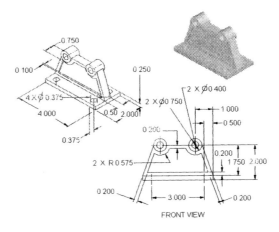

0.750

0 100

0 250

4 × ⌀ 0.375

2 × ⌀0 750

2 × ⌀0 400

4 000

0.50 2.000

1 000

0.500

0.375

0 200

2 × R 0 575

0.200

0.200

1 750 2.000

0 200

3.000

0 200

FRONT VIEW

CHAPTER 7

FILLETS, ROUNDS, AND CHAMFERS

INTRODUCTION AND OBJECTIVES

In this section, we will examine the features *Round (Fillet)* and *Chamfer*, which can be used to remove a sharp edge. These features remove material from a model.

Round and *Fillet* are essentially the same operation. One (fillet) removes material from an inside corner, while the other (round) removes material from an outside corner. The *Chamfer* option is used to bevel an edge.

In engineering practice, rounds are placed to remove sharp edges or at corners where the geometry leads to stress concentrations. For molded parts, the rounding or beveling of sharp corners allows the part to be removed from the mold easier.

The typical value for the radii of the round is 1/8″ (3 mm). Larger-radii rounds yield parts with better strength characteristics. However, it is not always possible to regenerate a part with larger-radii rounds due to possible interference between the round/fillet and other features. In general, it is best to place rounds/fillets on a model after all other features have been added.

If a part with a round/fillet fails to regenerate, the *Resolve* option may be used to investigate the failure. This option is discussed at some length in Chapter 9. Often the problem can be fixed by simply reducing the radius of the round/fillet or changing the order in which the rounds and/or fillet are placed.

The objectives of this chapter are to

1. Place rounds/fillets at an edge or between two surfaces
2. Place an edge chamfer

7.1 THE USE OF ROUNDS AND FILLETS

A *Round* is an *exterior* corner that has been given a curvature. The curvature is indicated by a small arc with a given radius. A *Fillet* is an *interior* corner that has been given a curvature. Like rounds, the size of the fillet is dimensioned with a radius.

In Pro/E, rounds and fillets may have a constant radius along their length, or the radius may vary, to produce a tapered round.

These options are shown in Figure 7.1 and are selected from the *Sets* option during the definition of the round geometry. A variable round is created by adding more than one radius. This is accomplished by right-mouse-clicking in the cell, selecting Add Radius, and then entering the new value.

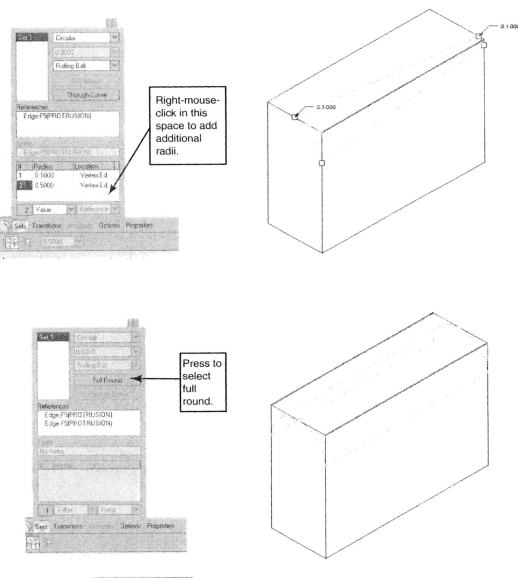

FIGURE 7.1 *For constructing a* round *or* fillet, *use the options in the dashboard.*

A *Full Round* is placed on planar surfaces bounded by two parallel edges or surfaces as shown in Figure 7.1. Select the edges and then press the *Full Round* button in the *Sets* option dialog box as shown.

Many parts contain multiple edges that must be rounded. If these edges require a round with the same radii, then the edges can all be rounded at the same time by selecting the edges to form a chain.

Often, parts with circular features have edges that must be rounded. Adding a round to a circular edge is as simple as adding a rounding to a linear edge. Since the circular edge is continuous around the part, it need be selected only once.

7.1.1 TUTORIAL 7.1: A ROUND AND FILLET FOR THE ANGLE BRACKET

This tutorial illustrates:

- Adding a round
- Adding a fillet

1. If the model "AngleBracket" is not already in memory, select and double-click on the file.
2. Select *Insert* and *Round* or simply .
3. Using Figure 7.2, select Edge A with your mouse.
4. Change the value of the round radius from the default to 0.125.
5. Press in the dashboard to build the model.
6. Press once again.
7. Now, using Figure 7.2 as a guide, select Edge B. You may wish to use your middle mouse button and rotate the model.
8. Change the value of the round radius from the default to 0.625.
9. Press in the dashboard to build the model.
10. Save the part by pressing .

7.1.2 TUTORIAL 7.2: EDGE ROUNDS FOR THE CONNECTING ARM

This tutorial illustrates:

- Adding a round to multiple edges

1. If the model "ConnectingArm" is not already in memory, select and double-click on the file.

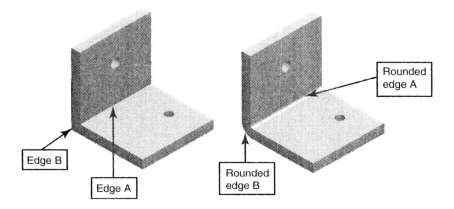

FIGURE 7.2 *The angle bracket with added round and fillet.*

Edge B

Edge A

Rounded edge A

Rounded edge B

FIGURE 7.3 *Round the edges on the rib.*

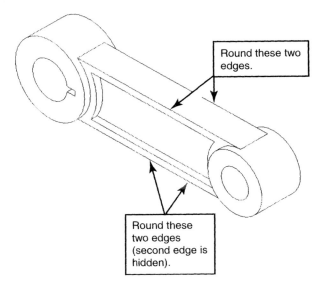

Round these two edges.

Round these two edges (second edge is hidden).

FIGURE 7.4 *The connecting arm with added rounds (compare with Figure 7.3).*

2. Select *Insert* and *Round* or simply ⬚.
3. With your mouse, click on the edges shown in Figure 7.3. There is a total of four edges and you may wish to rotate your model to access the hidden edges.
4. Enter a radius of 0.100.
5. Press ⬚ in the dashboard to build the model.
6. Save the part by pressing ⬚. The model with added rounds is shown in Figure 7.4.

7.1.3 TUTORIAL 7.3: A ROUND ON A CIRCULAR EDGE

This tutorial illustrates:

- Adding a round to circular edge

1. If the model "PipeFlange" is not already in memory, select ⬚ and double-click on the file.
2. Select ⬚.

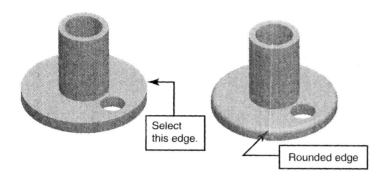

FIGURE 7.5 *The pipe flange with a 0.1-radius round.*

3. Using Figure 7.5, select the circular edge with your mouse.
4. Change the value of the round radius from the default to 0.100.
5. Press in the dashboard to build the model.
6. Save the part by pressing .

7.1.4 TUTORIAL 7.4: A PART WITH SURFACE TO SURFACE ROUNDS

This tutorial illustrates:

- Adding a surface to surface round

1. If the model "PipeFlange" is not already in memory, select and double-click on the file.
2. Select .
3. Using Figure 7.6, select the surface of the long cylinder.
4. Press the control (CTRL) key on your keyboard and then click on the surface of the disk.
5. Change the value of the round radius from the default to 0.050.
6. Press in the dashboard to build the model.
7. Save the part by pressing .

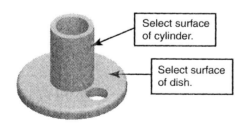

Select surface of cylinder.

Select surface of dish.

FIGURE 7.6 *The pipe flange with the added surface to surface round.*

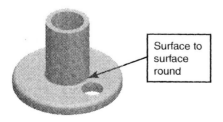

Surface to surface round

FIGURE 7.7 *The different ways to define a chamfer in Pro/E.*

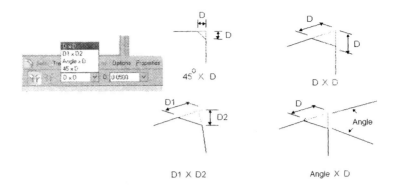

7.2 CHAMFERS

Sharp corners can also be removed by beveling the corner; such a bevel is called a *chamfer*. Chamfers are used at the ends of cylinders as well as on an edge. A chamfer that occurs on a corner is called a corner chamfer. Chamfers are specified by giving a linear and an angular dimension.

In Pro/E, for corners that are formed by surfaces at 90°, a feature called *45 × D* is available. For such a feature, the user need only give the linear dimension (the value D). For edges that are not at 90°, there are several options available. The first of these is the *D × D* chamfer. For this option, a chamfer is created by specifying the distance, D, from the edge along each surface. If the distance along each surface needs to be different, then use the *D1 × D2* option. Of course, you will need to give the program the values D1 and D2. Finally, the chamfer can be defined with an angle and linear distance. This is the *Angle × D* option. The options for dimensioning a chamfer are shown in Figure 7.7.

7.2.1 TUTORIAL 7.5: ADDING A CHAMFER TO THE ADJUSTMENT SCREW

This tutorial illustrates:

- Adding a chamfer to a part

1. If the model "AdjustmentScrew" is not already in memory, select and double-click on the file.
2. Select *Insert* and *Chamfer* or simply .
3. Select the edge shown in Figure 7.8.

FIGURE 7.8 *Adjustment screw with one end chamfered.*

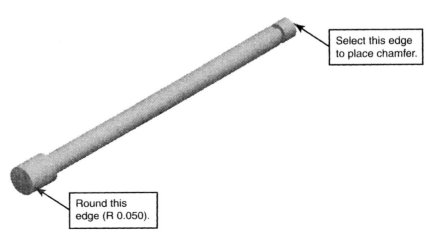

4. Change to 45 × D and enter a value of 0.050 for D in the appropriate cell.
5. Press ☑ in the dashboard to build the model.
6. Let us also add a round to the model. Select ▨.
7. With your mouse, pick the edge for the round shown in Figure 7.8.
8. Enter a value of 0.050 for the round radius.
9. Press ☑ in the dashboard to build the model.
10. Save the part by pressing ▨.

7.3 SUMMARY

In engineering practice, sharp edges and corners must be rounded or beveled to reduce stress concentrations. *Round* and *Chamfer* are the Pro/E options for removing material at model corners, edges, and between surfaces.

Because the placement of these features leads to problems in regeneration, it is best to place rounds, fillets, and chamfers after all other features have been located on the model. If the software is unable to regenerate the model, often the problem can be remedied by simply reducing the radius of the round/fillet or changing the order in which the rounds and/or fillets are placed.

7.3.1 STEPS FOR ADDING A ROUND OR FILLET

The general steps for placing a round or fillet are identical. The steps are:

1. Select *Insert* and *Round* or simply ▨.
2. Select the edge or chain of edges as necessary. If you are generating a surface to surface round, press the CTRL key to select each additional surface.
3. Enter the value for the round radius.
4. Press ☑ in the dashboard to build the model.

7.3.2 STEPS FOR ADDING A CHAMFER

In general, the procedure for adding a chamfer to a model is as follows:

1. Select *Insert* and *Chamfer* or simply ▨.
2. Pick the option for dimensioning the chamfer. For example, you may select *45 × D*.
3. Enter the value for D (or D1 and D2, if D1 × D2 is chosen).
4. Press ☑ to build the model.

7.4 ADDITIONAL EXERCISES

7.1 Complete the model of the cement trowel (also see Exercise 3.5) using the geometry in Figure 7.9.

FIGURE 7.8

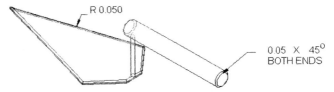

7.2 Add the round to the model of the side guide (also see Exercises 3.8, 4.10, and 5.2). Use Figure 7.10 as a reference.

FIGURE 7.10

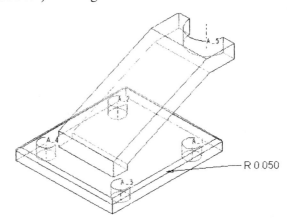

7.3 Add the chamfer to the model of the control handle (also see Exercises 3.6 and 4.8). The geometry of the chamfer is given in Figure 7.11.

FIGURE 7.11

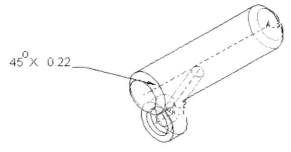

7.4 As shown in Figure 7.12, add the chamfer to the bushing (also see Exercises 3.2 and 4.3).

FIGURE 7.12

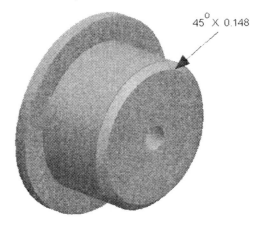

7.5 Add the chamfer to the model of the pipe clamp top (also see Exercise 5.7). Use Figure 7.13 for reference.

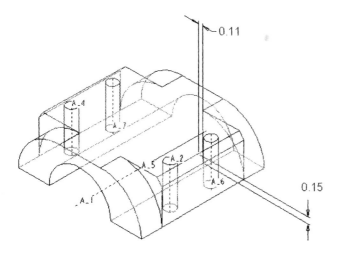

FIGURE 7.13

7.6 Add the chamfer to the model of the pipe clamp bottom (also see Exercises 5.8 and 7.2). Use Figure 7.14 for reference.

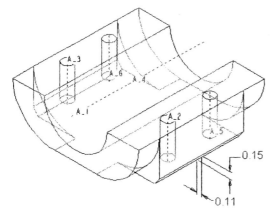

FIGURE 7.14

7.7 Add the round and chamfer to the model of the side support (also see Exercises 3.10 and 4.7). The geometry of the round and chamfer is given in Figure 7.15.

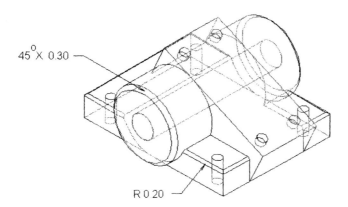

FIGURE 7.15

7.8 Construct the guide plate. Use Figure 7.16 for reference.

FIGURE 7.16

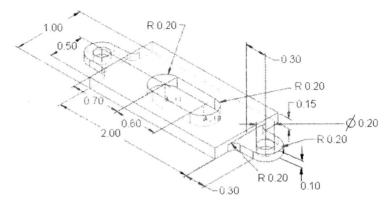

7.9 For the single bearing bracket (also see Exercise 6.12) in Figure 7.17, add a round to base and edges of the rib of radius 0.050. Also, add a round between the base and rib of radius 0.100.

FIGURE 7.17

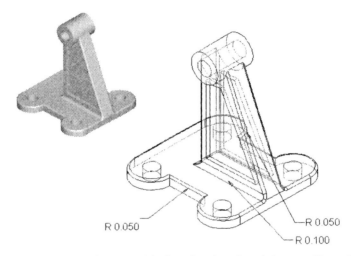

7.10 For the model of the double bearing bracket (also see Exercise 6.13) shown in Figure 7.18, add rounds to the rib, base edge, and the surface between the rib and the base. The value of the rounds is R 0.050.

FIGURE 7.18

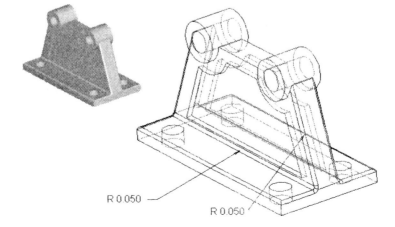

7.11 Round the edge of the clamp base (Exercise 4.13). Use Figure 7.19
      as a reference.

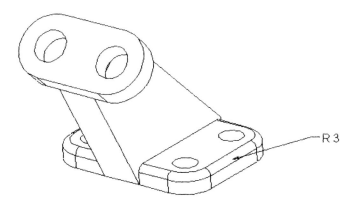

FIGURE 7.19

R 3

7.12 Add the surface to surface round to the slotted support (Exercise
      4.12). The geometry of the round is shown in Figure 7.20.

FIGURE 7.20

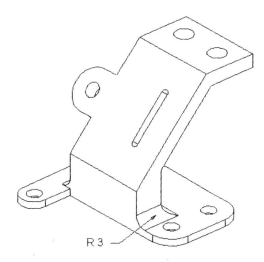

R 3

CHAPTER 8

PATTERN, COPY, GROUP, AND UDF

INTRODUCTION AND OBJECTIVES

Many of the models that we have constructed contain multiple copies of the same feature. So far, you have been asked to place all the copies individually, using the fundamental methods outlined for construction of a feature. You may have wondered if there is an easier way of placing these features.

Furthermore, some models that we have examined have features that are symmetric about a plane. You may ask if the symmetry can be used to place the feature.

You may have also noticed that some parts have features that are contained in a separate part. You may have wondered if there is some way to take the feature from one part and incorporate it into another part.

The answer to all these queries is, of course, yes. Multiple copies of a feature may be generated by using the *Copy* or *Pattern* options. Several features may be collected into a single entity using the *Group* option and then copied. Symmetry about a plane may be exploited to complete a part by using the *Mirror* option. Finally, a feature from a given part may be stored as a User-Defined Feature (UDF) and retrieved into a new part.

The objectives of the chapter are to

1. Construct copies of an original feature by using the *Copy* or the *Pattern* option
2. Make a copy of a feature using *Copy*
3. Construct rotational patterns of sketched features by creating datums on the fly
4. Place additional features on patterned features by using the *Reference Pattern* option
5. Collect multiple features into a single-named feature using the *Group* option and copy the group
6. Create user-defined features (*UDFs*) and load a UDF into a new part

8.1   USING THE *PATTERN* OPTION

One option that can be used to more easily place features is the *Pattern* option. In Chapter 4, we were faced with placing multiple holes that were identical except for an increment among their centers. The holes were placed individually, but they could have been placed by simply making a copy of the original hole and locating the new hole by incrementing from the original hole. This is how the *Pattern* option works.

A pattern can be one of three types as shown in Figure 8.1. In creating *Identical* patterns, Pro/E makes some assumptions. These assumptions are:

1. All the pattern instances are the same size.
2. The patterns are to be placed on the same surface as the original.
3. The only intersection between the patterns is with the placement surface.

The last assumption implies that the pattern(s) cannot intersect another feature or pattern. Because of these assumptions, *Identical* patterns are generated faster than all the other types of patterns. If the first two assumptions are relaxed, then a *Variable Pattern* is generated. A *General Pattern* has no assumptions, so it becomes the slowest pattern to generate.

The pipe flange from Chapter 4 is incomplete at this point and is an example of a part with a radial pattern. What remains to complete the model is the addition of three more holes along a bolt circle. Recall that in locating the original hole, the *Diameter* option was used with a value of 0° from datum plane Front. This angular value will serve as our increment. In general, if N holes are to be located on a bolt circle, then the holes are $360°/N$ apart. In this case, there are to be four holes; thus the holes will be 90° apart.

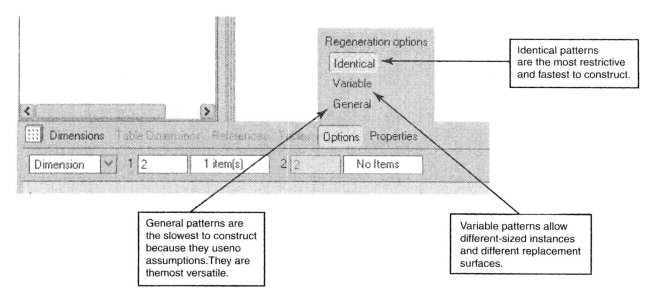

FIGURE 8.1 *A* Pattern *is constructed using the options from the dashboard and may be one of three types.*

The pipe flange has an existing radial dimension that may be used to pattern a feature. This will not always be the case. For features that do not have built-in radial dimensions, a datum needs to be created during construction of the pattern. This is called creating the datum "on the fly." Patterns that fall in this group are called "Rotational Patterns."

Datum planes created "on the fly" appear only during the construction of the feature. The datum will not appear on the model after the pattern feature has been built.

As will be illustrated in Tutorial 8.4, often the datum plane that is created "on the fly" cannot be used to sketch the cross section of the feature. In such cases an additional datum plane must be created.

After a feature has been patterned, it can be modified or additional features can be associated with the pattern. For example, the original instance of a pattern may be cut and the cut placed on the other instances by simply making a pattern of the cut and then choosing the *Reference* option.

8.1.1 TUTORIAL 8.1: A PATTERN WITH A LINEAR INCREMENT

How To Pattern w/

This tutorial illustrates:

- Generating an *Identical* linear pattern

1. If the model "PlateWithCounterboreHoles" is not already in memory, select ☞ and double-click on the file.
2. In the model tree, select the hole feature (this is the last item in the list. Press your right mouse button and select *Pattern* (or select *Edit* and *Pattern*).
3. Select *Dimensions* in the dashboard (see Figure 8.2).
4. Note the dimensions of the hole feature. Click on the 0.75 dimension. In the cell, enter 10.50 as shown in Figure 8.2. Hit *return*.
5. Click on the cell below *Direction 2* as shown in Figure 8.2.
6. Click on the 0.500 dimension.
7. Enter an increment of 4.00 as shown in Figure 8.2.
8. Press *Dimensions* to close the option.
9. Enter 2 for the number of instances in both directions as shown in Figure 8.2.
10. Press ☑ in the dashboard to build the model.
11. Save the part by pressing ⊞. The plate with the additional holes is shown in Figure 8.3.

8.1.2 TUTORIAL 8.2: A PATTERN WITH A RADIAL INCREMENT

This tutorial illustrates:

- Generating a pattern with a radial increment.

1. If the model "PipeFlange" is not already in memory, select ☞ and double-click on the file.
2. Click on the hole in the model tree. See Figure 8.4.

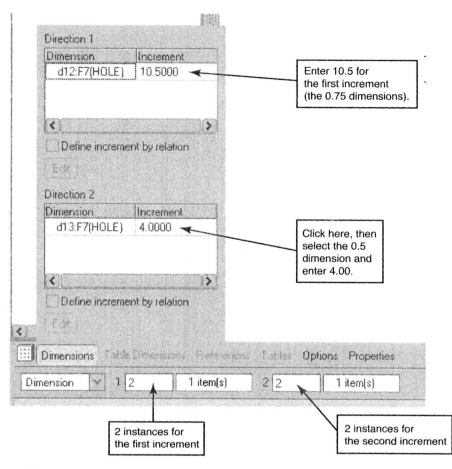

FIGURE 8.2 *Enter the increment in the appropriate cells.*

Enter 10.5 for the first increment (the 0.75 dimensions).

Click here, then select the 0.5 dimension and enter 4.00.

2 instances for the first increment

2 instances for the second increment

3. Press your right mouse button and select *Pattern* from the pop-up menu. Or, select *Edit* and *Pattern*.

4. Select *Dimensions* in the dashboard (see Figure 8.5).

5. Note the dimensions of the hole feature. Click on the 0.00° dimension. In the cell, enter 360/4 as shown in Figure 8.5. Hit *return*.

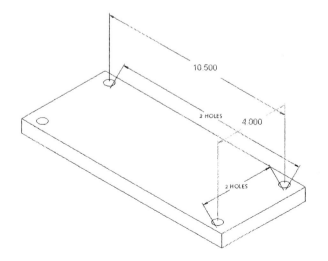

FIGURE 8.3 *The plate from Chapter 4 with all four holes placed.*

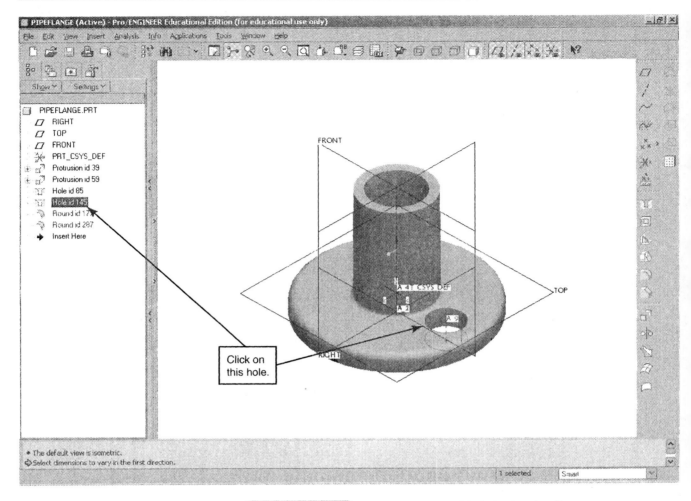

FIGURE 8.4 *Select the diametral hole in the pipe flange.*

6. Deselect *Dimensions* and enter four instances as shown in Figure 8.5.

7. Build the feature by selecting 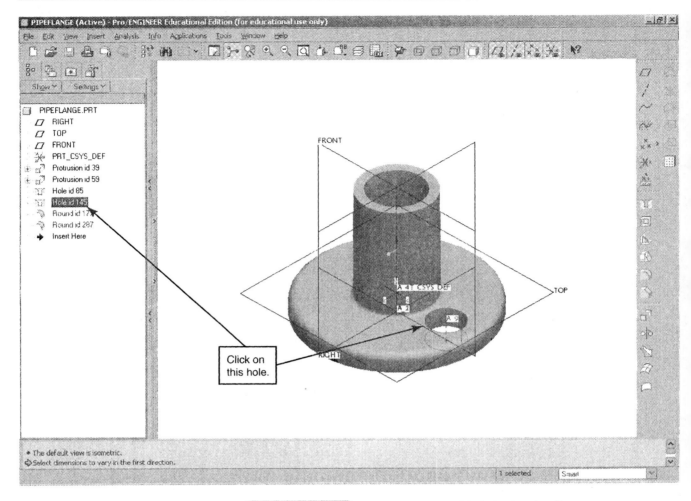. The completed model is shown in Figure 8.6.

8.1.3 TUTORIAL 8.3: A NUT WITH A ROTATIONAL PATTERN

This tutorial illustrates:

- Creating a rotational pattern
- Creating a datum plane "on the fly"

1. Consider the model shown in Figure 8.7. Then, select 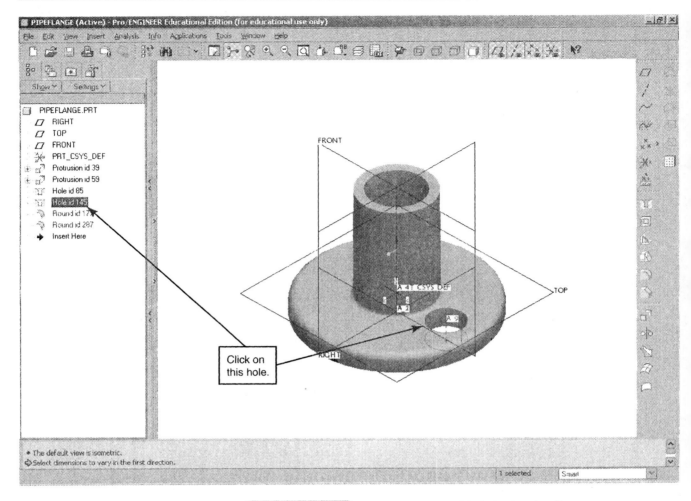.

2. Make sure that the *Use default template* option is checked.

3. Enter the name "Nut" in the *Name* field.

4. Select *OK*.

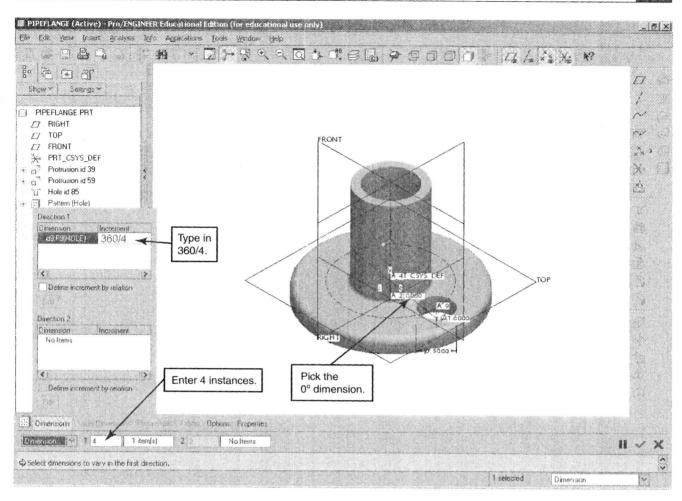

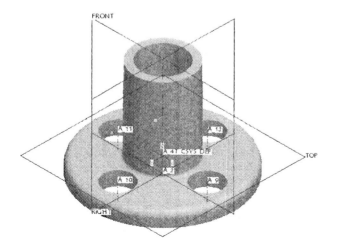

FIGURE 8.6 *The pipe flange with all four holes on the bolt circle.*

FIGURE 8.7 *The completed model of the nut and the model with just the cut.*

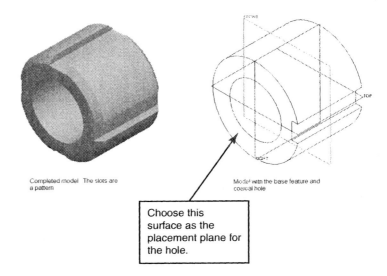

Completed model The slots are a pattern

Model with the base feature and coaxial hole

Choose this surface as the placement plane for the hole.

5. Select ▨.

6. Choose ▨ in the dashboard.

7. With your mouse, choose datum plane Front as the sketching plane.

8. If necessary, use ▨ for the *Orientation* field and to "right." Then, click on datum plane Right.

9. Again, the arrow should point into the screen. Use *Flip* if it does not.

10. Select ▨.

11. Sketch a circle whose center is aligned with the datum planes.

12. Double-click on the diameter dimension and change the value to 0.500.

13. Select Blind ▨ and enter a depth of 3/8″.

14. Build the section by pressing ▨.

15. Press ▨ in the dashboard to build the model.

16. Now, we may create the feature to be patterned.

17. Select ▨.

18. Choose ▨ and enter a depth of 0.100.

19. Choose ▨.

20. We need to create a datum for use in the rotational pattern. So, select ▨. This will pause the extrude process and allow you to create a datum "on the fly."

21. Click on the axis running through the cylinder (either A_1 or A_2) as shown in Figure 8.8.

22. Press the CTRL key and click on datum plane Top. Leave the rotation as 0.0000. Choose the *OK* button.

23. Resume the paused process by pressing ▨.

24. Select ▨.

25. Click on the new datum plane (DTM1) as the sketching plane. Select the surface shown in Figure 8.9 as the "bottom." Note the direction of the arrow. Flip, if necessary.

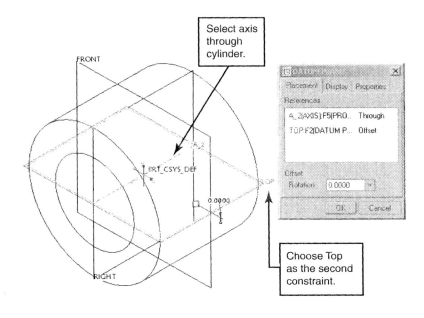

FIGURE 8.8 *Create a datum plane with an angular reference. This datum plane will be used to pattern the cut.*

26. Press the *Sketch* button [icon].
27. Dimension references need to be chosen; pick the edges shown in Figure 8.10. Press *Close* to close the *References* box.
28. Use the Rectangle tool [icon] and sketch the geometry shown in Figure 8.11.
29. Modify the dimension to 0.040 by double-clicking on the dimension on the screen.
30. Build the section by pressing [icon].
31. Press [icon] in the dashboard to build the model. The model with the slot is shown in Figure 8.12.

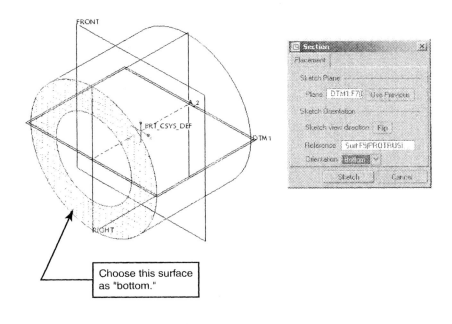

FIGURE 8.9 *Choose the shown surface to reorient the model in the Sketcher.*

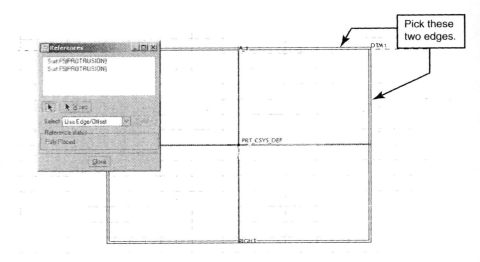

Pick these two edges.

FIGURE 8.10 *Pick the shown edges. These edges will be used as references to dimension the sketch.*

32. Now, let us create the rest of the slots by using a pattern.

33. In the model tree, select the Group feature. (This is the last item in the list. It is a group because it contains DTM1 as well as the cut.) Press your right mouse button and select *Pattern* (or select *Edit* and *Pattern*).

34. Select *Dimensions* in the Dashboard.

35. Note the dimensions of feature. Click on the 0.000° dimension. In the cell, enter 360/4 as shown in Figure 8.13. Hit *return*.

36. Press *Dimensions* to close the option.

FIGURE 8.11 *The nut with the dimensioned slot in the Sketcher.*

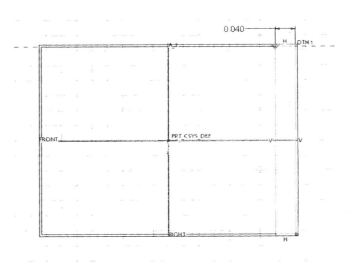

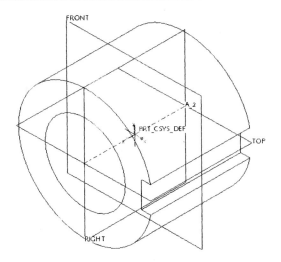

FIGURE 8.12 *The nut with one slot.*

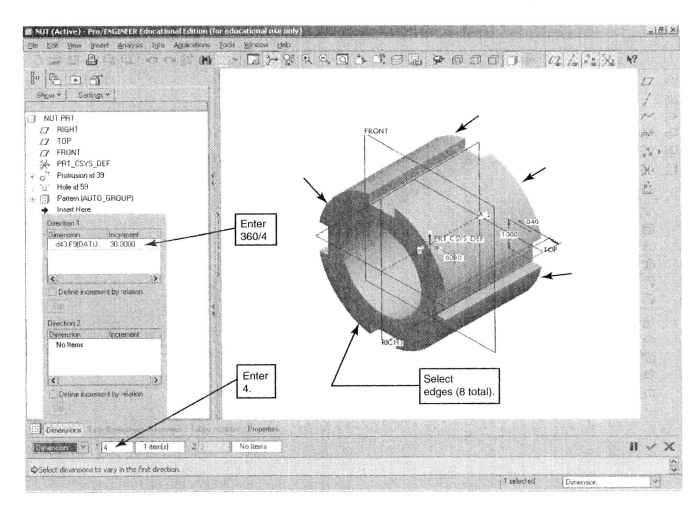

FIGURE 8.13 *Enter an increment of 90 to pattern the slot.*

37. Enter 4 for the number of instances as shown in Figure 8.13.

38. Press 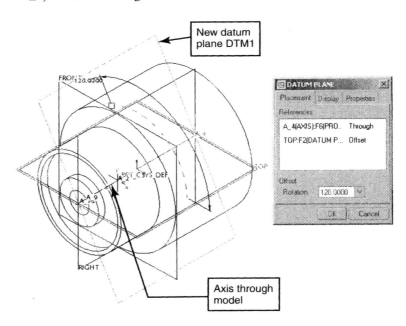 in the dashboard to build the model. The model at this point is shown in Figure 8.13.

39. Let us complete the model of the nut by adding chamfers at the ends. Select .

40. Select *Insert* and *Chamfer* or simply .

41. Select the eight edges shown in Figure 8.13.

42. Change to 45 × D and enter a value of 0.030 for D in the appropriate cell.

43. Press in the dashboard to build the model.

44. Save the part by pressing .

8.1.4 TUTORIAL 8.4: BLADES FOR THE PROPELLER

This tutorial illustrates:

- Creating a rotational pattern with advanced geometry
- Creating a datum plane on the fly
- Creating a sketching plane on the fly

1. If the model "Propeller" is not already in memory, select and double-click on the file.

2. Select .

3. Select . Enter a depth of 140-(65/2).

4. Choose . Enter a thickness of 10.

5. We need to create two datum planes, one for use in the rotational pattern and one as a sketching plane. So, select . This will pause the extrude process and allow you to create a datum "on the fly."

6. Click on the axis running through the cylinder (either A_1 or A_2) as shown in Figure 8.14.

FIGURE 8.14 *Construct a datum plane through the hub of the propeller.*

New datum plane DTM1

Axis through model

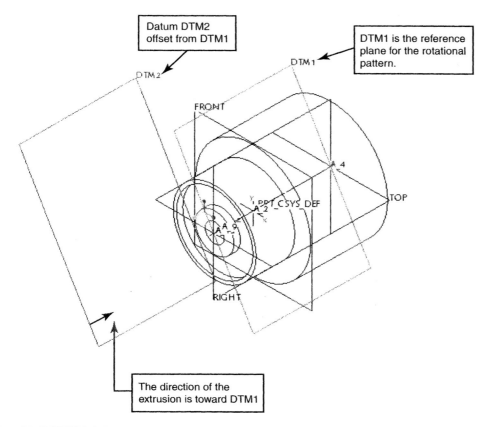

Datum DTM2 offset from DTM1

DTM1 is the reference plane for the rotational pattern.

The direction of the extrusion is toward DTM1

FIGURE 8.15 *Add a second datum plane to the model of the propeller.*

7. Press the CTRL key and click on datum plane Top. For the rotation enter 360/3. Choose the *OK* button.

8. Select ◻ once again. Click on datum plane DTM1. For the offset value enter 140. Pick *OK*. Note the new datum plane (DTM2) as shown in Figure 8.15.

9. Resume the paused process by pressing ▶.

10. Press ◻.

11. Click on datum plane DTM2 as the sketching plane.

12. Datum plane Front is orthogonal to DTM2, so click on datum plane Front and then select "bottom." Note the direction of the extrusion in Figure 8.15. Use *Flip,* if necessary. Press ▬.

13. Click on datum plane Front and an axis running through the center of the model as the references (see Figure 8.16).

14. Use ◥ and sketch a diagonal line as shown in Figure 8.17.

15. Use ⊟ and dimension as shown in Figure 8.17.

16. The values of the dimensions are driven by relations. Select *Tools* and *Relations*.

17. Enter the relations as shown in Figure 8.17. Note that your parameter names may be different from those in the figure. Then, select *OK*.

FIGURE 8. 16 *Choose the axis and datum plane Front as dimensioning references.*

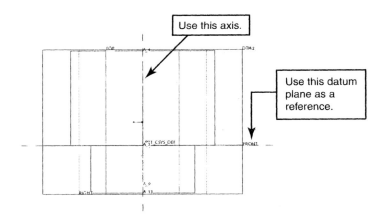

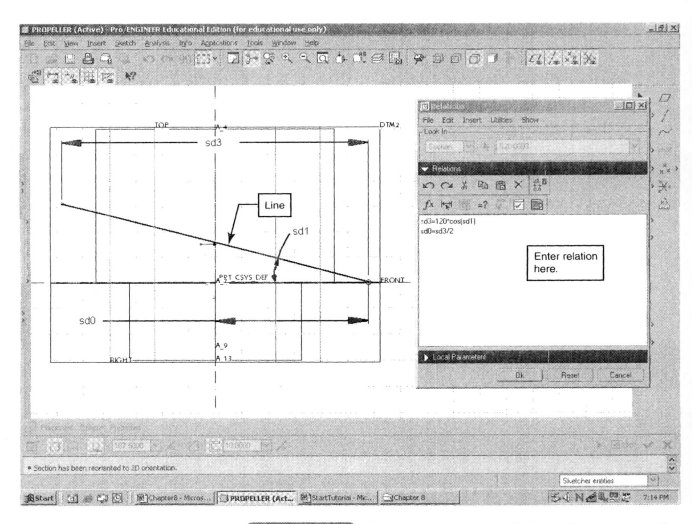

FIGURE 8. 17 *The model is shown in the Sketcher with the geometry of the first blade.*

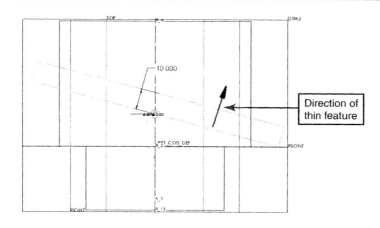

FIGURE 8.18 *Because the protrusion is constructed using a thin feature, a fill direction must be prescribed.*

18. Double-click on the angular dimension (sd1 in Figure 8.17) and enter 15.

19. Build the section by pressing 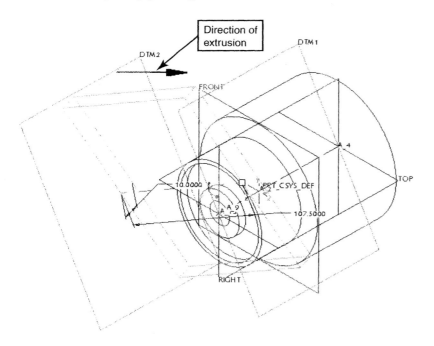.

20. Note the direction of the thin feature in Figure 8.18. Use next to to change the direction.

21. Use . Note the direction of the extrusion in Figure 8.19. Use next to to change the direction.

22. Build the feature by pressing .

23. Now, pattern the blade. Select the feature in the model tree (it will be the last feature in the list). Right-mouse-click and select *Pattern*.

24. Click on the 120° dimension. Enter 3 instances. Build the feature by selecting .

25. Save the part by pressing .

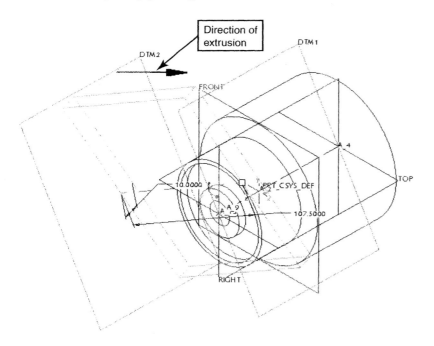

FIGURE 8.19 *After protruding the section shown in Figure 8.17, a rectangular protrusion is created.*

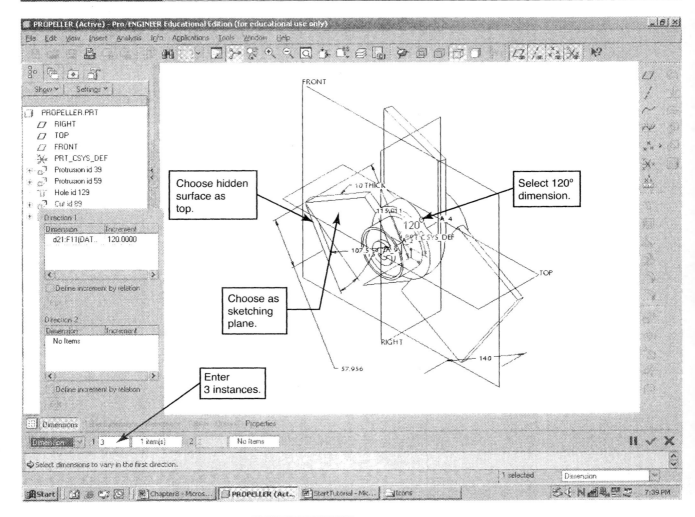

FIGURE 8.20 *Create the two other blades by making a pattern of the first.*

8.1.5 TUTORIAL 8.5: ADDITIONAL FEATURES FOR THE BLADES

This tutorial illustrates:

* Adding additional features to a pattern

1. We will now complete the geometry of the blades by using *Cut* and *Round*. If the model "Propeller" is not already in memory, select 📁 and double-click on the file.
2. Select 📎.
3. Choose ⬚ with a depth of 10 and ⬚.
4. Press 🖾.
5. Click on the blade surface shown in Figure 8.20 as the sketching plane.
6. Click on the hidden surface shown in Figure 8.20 and select "top" as the orientation.

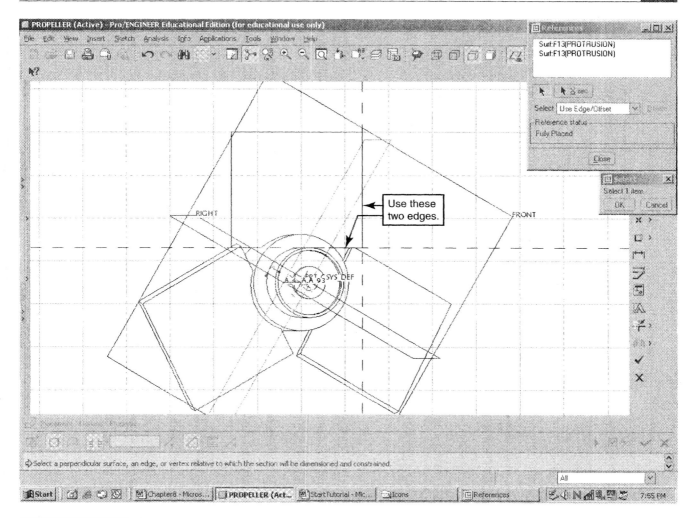

FIGURE 8.21 *The model of the propeller is shown reoriented in the Sketcher.*

7. Click on the two edges shown in Figure 8.21 to select the dimensioning references.

8. Use and and sketch the geometry shown in Figure 8.22.

9. Use and dimension the sketch. Use and change the dimensional values to the ones given in Figure 8.22.

10. Build the section by pressing .

11. Add the feature to the rest of the pattern. Select the cut in the model tree. Right-mouse-click and select *pattern*.

12. Make sure the pattern type is set to *Reference* as shown in Figure 8.23.

13. Build the section by pressing .

14. Now, repeat steps 2 through 13 for a cut on the opposite side of the blade (see Figure 8.24). Use the same sketching and orientation plane. The model with the cuts is shown in Figure 8.25.

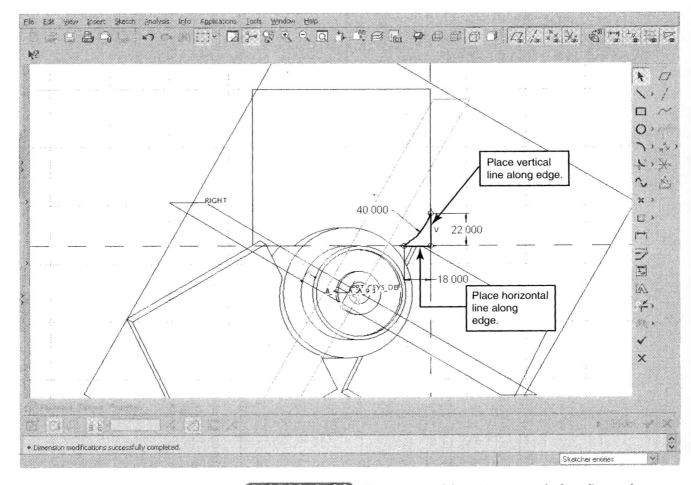

FIGURE 8.22 *The geometry of the cut is composed of two lines and an arc.*

15. Use and round the edges of the blades as shown in Figure 8.25. There are six edges. Select all six edges and then enter a round of 30. Build the feature.

16. Finally, complete the model by adding eight through holes on a 45-mm-diameter bolt circle. Select ▣.

FIGURE 8.23 *The cut may be added to the other instances in the pattern by using the* Reference *option.*

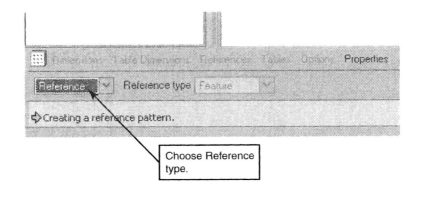

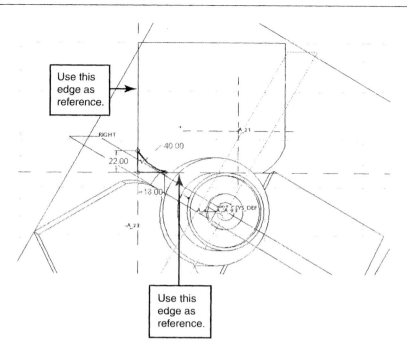

FIGURE 8.24 *Generate a cut to the other side of the blade.*

17. Select *Straight, Simple,* and *Through all* options.

18. Enter a diameter of 8.000.

19. Now click on the placement surface shown in Figure 8.25.

20. Press the *Placement* option and change to *Diameter.* Consider Figure 8.26.

21. Click in the Secondary references field and then on datum Top. Set the angular dimension in this case to 0.000°.

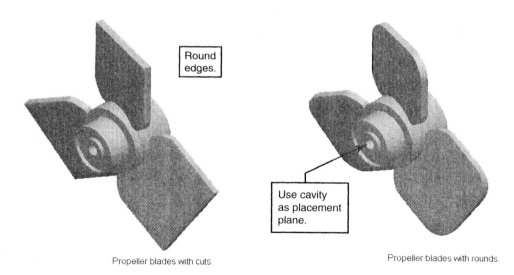

Propeller blades with cuts.

Propeller blades with rounds.

FIGURE 8.25 *The model before and after adding the cuts on the lower edges of the blades.*

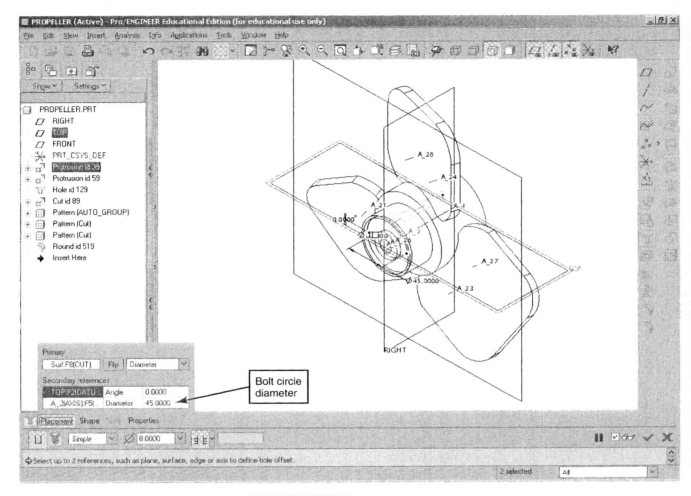

FIGURE 8.26 *Add a coaxial hole to the hub of the propeller.*

22. Note the diameter position handle. Drag this handle until it intersects with axis A_2. Enter a value of 45.000 for the bolt circle diameter.

23. Build the feature by pressing ☑.

24. Pattern the hole. Select the hole in the model tree. Right-click and select *Pattern*.

25. Select *Dimensions* in the dashboard.

26. Note the dimensions of the hole feature. Click on the 0.00° dimension. In the cell, enter 360/8. Hit *return*.

27. Deselect *Dimensions* and enter eight instances.

28. Build the feature by selecting ☑.

29. Save the part by pressing ☐. The model is shown in Figure 8.27.

FIGURE 8.27 *The completed propeller shown with the coaxial hole.*

8.2 USING THE *GROUP* OPTION

The *Group* option is a useful utility that allows the user to gather several features into a single group. A name is given to the group. The group can be treated as a feature.

The *Group* option is helpful when copying multiple features because the features can be collected into a single-named group and then copied as a single entity. The group option is accessed via the path *Feature* and *Group*. The *GROUP* menu, shown in Figure 8.28, contains the options for constructing, and to some extent, manipulating the group. Note the *Pattern* option within this menu. Grouped features cannot be patterned with the *Pattern* option discussed in Section 8.1. Instead, the creators of Pro/E have provided a *Pattern* option within the *GROUP* menu that can be used for the same purpose. You may ungroup a set of features by right-clicking on the group in the model tree and pressing the right mouse button and selecting *Ungroup*.

A group of features constructed by the user within the part is called a *Local Group*. The option *From UDF Lib* allows the user to retrieve features that have been grouped and stored in a library. This allows the user to place common features in different parts.

8.2.1 TUTORIAL 8.6: A VIBRATION ISOLATOR PAD

This tutorial illustrates:

- Constructing several features and then placing the features in a group

1. The model to be constructed is shown in Figure 8.29. In this tutorial we will create a group containing one boss and coaxial hole. Select ▢.
2. Make sure that the *Use default template* option is checked.
3. Enter the name "VibrationPad" in the *Name* field.
4. Select *OK*.

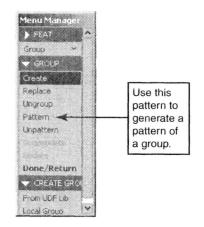

Use this pattern to generate a pattern of a group.

FIGURE 8.28 *The* GROUP *menu.*

FIGURE 8.29 *The vibration isolator pad is constructed with two protrusions and a hole. The first protrusion is used to construct the triangular base feature. The second protrusion forms the boss. After a coaxial hole is added to the boss, the two can be grouped and copied.*

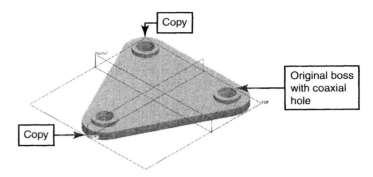

5. Select ⊞.
6. Choose the *Sketch* button ▨ in the dashboard.
7. With your mouse, choose datum plane Top as the sketching plane.
8. If necessary, use ▾ for the *Orientation* field and to "top." Then, click on datum plane Front.
9. Again, the arrow should point into the screen. Use *Flip* if it does not.
10. Select the *Sketch* button ▬.
11. Sketch the geometry shown in Figure 8.30 using ▧ and ◗. Dimension with ⊟ and modify the values to the ones shown in Figure 8.30.
12. Use ✓.
13. Use ⬚ and enter a depth of 0.2. The direction of the extrusion is up (Figure 8.31).

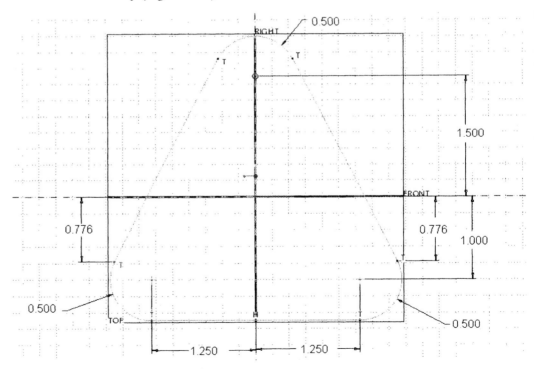

FIGURE 8.30 *Construct the triangular base feature by sketching the section shown.*

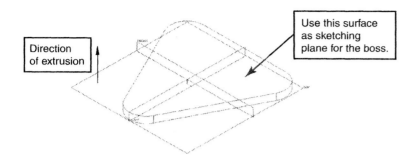

Direction of extrusion

Use this surface as sketching plane for the boss.

FIGURE 8.31 *Extrude the section as shown.*

14. Build the feature by selecting ☑.
15. Select ⊡.
16. Choose ▨ in the dashboard.
17. With your mouse, choose the surface shown in Figure 8.31 as the sketching plane. Use ▨ for the *Orientation* field and change to "bottom." Then, click on datum plane Front.
18. Sketch the geometry of the boss by using the *Circle* tool.
19. As necessary, dimension the feature so that it corresponds to Figure 8.32. Modify the dimensions to the values given in Figure 8.32 using ▨.
20. Use ☑.
21. Again extrude up by using the *Blind* option. The depth is 0.100. Select ▨ and enter a thickness of 0.100. The direction of the fill is inward.

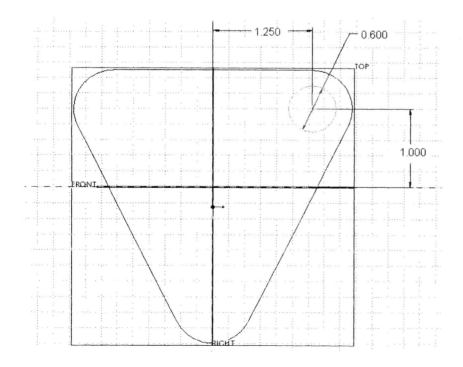

FIGURE 8.32 *The second protrusion is constructed by sketching a circle as shown.*

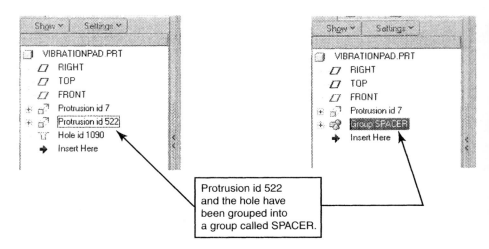

FIGURE 8.33 *By using the* Group *option, the boss and coaxial hole may be grouped.*

22. Build the feature by selecting ▩.

23. Select ▩. Use *Straight, Simple,* and *Through all.* Enter a diameter of 3/8″.

24. Click on the axis running through the boss. Then, press *Placement* and click in the *Secondary references* field. Click on the same surface as the sketching surface for the boss (Figure 8.31). Build the feature.

25. We are now ready to group the boss and coaxial hole into a single group feature. Select *Edit, Feature operations,* and *Group.* Then, select *Local Group.* Enter the name "Spacer" in the name field.

26. Select the protrusion and the hole in the model tree (Figure 8.33). Press the CTRL key to pick the second feature. Select *Done.* Note the group called SPACER in the model tree. Choose *Done/Return* and *Done.*

27. Save the part by pressing ▣.

8.3 THE *COPY* OPTION

The *Copy* option (Figure 8.34) is another useful option from the *FEAT* menu that may be used to increase model construction efficiency. The *Copy* option allows the user to make additional copies of one or more features by varying the dimensions and/or references describing the location of the feature. Features may be copied between the same or different models.

Features may be copied from a different part or assembly into the current part by using new references. When doing so, you must use the new references (*New Refs* option). Pro/E will prompt you to provide the proper information for the new references.

The options available when using the *Copy* command are shown in Figure 8.34. The *Mirror* option copies one or more features by using a given plane as a mirror. The *All Feat* option copies *all* geometry by performing a mirroring operation.

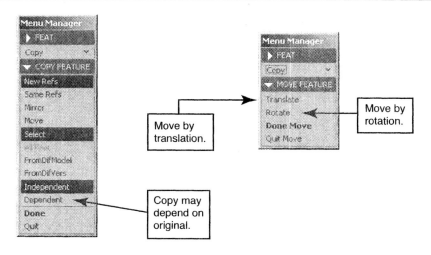

FIGURE 8.34 *The* COPY FEATURE *menu contains the options available for copying a desired feature.*

The *Independent* and *Dependent* options allow the user to prescribe whether the copy should depend on the original feature. Use the *Dependent* option if you want the copy of the feature to be automatically updated when the original feature is updated.

The *Move* option provides the user with the ability to make a copy of a given feature by performing a simple translation or rotation. Translation of the feature must be along a desired direction and an offset distance. When using the *Rotation* option, the user must define the rotation direction as well as the rotation angle. The translation or rotation is defined by selecting a plane (*Plane*), curve, edge, or axis (*Crv/Edg/Axis*), or coordinate axis (*Csys*).

8.3.1 TUTORIAL 8.7: COPYING A CUT BY USING MIRROR

This tutorial illustrates:

- Copying a cut feature by using *Copy* and *Mirror*

1. Retrieve the model "ConnectingArm" if it is not already in memory, select 🖼, and double-click on the file.
2. Select *Edit, Feature operations,* and *Copy.*
3. Choose *Mirror, Select,* and *Dependent.* Then click on the cut in the rib.
4. Choose *Done.* Click on datum plane Front, the mirror plane.
5. The software will add the feature. Select *Done.*
6. Save the part by pressing 🖻.

8.3.2 TUTORIAL 8.8: COPYING THE GROUP SPACER BY USING MIRROR

This tutorial illustrates:

- Copying a feature by using *Copy* and *Mirror*

1. If the model "VibrationPad" is not already in memory, select 🖼 and double-click on the file.
2. Select *Edit, Feature operations,* and *Copy.*
3. Choose *Mirror, Select,* and *Dependent.* Then click on the group Spacer.

FIGURE 8.35 *The model with a copy of the group.*

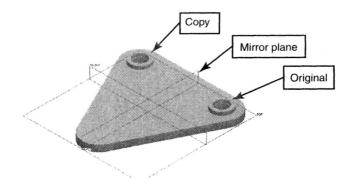

4. Choose *Done*. Click on datum plane Right, the mirror plane.

5. The software will add the feature as shown in Figure 8.35. Select *Done*.

6. Save the part by pressing 🖫.

8.3.3 TUTORIAL 8.9: A SECOND COPY OF THE SPACER

This tutorial illustrates:

- Copying a feature by using *Same Refs*

1. If the model "VibrationPad" is not already in memory, select 🖆 and double-click on the file.

2. Select *Edit, Feature operations,* and *Copy*.

3. Choose *Same Refs, Select,* and *Dependent*. Then, click on the group Spacer (the original) and pick *Done*.

4. The new copy will be created by simply changing the references of the original. In particular, the distance from datum plane Right and Front (see Figure 8.36).

5. Scroll through the list of dimensions in *GP VAR DIMS* menu. Note that as the mouse is placed over the dimension in the menu, the dimension is highlighted on the model. Click on the 1.250 and 1.000 dimensions.

6. Select *Done* in the *GP VAR DIMS* menu.

7. The software will now prompt for updated values of the chosen dimensions. For the 1.25 dimension, enter an updated value of 0.000.

8. Enter -1.500 for the 1.00 (the negative is used because of the direction from datum plane Front). Select *OK* and *Done*.

9. Compare your model to the one in Figure 8.29. Save the part.

8.3.4 TUTORIAL 8.10: ADDING A HOLE IN ANGLEBRACKET BY TRANSLATION

This tutorial illustrates:

- Copying a feature by using *Move*

1. This feature may also be created by using a pattern, but we will use *Copy*. Retrieve the model "AngleBracket" by selecting 🖆 and double-clicking on the file.

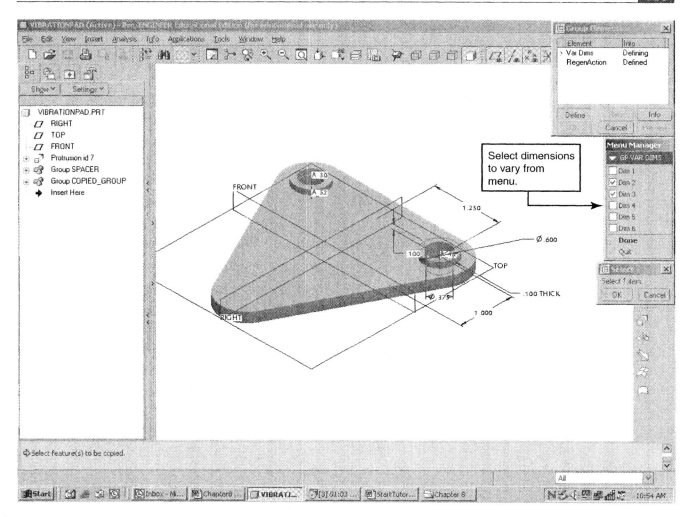

FIGURE 8.36 *The second copy may be constructed by using the same references as the original and changing the dimensional values.*

2. Select *Edit, Feature operations,* and *Copy.*

3. Choose *Move, Select,* and *Dependent.* Then click on the hole shown in Figure 8.37. Select *Done.*

4. Choose *Translate* and *Plane.* Then, pick on the surface shown in Figure 8.37. Note the direction of the offset in Figure 8.37. Select *Flip,* if it is necessary to change the direction. Then, choose *OK.*

5. The increment between the original and the copy is 2.00. Enter this value in the cell. Then choose *Done Move.*

6. There is no need to vary any of the copy dimensions, so select *Done* in the *GP VAR DIMS* menu.

7. Select *OK* and *Done.* This will create the copy. Save the part.

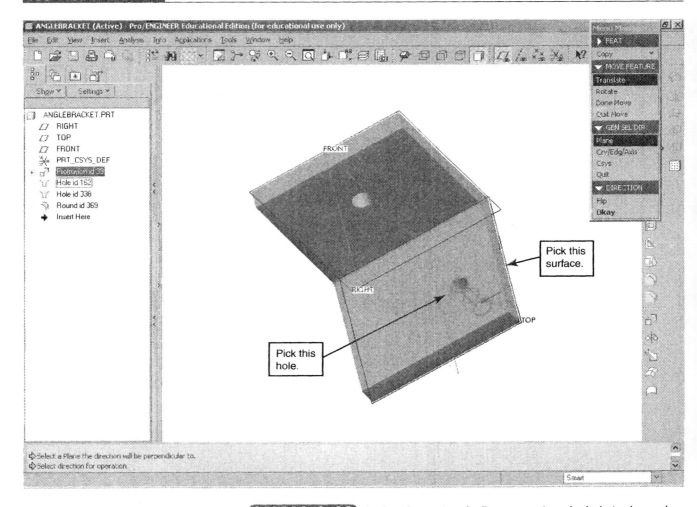

Rather than using the Pattern *option, the hole in the angle bracket can be constructed by using* Copy *and* Translate.

8.3.5 TUTORIAL 8.11: USING COPY AND ROTATE

This tutorial illustrates:

- Generating a feature by copying and rotating an existing feature

1. Retrieve the model "EmergencyLightHolder" by selecting 📑 and double-clicking on the file.
2. Select *Edit, Feature operations,* and *Copy.*
3. Choose *Move, Select,* and *Dependent.* Then click on the tab shown in Figure 8.38. Select *Done.*
4. Select *Rotate* and *Crv/Edg/Axis.* Pick the axis in Figure 8.38.
5. Note the direction of the operation. Compare the direction in your screen and that in Figure 8.39. If they are not the same, select *Flip.* Otherwise, choose *OK.*
6. Enter 90 in the cell and hit *return.*

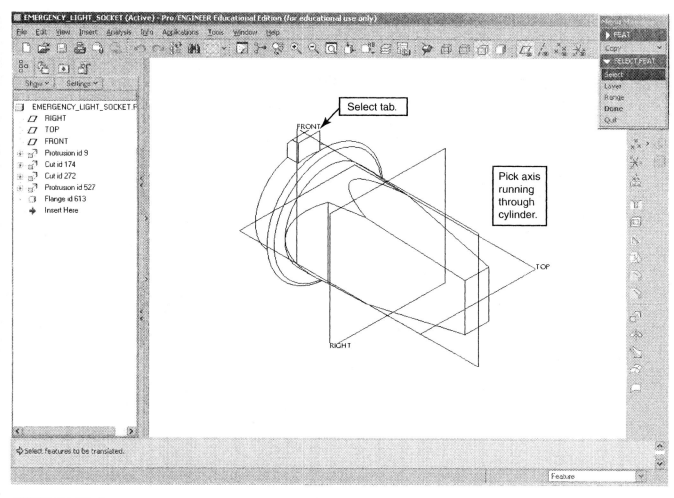

FIGURE 8.38 *An additional alignment tab may be added by using* Copy *and* Rotate.

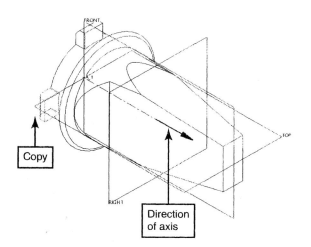

FIGURE 8.39 *Note the copy.*

7. Select *Done Move* and *Done,* as there is no need to vary the dimensions. Then, choose *OK* and *Done.*
8. The software will create the feature as in Figure 8.39.
9. Save the part.

8.4 USER-DEFINED FEATURES

For features that occur on many parts, Pro/E allows the user to save the feature or features as a User-Defined Feature (*UDF*). A library containing user-defined features is known as a *UDF Library*. The software will place a UDF in the working directory unless instructed otherwise. Most organizations will set up a special folder to be used as a UDF library; check with your system administrator to see if a UDF library folder has been established.

When creating a UDF, the feature may be saved with or without the reference part on which it was originally constructed. Saving the UDF with its reference part is useful in illustrating how the UDF was originally created. Of course, storing the reference part with the UDF will increase the memory required for storage.

When adding a UDF to a new model, the group may be one of four different types. These types are as follows:

1. *Stand Alone.* All the information describing the UDF is associated and copied with the UDF when the UDF is loaded into a new part. This requires more storage space. Furthermore, any changes made to the reference part will not show up in the UDF.

2. *Subordinate.* A subroutine UDF retrieves its values from the original reference part during execution. Therefore, the reference part must also be present. Use of the *Subordinate* option will require smaller memory allocations than the *Stand Alone* option. Furthermore, any changes made to the reference part will automatically be updated in the UDF.

3. *Independent Groups.* The use of this option allows the user to construct UDF groups that are independent to any changes in the UDF.

4. *UDF Driven Groups.* In some cases, you may wish to have changes in the reference part automatically updated in the UDF group. In such cases, use the *UDF Driven* instead of the *Independent* option.

User-defined features are created by selecting the appropriate features. If a pattern is chosen, then only one member of the pattern need be selected. In defining the UDF, the dimensions and placement reference must also be selected. The dimensions and placement reference may be defined to be variable. This allows the user to change the dimensions and/or placement reference when loading the UDF into a new part. In order to create variable dimensions and/or placement references, select the *Var Dims* option when creating the UDF.

In some cases, you may wish to create a UDF that refers to an entire family of features. For example, you may wish to create a UDF of a cap screw. You may construct a family of cap screws wherein different dimensions such as the thread pitch or screw diameter vary by a given amount.

In order to create a family, select the *Family Table* option during the creation of the UDF and select the dimensions to add to the table. Pro/E requires that a symbol be attached to each dimension.

When loading a UDF, the software allows the user to display the dimensions in three different ways: *Normal, Read Only,* and *Blank.* These options are characterized as:

1. *Normal.* The dimensions are defined in the manner traditional to Pro/E. That is, the user may modify the dimensions.

2. *Read Only.* The software will display the dimensions, but they cannot be modified.

3. *Blank.* Dimensions will not be displayed, nor can they be modified.

8.4.1 TUTORIAL 8.12: CREATING A UDF

This tutorial illustrates:

• Creating a user-defined feature

1. Retrieve the model "PlateWithCounterboreHoles" by selecting ⬚ and double-clicking on the file.
2. Select ⬚.
3. Select *Straight, Simple,* and *Through all* options. Enter a diameter of 0.500.
4. Click on the surface shown in Figure 8.40. Drag the reference handles to the edges shown in the figure. Double-click on the values and enter the numbers shown in the figure.
5. Build the feature by pressing ⬚.
6. Now pattern the hole. Select the hole feature. Press your right mouse button and select *Pattern* (or select *Edit* and *Pattern*).
7. Select *Dimensions* in the dashboard.
8. Click on the 2.650 dimension. In the cell, enter 6.700 (see Figure 8.41). Hit *return.*

FIGURE 8.40 *The plate with an added hole.*

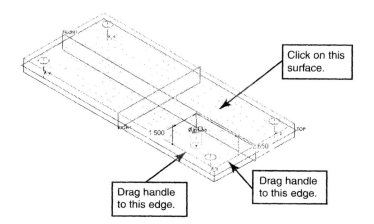

Click on this surface.

Drag handle to this edge.

Drag handle to this edge.

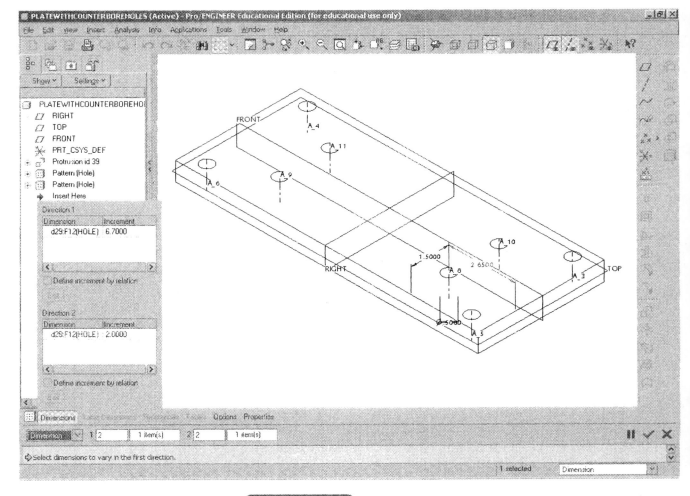

FIGURE 8.41 *Construct the rest of the holes by using a pattern and incrementing as shown.*

9. Click on the cell below *Direction 2*. Click on the 1.500 dimension.

10. Enter an increment of 2.000.

11. Press *Dimensions* to close the option.

12. Enter 2 for the number of instances in both directions.

13. Press ☑ in the dashboard to build the model.

14. Choose *Tools, UDF Library,* and *Create.*

15. Enter the name "Four_Holes" and hit *return.*

16. Select *Stand Alone* and *Done.*

17. In this case, the reference part is illustrative and simple, so answer *Yes* to the query to include the reference part.

18. Click on the first instance of the hole pattern as shown in Figure 8.42.

19. Select *Done* and *Done/Return.*

20. Choose *Single* and *Done/Return.*

21. The software will now ask for prompts for the various references. The feature has three: the placement surface and two linear

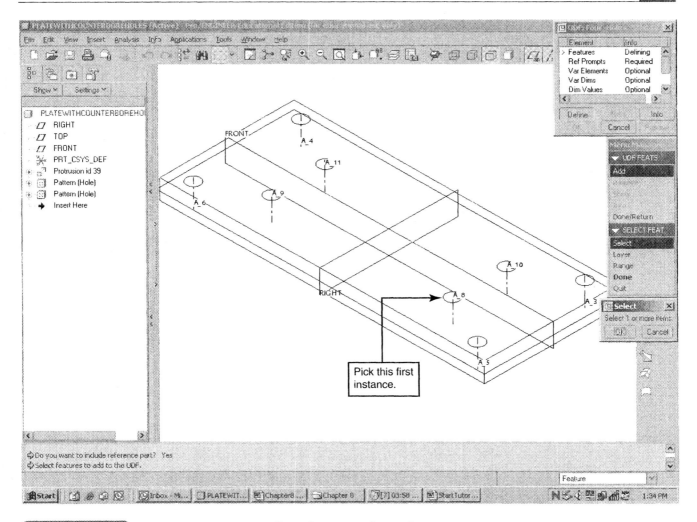

Pick this first instance.

FIGURE 8.42 *The* UDF FEATS *menu allows the user to choose the features to be included in the UDF.*

references. Enter: Placement_Plane for the first prompt as shown in Figure 8.43. Hit *return*.

22. Enter: First_Reference and hit *return*.

23. Enter: Second_Reference and hit *return*.

24. The prompts may be changed at this point. We are not going to do this; however, we will add prompts for the values of the feature dimensions such as diameter. Select *Var Dims* by double-clicking on the option (see Figure 8.43).

25. Click on the following dimensions: the 0.500, 1.500, 2.65, 2.000, and 6.700. Select *Done/Return* twice.

26. Now, enter text prompts for each dimension as it is highlighted (hit *return* after typing each prompt). For 0.5, type: "Enter the diameter of the hole." For 1.500, type: "Enter the distance from the first reference." For 2.65, type: "Enter the distance from the second reference." For 2.000, type: "Enter the increment from the first reference." For 6.700, type: "Enter the increment from the second reference."

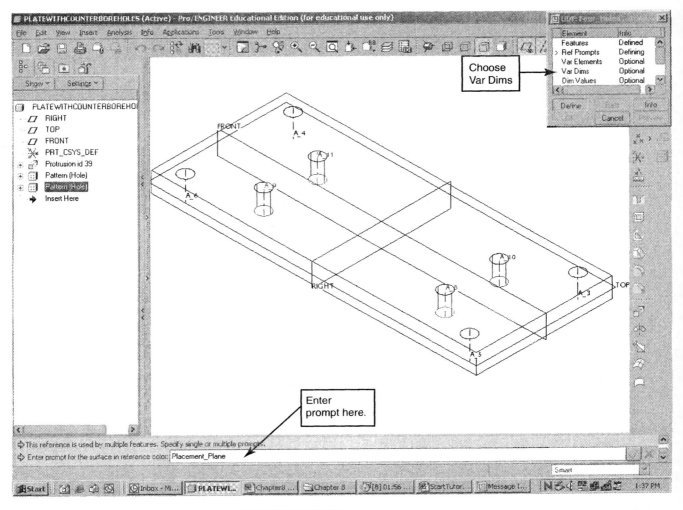

FIGURE 8.43 *The software allows the user to enter descriptive text called* Ref Prompts. *These* Ref Prompts *are used when loading the UDF into a new part.*

27. Select *OK*. The software will create and write the UDF to the library.
28. Save the model by using ⊡.

8.4.2 TUTORIAL 8.13: ADDING A UDF TO A PART

This tutorial illustrates:

- Adding a user-defined feature to a part

1. First construct a new part. Select ▯.
2. Make sure that the *Use default template* option is checked.
3. Enter the name "SquarePlate" in the *Name* field.
4. Select *OK*.
5. Select *Insert* and *Extrude* (or simply ▱).

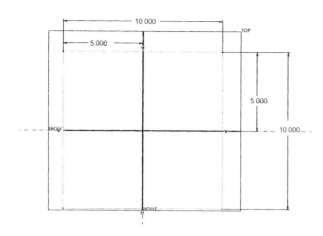

FIGURE 8.44 *Use this geometry to construct the base feature of the plate.*

6. Choose the *Sketch* button in the dashboard.

7. With your mouse, choose datum plane Top as the sketching plane and datum plane Front as "bottom." Again, the arrow should point into the screen. Use *Flip* if it does not.

8. Select the *Sketch* button.

9. Sketch a rectangular section by using.

10. Dimension and modify the section as shown in Figure 8.44.

11. Build the section by pressing.

12. Choose and enter a depth of 0.25. The direction of the extrusion is up.

13. Press in the dashboard to build the model.

14. Now, select Insert and User-Defined Feature. Find the file "Four_Holes.gph" and double-click on it.

15. Choose *Yes* for the query "Retrieve reference part."

16. The software will open a second window with the reference part as shown in Figure 8.45.

17. Choose *Independent* and *Done*.

18. The user may scale the UDF. This is useful if the model has different units from the reference part. This is not the case here. So, select *Same Dims* and *Done*.

19. As the software cycles through the prompts, respond with the values shown in Table 8.1.

TABLE 8.1 VALUES FOR THE PROMPTS

PROMPT	VALUE
Enter the diameter of the hole.	0.500
Enter distance from first reference.	9.000
Enter the distance from the second reference.	1.000
Enter the increment along the first reference.	8.000
Enter the increment along the second reference.	8.000

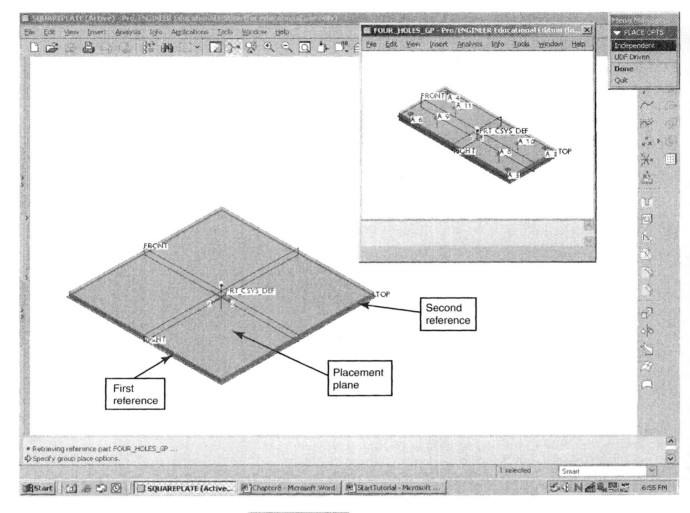

FIGURE 8.45 *The user may select the references when importing the UDF.*

20. Select *Normal* and *Done*.
21. The software will prompt for the references. Click on the references when prompted. Use Figure 8.45 as a guide.
22. Use *Redefine*, if you made a mistake. Select *Done* when finished.
23. Save the part using 🖫 . The part is shown in Figure 8.46.

8.5 SUMMARY

In this chapter, we considered options that increase model construction efficiency. These options are the *Pattern, Group, Copy,* and *UDF Library*.

FIGURE 8.46 *The plate is shown with the added UDF.*

8.5.1 REVIEW AND STEPS FOR THE *PATTERN* OPTION

The *Pattern* option allows the user to create additional features from an original if an increment separates the instances. The original is called the pattern leader. The *Pattern* option will pattern only a single feature. After a pattern has been created, the *Ref Pattern* option may be used to place additional features on the pattern.

There are three types of patterns. The first, the *Identical* pattern, has stringent requirements. If a pattern leader intersects another feature or another pattern, it cannot be patterned using the *Identical* option. The *Identical* pattern is the fastest to generate.

If the pattern is to vary in size or is to be placed on a different placement surface from the leader, then it may be constructed by using the *Varying* option. The *Varying Pattern* is the second type and the second fastest to generate.

The *General* Pattern is the slowest to create, but uses no assumptions. It allows features to be patterned that intersect another feature or pattern.

The general steps for constructing a pattern are the following:

1. In the model tree, select the feature. Press your right mouse button and select *Pattern* (or select *Edit* and *Pattern*).
2. Select *Dimensions* in the dashboard.
3. Click the dimension in the first direction. In the cell, enter the increment. Hit *return*.
4. If there is a second direction, click on the cell below *Direction 2*. Choose the corresponding dimension. Enter the increment in the cell.
5. Enter the number of instances in both directions.
6. Press ▨ in the dashboard to build the model.

For a *Rotational Pattern*, an angular dimension must exist prior to the construction of the pattern. This angular dimension may be added to the model during the construction of the feature by using ▨.

8.5.2 REVIEW AND STEPS FOR THE *GROUP* OPTION

The *Group* option is used to collect multiple features into a single-named entity. After a group is created, it may be treated as a single feature. If the group is to be patterned, then the *Group Pattern* option must be used. Grouped features cannot be patterned using the *Pattern* option summarized in Section 8.4.1.

The general steps for creating a group are:

1. Select *Edit, Feature operations,* and *Group*.
2. Then, select *Local Group*. Enter the name of the group in the name field.
3. Select the features to be grouped. Press the CTRL key to pick each additional feature.
4. Select *Done*.
5. Choose *Done/Return* and *Done*.

8.5.3 REVIEW AND STEPS FOR THE *COPY* OPTION

The *Copy* option may be used to copy features. The *Copy Mirror* option allows the user to copy individual features or a group from a mirror plane.

In addition, features may be copied by changing one or more of the dimensions of the original feature. In such cases, either the same references are used (*Same Refs* option) or new references are determined (*New Refs* option).

The steps for creating a copy of a feature or group using the *Copy Mirror* option are:

1. Select *Edit, Feature operations*, and *Copy*.
2. Choose *Mirror*.
3. Choose *All Feat* to mirror copy all features or *Select* to copy only certain features.
4. Choose *Independent* or *Dependent*.
5. Then click on the features.
6. Choose *Done*.
7. Click on the mirror plane.
8. Select *Done*.

The sequence of steps for constructing a copy using the *Same Refs* or *New Refs* option is:

1. Select *Edit, Feature operations*, and *Copy*.
2. Choose *Same Refs* and *New Refs*.
3. Choose *All Feat* to mirror copy all features or *Select* to copy only certain features.
4. Choose *Independent* or *Dependent*.
5. Then click on the feature and pick *Done*.
6. Scroll through the list of dimensions in *GP VAR DIMS* menu. Click on the desired dimensions.
7. Select *Done* in the *GP VAR DIMS* menu.
8. The software will now prompt for updated values of the chosen dimensions.
9. Select *OK* and *Done*.

The sequence of steps for constructing a copy using the *Move* and *Translate* option is:

1. Select *Edit, Feature operations*, and *Copy*.
2. Choose *All Feat* to mirror copy all features or *Select* to copy only certain features.
3. Choose *Independent* or *Dependent*.
4. Then click on the feature and pick *Done*.
5. Choose *Translate* and *Plane, Crv/Edg/Axis,* or *Csys*.
6. Then, pick the plane, curve, edge, axis of coordinate system.
7. Then, choose *OK*.
8. Enter any increment value. Then choose *Done Move*.
9. Vary the dimensions, if desired, by using the *GP VAR DIMS* menu.

10. Select *Done*.

11. Select *OK* and *Done*.

The sequence of steps for constructing a copy using the *Move* and *Rotate* option is:

1. Select *Edit, Feature operations,* and *Copy*.

2. Choose *All Feat* to mirror copy all features or *Select* to copy only certain features.

3. Choose *Independent* or *Dependent*.

4. Then click on the feature and pick *Done*.

5. Choose *Rotate* and *Plane, Crv/Edg/Axis,* or *Csys*.

6. Then, pick the plane, curve, edge, axis of coordinate system.

7. Choose *OK*.

8. Enter the rotation angle. Then choose *Done Move*.

9. Vary the dimensions, if desired, by using the *GP VAR DIMS* menu.

10. Select *Done*.

11. Select *OK* and *Done*.

8.5.4 Review and steps for creating and placing a UDF

A user-defined feature (UDF) may be created and placed in a UDF library. The benefit in creating a UDF is that common features occurring on multiple parts need be created only once.

User-defined features are created and placed in a UDF library using the *Tools* and *UDF Library* option. They may be retrieved and placed on a new part with *Insert* and *User-Defined Feature*.

The general steps for creating a UDF are:

1. Choose *Tools, UDF Library,* and *Create*.

2. Enter the name of the UDF and hit *return*.

3. Select *Stand Alone* or *Subordinate* and *Done*.

4. Click on the features to add to the UDF.

5. Select *Done* and *Done/Return*.

6. Enter text for the prompts.

7. If desired, select *Var Dims* to allow the dimensions to be varied when the UDF is loaded in a new part. Then enter text prompts for each dimension as it is highlighted.

8. Select *OK*.

A UDF may be retrieved from the UDF library and placed on a new part as a group. Of course, you must have permission to access the folder. In general, a UDF may be retrieved as follows:

1. Choose *Insert* and *User-Defined Feature*.

2. Select *Independent* or *UDF Driven*.

3. Choose the placement scale: *Same Size, Same Dims,* or *User Scale*.

4. The software will prompt if there are any variable dimensions. Respond with the appropriate values.

5. Select the *Display* option: *Normal, Read Only,* or *Blank*.

6. Select the placement references when prompted to do so.

7. The software will place the UDF. Use *Redefine* to modify the UDF; otherwise choose *Done*.

8.6   ADDITIONAL EXERCISES

8.1 Use Figure 8.47 for the model of the pipe clamp top (see also Exercise 5.7). Complete the model using *Copy, Mirror* and *All Feat*.

8.2 Using Figure 8.48, complete the model of the pipe clamp bottom (see also Exercise 5.8) using *Copy, Mirror,* and *All Feat*.

8.3 Add the holes on the part of the side support (see also Exercises 3.10 and 4.6), using the *Pattern* option and Figure 8.49

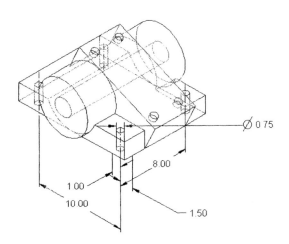

8.4 Add the holes to the two inclined planes on the double side support (see also Exercises 3.11 and 4.10). Use Figure 8.50 for reference.

FIGURE 8.50

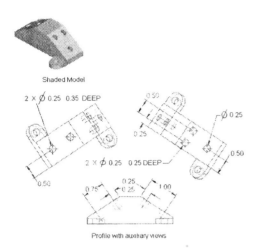

Shaded Model

2 X Ø 0.25 0.35 DEEP

Ø 0.25

0.50

2 X Ø 0.25 0.25 DEEP

0.50

Profile with auxiliary views

8.5 Construct the gasket by using a single protrusion and the *Copy, Mirror,* and *All Feat* options. Use Figure 8.51 as a reference.

FIGURE 8.51

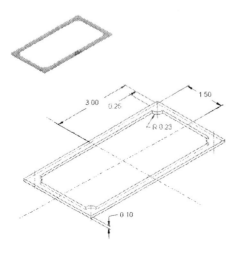

3.00   0.25   1.50   R 0.23   0.10

8.6 Construct the wing plate using a single protrusion and the *Copy, Mirror,* and *All Feat* options. Consult Figure 8.52 for the geometry of the model.

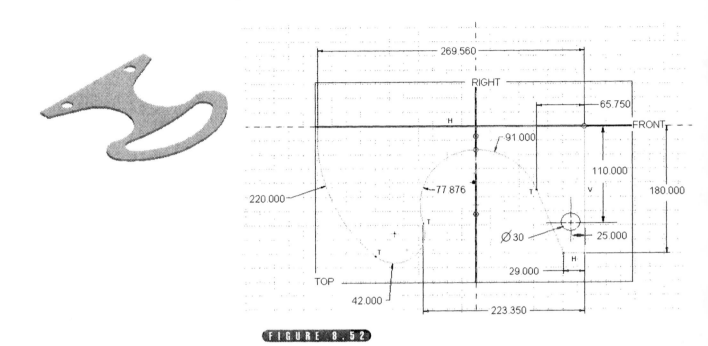

FIGURE 8.52

8.7 Create the model of the skew plate. Use the *Pattern* and the *Copy* options to place the holes. Notice this is a metric part. The geometry of the part is given in Figure 8.53.

FIGURE 8.53

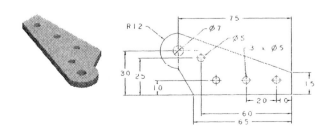

8.8 Construct the ratchet, using the *Pattern* option to place the teeth. The geometry of the part is given in Figure 8.54.

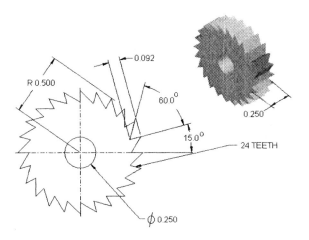

FIGURE 8.54

R 0.500
0.092
60.0°
0.250
15.0°
24 TEETH
Ø 0.250

8.9 Use the *Pattern* option to add the teeth to the index. The geometry is given in Figure 8.55.

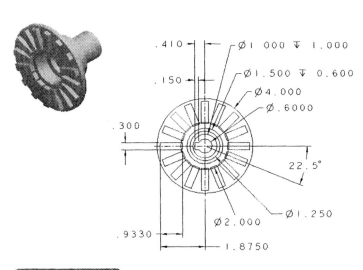
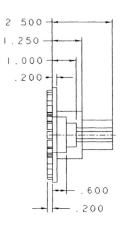

.410
.150
Ø1.000 ∇ 1.000
Ø1.500 ∇ 0.600
Ø4.000
Ø.6000
.300
22.5°
.9330
Ø2.000
Ø1.250
1.8750

2.500
1.250
1.000
.200
.600
.200

FIGURE 8.55

8.10 Construct the model of the hinge. *Hint:* Create one half of the model with one tooth. Pattern the tooth using the 5.000 dimension. Then use *Copy, Mirror,* and *All Feat* options to create the remaining half of the model. Use Figure 8.56.

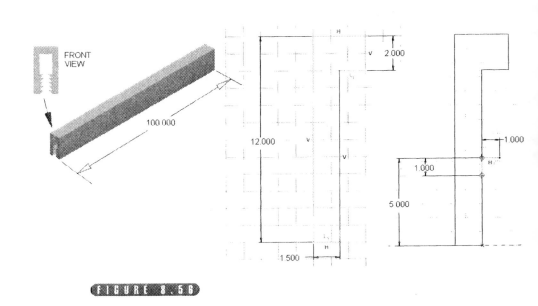

FIGURE 8.56

8.11 Construct the pinion gear. Use Figure 8.57.

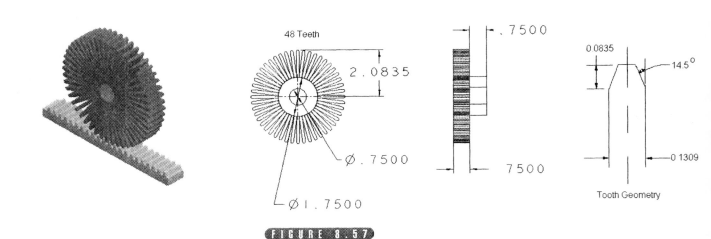

FIGURE 8.57

8.12 Construct the rack. Use Figure 8.58.

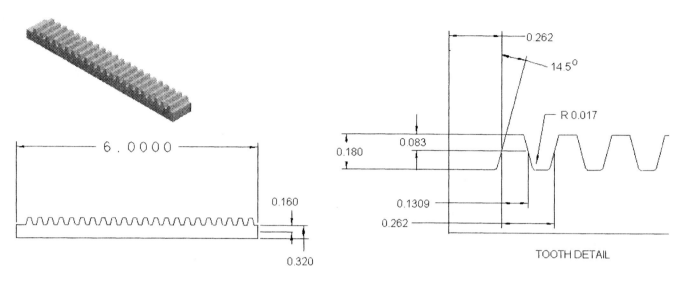

FIGURE 8.58

8.13 Create the model of the spur gear using Figure 8.59 as a guide.

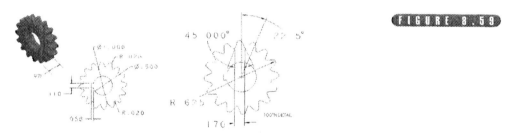

FIGURE 8.59

8.14 Construct the post hinge. *Hint:* The *Group* option is helpful in completing this part. Use Figure 8.60 as a reference.

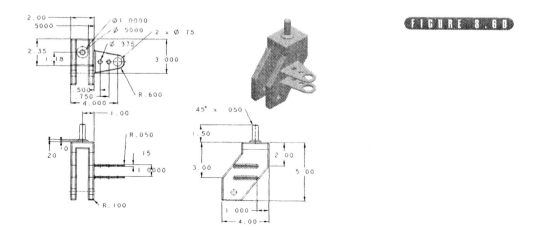

FIGURE 8.60

CHAPTER █9█

REORDER, REFERENCES, SUPPRESS, AND LAYER

INTRODUCTION AND OBJECTIVES

The *Reorder, References,* and *Suppress* options allow the user to control th
generating and displaying of features. The *Resume* option restores the visi
bility of suppressed features.

A part, drawing, or assembly may contain multiple levels. These level
are called layers in Pro/E. Often different types of features are placed o
different layers. Features in a layer are treated as a group. Operations ma
be performed on the group. For example, say it is desired to turn off the dis
play of all the rounds in a part. The rounds may be placed on a layer calle
"All_Rounds" and then the visibility of this layer turned off. Obviously, thi
saves time, because without the use of layers, the visibility of each roun
would have to be deactivated. With names given by the user, layers allo
the user to group features in a logical manner.

The last option that will be considered in this chapter is *Resolve*. Thi
option does not appear on any of the standard menus. It becomes activ
only after a failure occurs in feature construction. The option allows th
user to diagnose and fix the failure.

The objectives of this chapter are to

1. Attain experience in creating *Layers,* and add features to a layer
2. Obtain proficiency in using the feature management options *Reroute, Reorder, Suppress,* and *Resume*
3. Diagnose and fix a failure in creating a feature with the use of *Resolve*

█9.1█ THE *REORDER* AND *REFERENCES* OPTIONS

The order in which a feature or set of features is created is very importan
It is possible that after working on a part, you discover that a feature shoul

have been added to the model before several other features. So what do you do? Do you delete all the features, add the new feature, and then reconstruct those that you delete? Of course not! The correct solution is to reorder the feature before the other features.

When performing this operation, the user must keep in mind the relationship between the features. If a feature is a child of another feature, then it cannot be reordered so that it appears in the regeneration list *before* its parent. A child may be made independent of its parent by using the *References* option. This option allows the user to define new sketching and dimensioning features.

Take, for example, the emergency light cover from Chapter 5. Recall that in constructing the cover, we used the *Shell* option to remove material from the base feature *after* holes had been located in the base feature. Because the shelling process was performed after the holes were added, the holes were shelled as well as the rest of the base, leaving bosses to be used in the assembling process (Figure 9.1(a)).

Suppose that it was our intention not to shell the holes; that is, the bosses are not desired. Can the model be modified with this change in mind? Yes, if the shelling process was performed before the holes were placed (Figure 9.1(b)). The way to do this is to reorder the shell.

9.1.1 TUTORIAL 9.1: AN EXAMPLE USING *REORDER*

This tutorial illustrates:

- Adding a standard hole to a part
- Adding a straight hole to a part

1. Retrieve the model "EmergencyLightBase." Select 📄 and double-click on the file.
2. Consider the model tree for this model, which has been reproduced in Figure 9.2.
3. Select the Shell feature as shown in Figure 9.2 and drag it up the list. Release your mouse in the list just before the Pattern (Hole) feature.
4. Note the updated model tree in Figure 9.3.
5. You do not need to save the model.

9.2 LAYERS AND FEATURES

In Pro/E, a *Layer* is used to blank—that is, hide—certain features during the construction of the part. Often datums, coordinate systems, and dimensions are placed on a different layer because they tend to clutter the drawing. If these features are placed on a separate layer from the rest of the model, then their display can be turned off by using *Blank Layer*. Rounds, fillets, and chamfers are often placed on different layers so that their visibility can be suppressed.

Layer options may be accessed by selecting the *Layer* tool button in the main toolbar 🗐. Doing so will replace the model tree with a tree listing the layers in the model (see Figure 9.4). A pop-up menu may be activated as shown by selecting one of the layers and pressing the right mouse button. Additional layers may be added to the model by either selecting *New Layer* from this pop-up menu or by selecting *Edit* ▬▬ and *New Layer*.

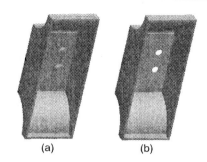

FIGURE 9.1 *Changing the order of the Shell feature leads to (b), where the part no longer has any bosses.*

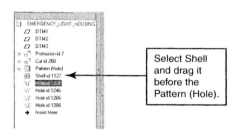

FIGURE 9.2 *The model tree of the part before changing the order of the Shell feature.*

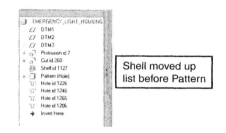

FIGURE 9.3 *The model tree of the part after changing the order of the Shell feature.*

FIGURE 9.4 *From the Layers Model Tree, the user may turn the visibility of different layers on and off.*

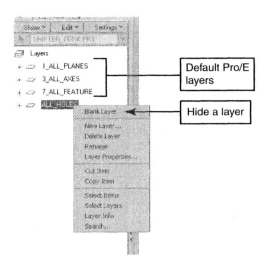

Features may be removed from a layer by selecting the layer in the tree, pressing the right mouse button, and choosing *Layer Properties*. Then, in the *Layer Property* box highlight the feature to be removed by clicking on it and press the *remove* button.

Solid features such as cuts, holes, and so on, may be placed in a layer. However, such features cannot be blanked; rather, the *Suppress* option is used to turn their visibility off. The features may be redisplayed using the *Resume* option.

9.2.1 TUTORIAL 9.2: CREATING A LAYER AND ADDING FEATURES TO THE LAYER

This tutorial illustrates:

- Creating a layer and adding features to the layer

1. Retrieve the model "ShifterFork." Select 🖼 and double-click on the file.
2. Select ⬜.
3. Select *Straight, Simple,* and *Through all* options.
4. Enter a diameter of 1.000.
5. With your mouse, click on the axis A_1 (see Figure 9.5).
6. Press the *Placement* option in the dashboard. Click in the Secondary references field and select the surface (for the first hole) in Figure 9.5.
7. Build the feature by pressing ✓.
8. Select ⬜.
9. Select *Straight, Simple,* and *Through all* options.
10. Enter a diameter of 1.125.
11. With your mouse click on the axis A_2 (see Figure 9.5).
12. Press the *Placement* option in the dashboard. Click in the Secondary references field and select the surface (for the second hole) in Figure 9.5.
13. Build the feature by pressing ✓.

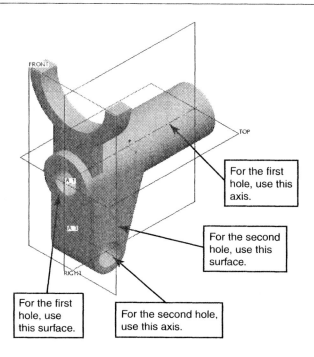

FIGURE 9.5 *The shifter fork with the added holes.*

For the first hole, use this axis.

For the second hole, use this surface.

For the first hole, use this surface.

For the second hole, use this axis.

14. Select ▨. In the layer tree, choose *Edit* ▬▬◀ and then *New Layer*.

15. In the *Layer Properties* box (Figure 9.6), enter the text "ALL_HOLES."

16. Click on the two holes. The features will be added to the layer. Note the entries in the *Layer Properties* box.

17. Select *OK*.

18. Save the part by pressing ▨.

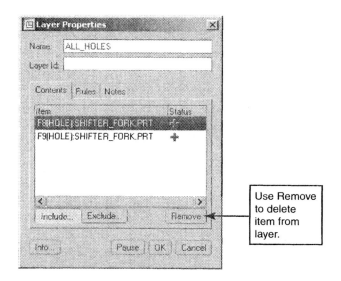

FIGURE 9.6 *When creating a layer, specific items can be chosen from the model.*

Use Remove to delete item from layer.

9.3 SUPPRESS AND RESUME

The *Suppress* option may be used to hide one or more features. The *Resume* option is used to restore any suppressed features. The *Suppress* option may be accessed from the model tree by highlighting the feature, pressing the right mouse button, and selecting *Suppress* from the pop-up menu. In addition, the *Suppress* option may be accessed via the path *Edit* and *Suppress*. Resume is accessible by the path *Edit* and *Resume*.

Suppressing a feature is helpful when the design intent is not clear. This often occurs when one particular feature interferes with another. The feature causing the interference can be suppressed and the model analyzed for the design intent. Later, the suppressed feature can be restored.

The *Suppress* option can also be used when a feature fails to regenerate. In such cases, by suppressing the troublesome feature, the scheme used in generation of the feature and its interaction with the model can be more easily addressed.

9.3.1 TUTORIAL 9.3: SUPPRESSING AND RESUMING FEATURES

This tutorial illustrates:

- Using *Suppress* to turn off the visibility of a feature
- Using *Resume* to turn on the visibility of suppressed features

1. If the model "ShifterFork" is not in memory, select 🖻 and double-click on the file.
2. Suppress the first protrusion in the list. Select the protrusion in the model tree, press your right mouse button, and select *Suppress*.
3. Select *OK*.
4. Now, resume the visibility of the suppressed features. Choose *Edit* and *Resume*.
5. Note that the only certain features may be resumed. Select *All*.

9.3.2 TUTORIAL 9.4: SUPPRESSING FEATURES IN A LAYER

This tutorial illustrates:

- Suppressing and resuming features by layer

1. If the model "ShifterFork" is not in memory, select 🖻 and double-click on the file.
2. Select *Feature operations* and *Suppress*.
3. Choose *Layer*. Click on the layer "ALL_HOLES."
4. Select *Done Sel* and *Done*.
5. Resume the suppressed features. Choose *Edit*, *Resume*, and *Last*.

9.4 THE *RESOLVE* OPTION

The *Resolve* option is used to diagnose and fix a failed regeneration. Note that a failed feature *must* be resolved before continuing any further.

Several options are available for resolving a failed feature. These options are:

1. *Undo Changes.* Remove the changes made to the model and return to the previous successful regeneration.
2. *Investigate.* Using the *Investigate* menu, attempt to find the cause of the failure. The *Investigate* menu contains the option *Show Ref.* This option shows the references of the failed feature.
3. *Fix Model.* Fix the model by using various options.
4. *Quick Fix.* Fix the model by redefining the elements of the failed feature.

9.4.1 TUTORIAL 9.5: RESOLVING A FAILED REGENERATION

This tutorial illustrates:

- Investigating a failed feature
- Using *Resolve* to fix a failed feature
- Resolving a failure by using *Undo Changes*
- Resolving a failure by using *Quick Fix*

1. Retrieve the model "EmergencyLightHolder." Select and double-click on the file.
2. Round the edges of the part. Select ▨.
3. Enter a round of 0.1000. Select the edges shown in Figure 9.7. *Remember:* Press the CTRL to pick additional edges.
4. Build the feature. Press ✔.
5. Now, select the round feature in the model tree. Press your right mouse button and choose *Edit.*
6. Click on the round value on the screen and change it to 0.150. Hit *return.*
7. Choose *Edit* and *Regenerate.* The feature will fail and the *Failure Diagnostics* box will appear (see Figure 9.8).
8. Select *Overview* to see the alternatives for fixing the failure. Choose *Close.*

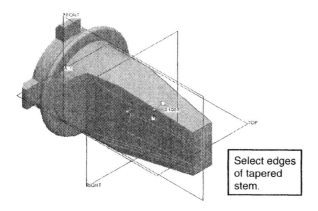

FIGURE 9.7 *Use the shown edges as the edges for the round.*

Select edges of tapered stem.

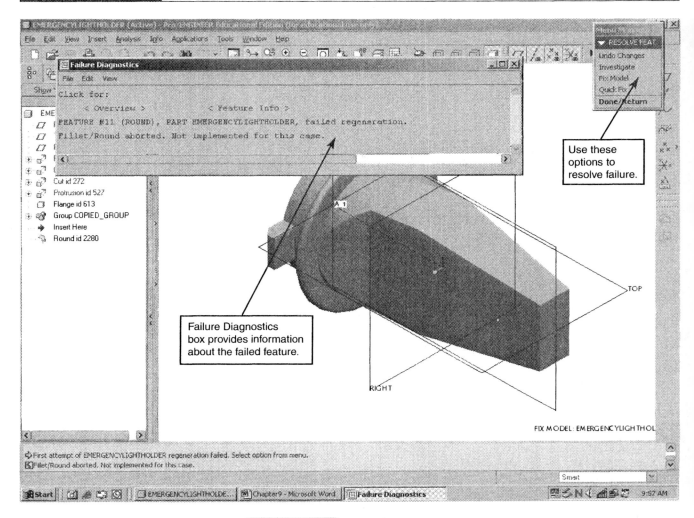

FIGURE 9.8 *The* RESOLVE FEAT *menu may be used to resolve a failed regeneration.*

9. Now, press *Investigate* and *List Changes* to see the changes that have been made to the model. Select *Close*.

10. Select *Show Ref.* Note that this feature has no children. This is important because any changes you make will affect the children.

11. Note the parents of this feature. Click on the parents to show them. Pressing 🔳 will list information about the feature and its parents. Select *Close*.

12. Choose *Undo Changes. Confirm.*

13. Once again, select the round feature in the model tree. Press your right mouse button and choose *Edit.*

14. Click on the round value on the screen and change it to 0.150. Hit *return.*

15. Choose *Edit* and *Regenerate.* The feature will fail.

16. Choose *Quick Fix.* Note the choices available.

17. Select *Suppress, Confirm,* and *Yes.* This will remove the feature from the generation list.

18. Select *Edit, Resume,* and *Last.* This will restore the failed feature and redisplay the *Failure Diagnostics* box.

19. Select *Redefine.* In the dashboard change the value of the round to 0.1 and hit *return.*

20. Press ✔. Choose *Yes.*

21. Save the part by selecting 🖫.

9.4.2 TUTORIAL 9.6: ANOTHER EXAMPLE WITH *RESOLVE*

This tutorial illustrates:

- Investigating a failed feature
- Using *Resolve* to fix a failed feature
- Resolving a failure by using *Quick Fix*
- Resolving a failure by changing the section of the feature

1. Retrieve the model "EmergencyLightHolder." Select 🖼 and double-click on the file (Figure 9.9).

2. Now, select the flange feature in the model tree. Press your right mouse button and choose *Edit.*

3. Click on the round value on the screen and change it from 0.024 to 0.050. Hit *return.*

4. Choose *Edit* and *Regenerate.* The feature will fail and the *Failure Diagnostics* box will appear.

5. Note the error message: "Could not intersect part with feature."

6. Choose *Investigate, Current Modl,* and *Failed Geom.*

7. Note the curves that have appeared on the model (see Figure 9.9). The curves indicate that the feature can be created only along the curves.

8. Select *Item Info.* Note the recommendation: "Redefine the feature so that it intersects the part."

9. Now, pick *Quick Fix.* Choose *Redefine, Confirm, Section, Done,* and *Sketch.*

10. Pick 🖱. Click on the two lines shown in Figure 9.10 and then press your *delete* key.

11. Draw two new longer lines as shown in Figure 9.11. Dimension the lines by using 🖴. Modify the value to that given in Figure 9.11.

12. Build the feature by pressing ✔.

13. Choose *Yes.*

14. DO NOT save changes made in this tutorial.

9.5 SUMMARY AND STEPS FOR USING THE OPTIONS IN THIS CHAPTER

Feature management is the subject of this chapter with regard to the options *Reorder, Reroute, Layer, Suppress, Resume,* and *Resolve.* Features may be placed on a different level within the model with the *Layer* option.

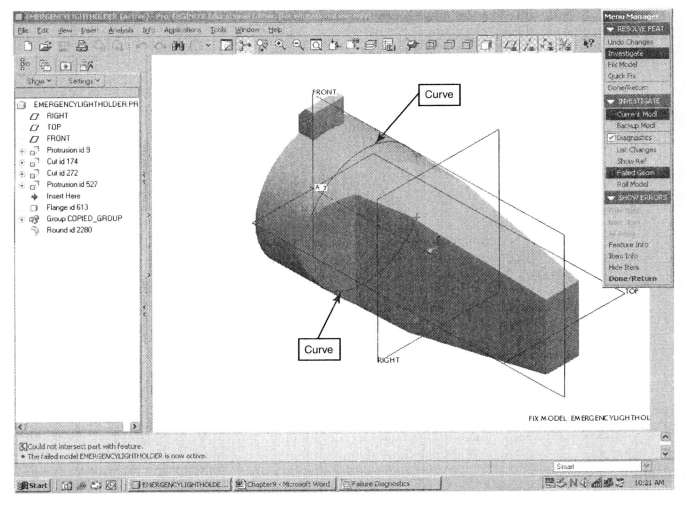

FIGURE 9.9 *The software highlights the failed feature.*

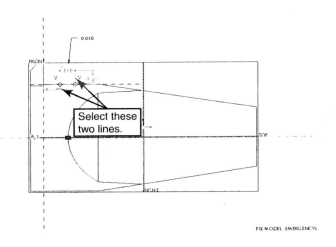

FIGURE 9.10 *Redraw the flange geometry with the entities shown.*

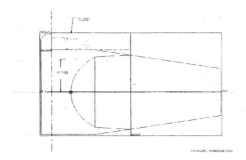

A feature or set of features may be removed from the generation list with the *Suppress* option and placed back on the list using the *Resume* option. The *Reorder* option may be used to change the order in which a feature is generated, with respect to the other features in the model. If a feature is a child of another feature, it may be moved ahead of its parent by using the *Reroute* option. Finally, the *Resolve* option may be used to correct a failed regeneration by diagnosing the failure.

The general sequence of steps to follow for reordering a feature is:

1. Select the feature in the model tree.
2. Drag the feature up or down the tree list as desired.

The following sequence of steps may be used to create a layer:

1. Select 🖺. In the layer tree, choose *Edit* ▭▭ and then *New Layer*.
2. In the *Layer Properties* box enter the name of the layer.
3. Click on features to be included in the layer.
4. Select *OK*.

After a layer has been created, features may be added to the layer using the following procedure:

1. Select 🖺.
2. In the layer tree, choose the desired layer.
3. Press the right mouse button and select *Layer Properties* from the pop-up menu.
4. Click on features to be included in the layer.
5. Select *OK*.

Solid features may be removed from the generation list by using the *Suppress* option. The features may be in a specific layer. In order to suppress one or more features use the following procedure:

1. Select the feature in the model tree.
2. Press your right mouse button and select *Suppress*.

If the desired features are contained in a layer, then use the following procedure:

1. Select Feature Operations and *Suppress*.
2. Choose *Layer*. Pick the name of the layer in the list.
3. Select *Done Sel*.
4. Choose *Done*.

Pro/E will regenerate the model without the suppressed features. The features may be placed back in the generation list by using the *Resume* option. In general, this option may be used as follows:

1. Select *Edit* and *Resume.*
2. Choose *All* or *Last* as needed.

In order to resume features in a layer, use the following steps:

1. Select *Feature operations, Resume,* and *Layer.*
2. Choose the layer by name from the list.
3. Choose *Done Sel* and *Done.*

The model will be regenerated and the suppressed feature restored.

9.6 ADDITIONAL EXERCISES

9.1 Create a layer called "Hole_Layer" for the "EmergencyLightHolder." Add a 0.250-diameter, 0.500-deep coaxial hole to the model. Then place on the hole in the layer. Use Figure 9.11 as a reference.

9.2 For the "ShifterFork" model (Figure 9.12), create two layers; call the first layer "All_Rounds" and the second layer "Cut."

9.3 Create 0.50-radii rounds, as shown in Figure 9.12, for the part "ShifterFork" and add these rounds to the layer "All_Rounds."

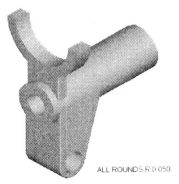

ALL ROUNDS R 0.050

9.4 Complete the support in the model "ShifterFork" using a 1″ wide, 2.25″ high rectangular cut. Add this feature to the layer "Cut." The geometry of the cut is shown in Figure 9.13.

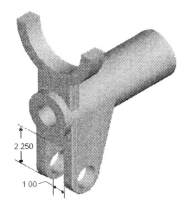

2.250

1.00

CHAPTER 10

THE DRAWING MODE

INTRODUCTION AND OBJECTIVES

So far, we have studied only options available to create models using Pro/E. However, it is often desired to obtain special views of the model in order to convey its features. This can be done in several ways. A standard method is the use of multiviews.

Typically, multiview drawings are obtained of the top, front, and side of the model. The views are fully dimensioned and tolerances are specified. If desired, a pictorial, such as an isometric, is often added to further clarify the features inherent in the model.

The module for creating multiviews in Pro/E is the *Drawing* mode. Similar to the approach used for part files, the software allows the user to create drawing files. A standard format may be loaded during the creating of the drawing file.

The objectives of this chapter are to

1. Become familiar with the drawing mode
2. Retrieve or create a format
3. Create a multiview drawing and add dimensions and notes to the views

10.1 DRAWINGS AND FORMATS

A drawing is a layout containing various views of a model. Most drawings contain a front, top, and right-hand side view of the model. Often, the drawing may contain a section of the model created by slicing the model with a plane and displaying one side of the cut model.

All drawing files are designated with the extension ".drw" by Pro/E. In order to create a drawing, select either *File* and then *New* or the icon. As shown in Figure 10.1, choose the drawing option and enter the name of the drawing.

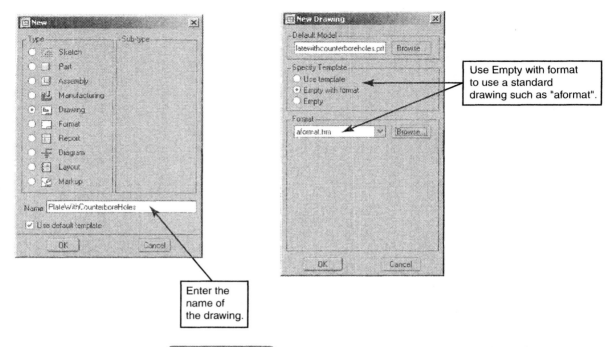

Enter the name of the drawing.

Use Empty with format to use a standard drawing such as "aformat".

FIGURE 10.1 *Create a drawing by using the* New *and* New Drawing *dialog boxes.*

The size and orientation of the drawing may be defined by the *New Drawing* dialog box illustrated in Figure 10.1. If desired, enter the name of the model in the cell provided. You can also give the name of the model when the view is added to the drawing. Select the desired paper size or retrieve a format using the *Empty with format* option. In general, we suggest that the *Landscape* orientation be used.

A format is nothing more than a boundary for the drawing, with spaces for your name, date, name of the model, etc. The Pro/E software is bundled with standard formats. Any of these formats may be chosen by simply selecting the paper size. However, a simple format contained in the file "Aformat.frm" is bundled with this book and may be downloaded from the Prentice Hall website (www.prenhall.com/rizza). This simple format is provided so that the user may complete the tutorials. The format is designed for A-size paper and *Landscape* orientation. You will need to make sure that you have access to this file.

Custom formats may be created with the *Format* mode. The geometry of the format may be created by sketching the entities. In order to create a format, choose *File* and *New*, or the ▢ icon. Then, choose the *Format* option, as shown in Figure 10.2. Give the format a name and select *OK*. Select the paper size and orientation.

In adding views to a drawing, the model must be defined, and Pro/E must have access to the model. If the part file is not located in the default directory, then the directory must be changed or the file moved.

When adding the first view, the drawing may be scaled, if desired. If no scale is given, then Pro/E scales the initial view so that additional views may be added to the drawing. The scale size is written at the bottom of the

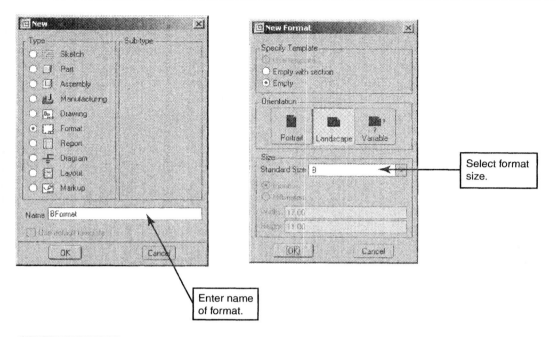

Create a format by using the New *and* New Format *dialog boxes.*

drawing window and may be modified by clicking on *Edit, Value,* and then on the scale with your mouse pointer.

The first view added to the drawing is a *General* view. Subsequent views may be *Projections, Auxiliary, Revolved,* and *Detailed* views. Most often, we will be using the *Projection* option to create additional views. A *Detailed* view shows a portion of the model. Detail views are helpful in highlighting special features inherent in the model. *Auxiliary* views are special views. Such views are usually used to illustrate incline or oblique surfaces. A *Revolved* view is a view created by revolving a section 90° around a cutting plane. Revolved views are helpful in illustrating cross sections.

10.1.1 TUTORIAL 10.1: CREATING A FORMAT

This tutorial illustrates:

- Construction of a standard format
- Changing drawing options

1. Begin the creation of this format by selecting *File* and *New* or 🗋.
2. In the *New* dialog box (Figure 10.2), choose the *Format* option and enter the name "BFormat."
3. Select the *OK* button.
4. In the *New Format* dialog box choose *Landscape* orientation, B size paper, and *Empty*.
5. Then, press the *OK* button. As shown in Figure 10.3, note that the paper size is indicated below this outline.

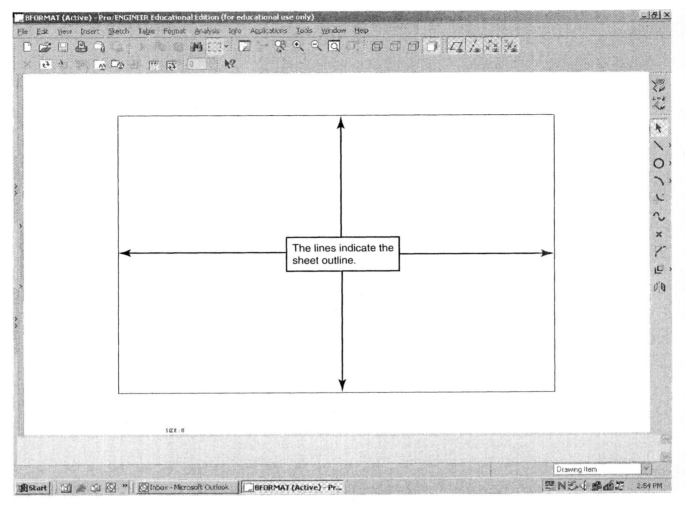

The lines indicate the sheet outline.

FIGURE 10.3 *Note the boundaries of the format.*

6. First, let us change some of the parameters describing the format. Select *File* and *Properties*.

7. The software will display the drawing configuration file for this format. Change *drawing_text_height* to 0.3 by double-clicking on the default value. Enter the new value and select ▭.

8. Repeat step 7 and change *draw_arrow_style* from *Closed* to *Filled*.

9. Select *Apply* button. Then, select *Close*.

10. Pick ▭.

11. Select *Sketch* and *Specify Absolute Coordinates* (see Figure 10.4).

12. Enter the *x* coordinate of the first point given in Table 10.1.

13. Enter the *y* coordinate of the first point given in Table 10.1.

14. Select ▭.

15. Repeat steps 11 through 14 for the remaining points in Table 10.1.

16. Add the text given in Table 10.2. Select *Insert* and *Note*.

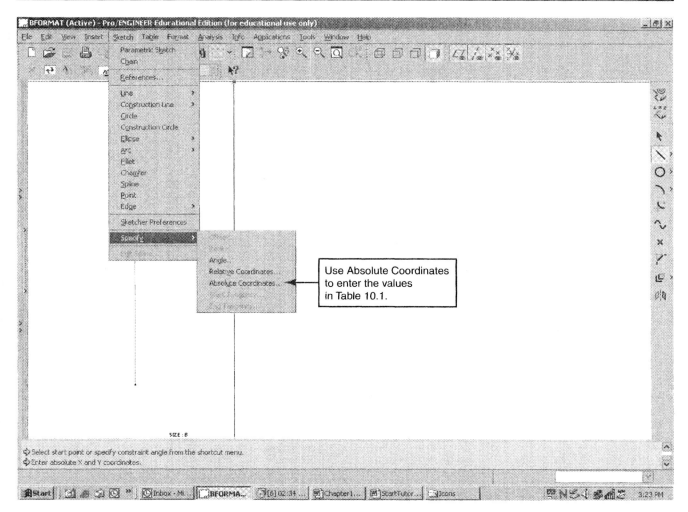

FIGURE 10.4 *Specify the location of the lines by using coordinates.*

17. In this case, select the defaults *No Leader, Horizontal, Standard,* and *Default* justification.

18. Select *Make Note.*

19. In this case, choose *Abs Coord.*

20. Enter the *x* coordinate of the text using Table 10.2.

21. Enter the *y* coordinate of the text using Table 10.2.

22. Enter the text and hit *return.*

23. Repeat steps 18 through 22 for the remaining text in Table 10.2.

24. Select *Done/Return.*

25. Add the name of your university or company to the 4" wide by 1" high affiliation space (see Figure 10.5). Since the size of this text varies with each organization, we cannot give a sequence of steps for this task. However, the text may be added by using *Abs Coord* option.

26. Save the format by using ⬚. The format is shown in Figure 10.6.

TABLE **10.1** COORDINATES OF THE LINES NEEDED TO COMPLETE THE FORMAT

LINE NO.	AXIS	FIRST POINT	SECOND POINT
1	x	0.5	0.5
	y	10.5	0.5
2	x	0.5	16.5
	y	10.5	10.5
3	x	16.5	16.5
	y	10.5	0.5
4	x	0.5	16.5
	y	0.5	0.5
5	x	0.5	16.5
	y	1.5	1.5
6	x	4.5	4.5
	y	1.5	0.5
7	x	10.5	10.5
	y	1.5	0.5
8	x	11	16
	y	1	1
9	x	5	10
	y	1	1

TABLE **10.2** TEXT AND THE LOCATION OF THE TEXT NEEDED TO COMPLETE THE FORMAT

TEXT NO.	AXIS	COORDINATES	TEXT
1	x	11	NAME:
	y	1.2	
2	x	11	DATE:
	y	0.8	
3	x	5	TITLE:
	y	1.2	

10.1.2 TUTORIAL 10.2: CREATING A DRAWING

This tutorial illustrates:

- Creating a drawing
- Adding views to a drawing
- Positioning views in a drawing
- Sketching geometry in a drawing

1. Select *File* and *New* or click on ▯.
2. Choose the drawing mode. Enter the name: "SquarePlatewith-Holes."

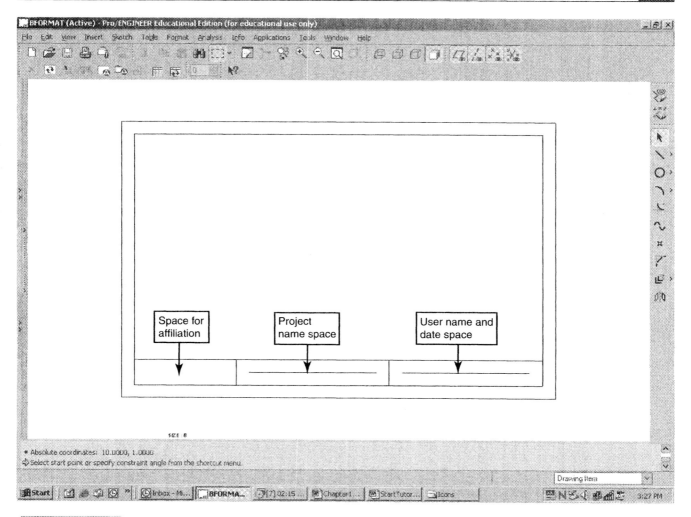

FIGURE 10.5 *The remaining lines are added to the format. This creates the spaces for the necessary information.*

3. Click on the *OK* button.

4. Use *Browse* or enter the model name (SquarePlatewithHoles.prt) and then select the *Empty with format* option.

5. Use *Browse* or enter "AFormat" in the cell provided.

6. Choose *OK*.

7. Change the display to isometric. Choose *Tools, Environment,* and change the default orientation to *Isometric.*

8. Select *OK* to save the change.

9. Add the first view. Choose 🔲.

10. Accept the defaults shown in Figure 10.7 (the *VIEW TYPE* menu) by selecting *Done.*

11. Click somewhere in the lower left-hand portion of the screen. Pro/E will place an isometric view of the model.

12. Reorient the model so that it appears as in Figure 10.7. Select datum plane Front as *Reference 1* and Top as *Reference 2.* Choose *Done.*

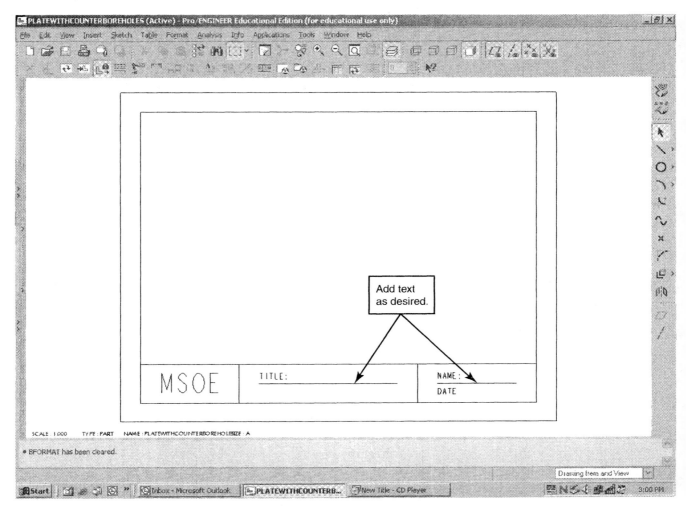

FIGURE 10.6 *The format is reproduced with the added text.*

13. Obviously, the view in Figure 10.7 needs to be rescaled. Select *Edit* and *Value*. Click on the scale (the value 0.167), and enter a value of 0.250. The resized view is shown in Figure 10.8.

14. Add a top view. Choose 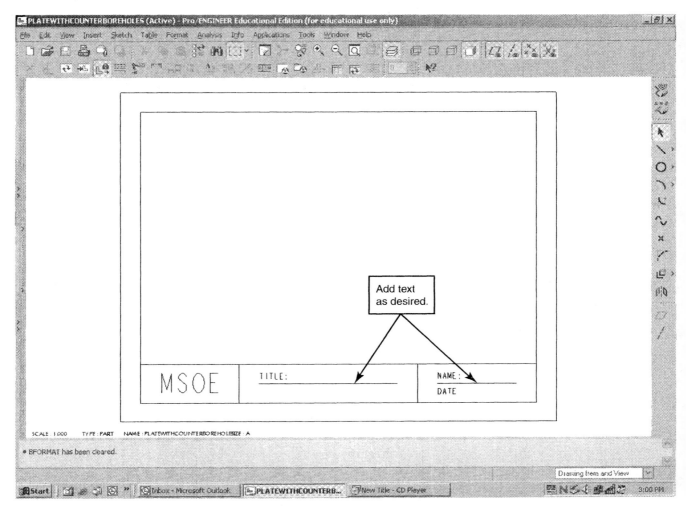.

15. Accept the defaults in the *VIEW TYPE* menu by selecting *Done*.

16. Click somewhere above the first view.

17. Choose .

18. Accept the defaults in the *VIEW TYPE* menu by selecting *Done*.

19. Click somewhere to the right of the first view.

20. A right-hand side view will be added. Your drawing should now appear as in Figure 10.9.

21. Pro/E will place an isometric view of the model.

22. The views might need to be repositioned. Make sure is not pressed. If it is, press it to allow dragging of the views.

23. Click on the first view and move it as desired. Place the view so that there are equal borders around all the views.

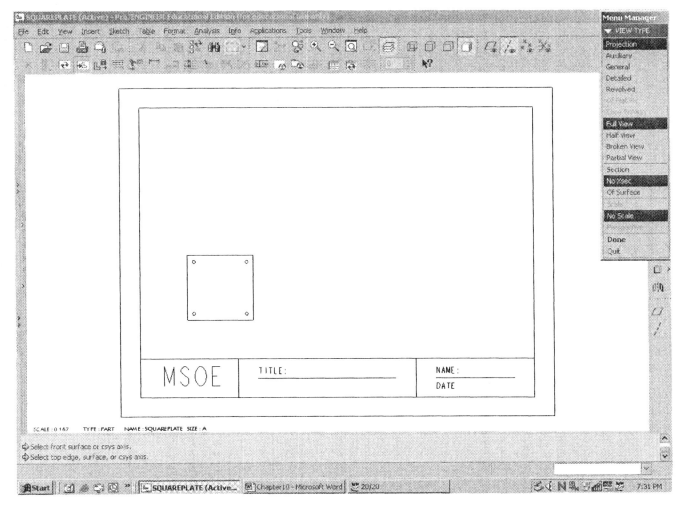

FIGURE 10.7 *The drawing with the first view added.*

24. Add an isometric view of the model. Select .

25. Choose *General* from the *VIEW TYPE*.

26. In comparison to the other views, this view will be too big to fit on the drawing. Therefore, we need to scale the view. Choose *Scale*.

27. Select *Done*.

28. Click in the upper right-hand side of the screen.

29. Enter a value of, say, 0.125 for the scale (this can always be changed).

30. Select *OK* from the *Orientation* dialog box.

31. Since the new view is of different scale from the rest, it is convenient to sketch a border around this view. Choose .

32. Then sketch a horizontal and vertical line, as shown in Figure 10.10. The lines may be deleted by clicking on them and then hitting the *delete* key.

33. Save the drawing by pressing .

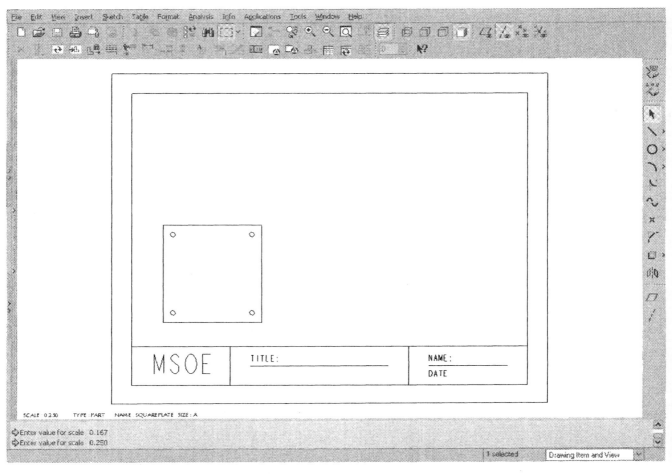

FIGURE 10.8 *Change the size of the drawing.*

10.2 ADDING DIMENSIONS AND TEXT TO A DRAWING

Dimensions that were used to construct a model (driven dimensions) can be displayed on the drawing by using *View* and *Show and Erase* or by selecting ▥. After selecting the *Show and Erase* command, the dialog box in Figure 10.11 appears. The buttons across the top of the box are from left to right: dimension, reference dimension, geometric tolerance, note, balloon, axis, symbol, surface finish, datum plane, and cosmetic feature. If you place your mouse pointer on any of these buttons and leave it there, the name of the button will be displayed.

Notice that any of these options can be activated per view, for each feature or for the entire part. When activating the option per feature, you will need to click on the desired feature. Selection of multiple features is allowed; simply click on each feature in succession.

Often when using the *Show* option in placing dimensions, Pro/E places the dimension in an inconvenient location and dimensions may be on top of or too close to another dimension. Luckily, by simply selecting the

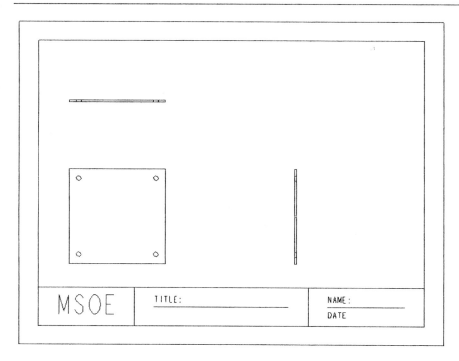

FIGURE 10.9 *Additional views may be created as projections of the first.*

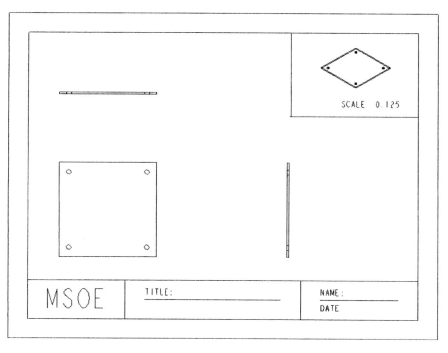

FIGURE 10.10 *An isometric pictorial may be added to the drawing. Lines may be sketched to separate this view, which is of different scale from the rest of the drawing.*

dimension with the mouse and dragging it to a new location, the user can move the text and dimension.

The *Erase* command can be used to hide any driven dimension. The corresponding dialog box is shown in Figure 10.11.

You can create your own dimensions by clicking on *Insert* and then on *Dimension*. The dimension is constructed by selecting the appropriate entities.

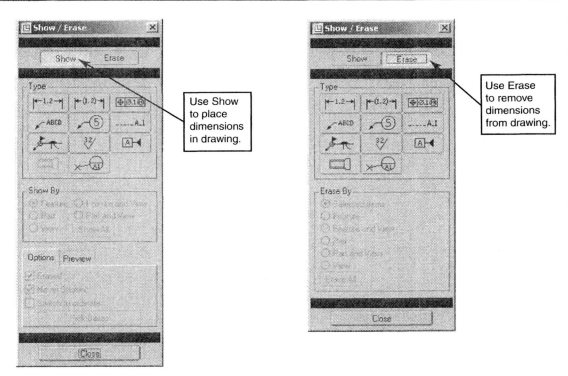

FIGURE 10.11 *Driven dimensions may be displayed or hidden by using the* Show *and* Erase *options,* respectively.

In placing the dimension values between the leaders, you may find the *Text Style* option (Figure 10.12) useful. By changing the height or width of the text, you can make the text fit between the leaders. The *Text Style* box can be used to change the appearance and orientation of any text on the drawing. The sequence for activating these options is *Format* and *Text Style*.

FIGURE 10.12 *The* Text Style *box is shown.*

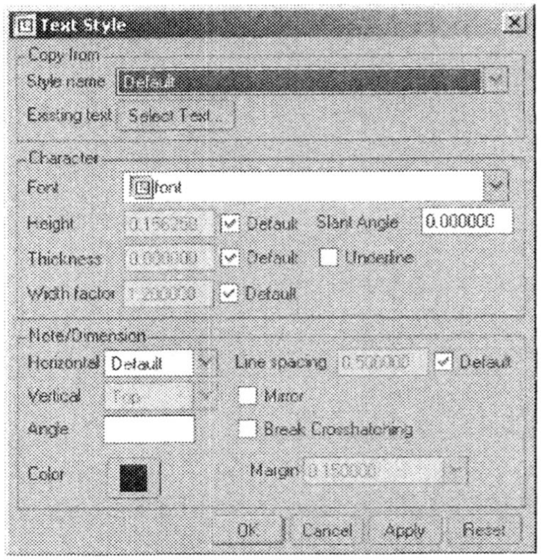

By selecting *Format* and *Arrow Style* the user can change the style of the arrowhead type.

Pro/E supports eight different arrowhead styles. These styles are:

1. *Default.* The default style depending on the type of dimension.

2. *Arrowhead.* A single triangle is drawn on the end of the line. Whether the arrow is Filled or Closed is controlled by the *draw_arrow_style* option in the drawing setup file.

3. *Dot.* A dot is placed at the end.

4. *Double Arrow.* A double triangle is drawn at the end of the line. This symbol is used to indicate a clipped dimension.

5. *Slash.* A slash is drawn at the line's end. The slash is constructed by drawing a line 45° to the dimension line.

6. *Integral.* An integral sign is used for the end of the dimension line.

7. *Box.* The line is completed with a rectangular box.

8. *None.* No ending is placed on the line.

It is possible to modify a part without leaving the drawing mode and restarting the part mode. By selecting *Edit* and *Dimension Scheme,* the user can redefine a selected feature.

10.2.1 TUTORIAL 10.3: DIMENSIONS AND TEXT FOR THE DRAWING OF THE PLATE

This tutorial illustrates:

- Adding dimensions to a drawing using *Show and Erase*
- Positioning dimensions and text
- Modifying dimension text
- Moving dimensions from one view to another

1. Retrieve the drawing "SquarePlatewithHoles" using ⬛.

2. Press ⬛ or select *View* and *Show and Erase.*

3. Press the *Dimension* button as shown in Figure 10.13. Also, select *Feature.*

4. Click on the plate in the front view.

5. Select *OK.*

6. We will keep the two 10.000 dimensions as well as the 2.000 dimensions, so choose *Sel to Keep* and click on these dimensions. Use the CRTL key to pick each additional dimension. Choose *OK* and *Close.*

7. Use ⬛ and click on each dimension one at a time. Drag it to the desired position. Use Figure 10.14 as a reference.

8. Press ⬛. Press the *Dimension* button and *Feature* option.

9. Click on one of the holes shown in Figure 10.15. Select *OK.*

10. We will keep all the dimensions. Select *Accept All.* Then choose *Close.*

11. Move the diameter dimension away from the view as shown in Figure 10.16. Select the 1.000 dimension in the top view. Press your right mouse button and select *Move Item to View.* Click on the front view.

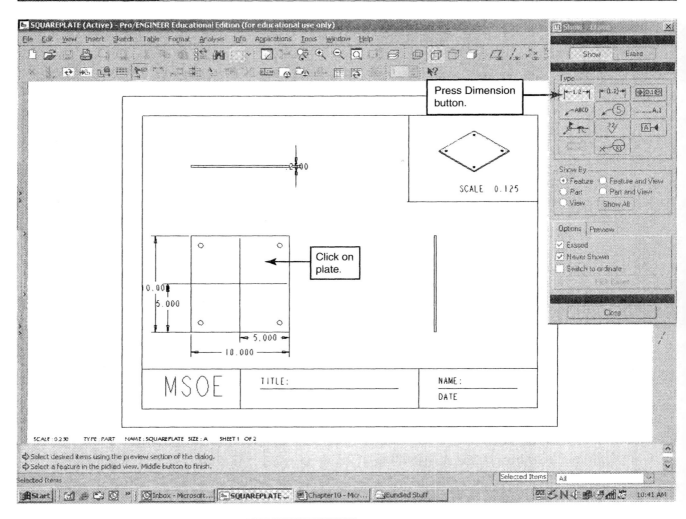

FIGURE 10.13 *The drawing is shown after displaying the dimensions of the plate.*

12. In the front view, drag the 1.000 dimension to position (see Figure 10.16).

13. Click on the 1.000 dimension in the right-hand view. Press your right mouse button and select *Move Item to View*. Click on the front view.

14. Drag this dimension into position. Use Figure 10.17 as a reference.

15. Press █. Press the *Axis* button. Deselect the *Dimension* button. Press the *Show All* button. Choose *Yes, Accept All,* and *Close*.

16. All the axes will appear. Turn off the display of their labels by deselecting ▟.

17. With the mouse pointer, click on the diameter dimension. Press your right mouse button and select *Flip Arrows*.

18. Grab the handle in Figure 10.18 and move the text to the left of the leader as in Figure 10.19.

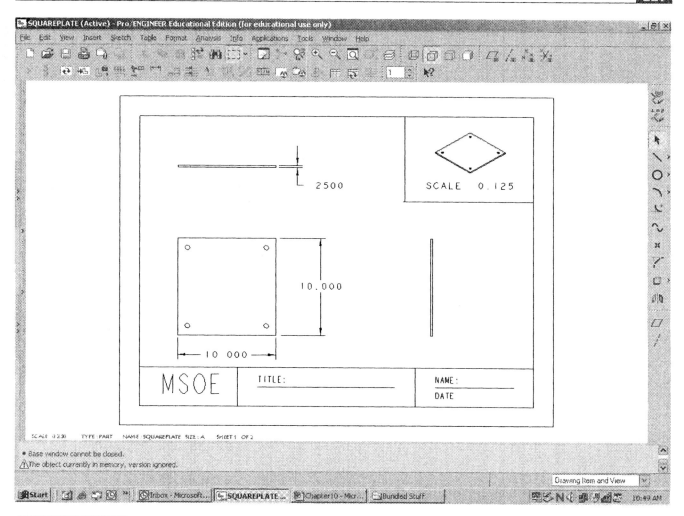

FIGURE 10.14 *Move the dimensions to roughly the same positions shown.*

19. Select *Insert* and *Note*.

20. Select the defaults No Leader, Horizontal, Standard, and Default justification.

21. Select *Make Note*.

22. Click somewhere near the lower right-hand corner of the drawing.

23. Type in: SCALE 1=4. Hit *return* twice. Position the text as shown in Figure 10.19.

24. Repeat the process and add your name, title, date, and affiliation as shown in Figure 10.19. If you need to modify the style of the text—in particular, the size—double-click on the text and select *Text Style*.

25. There are four holes, so double-click on the diameter dimension. Select the Dimension Text tab and enter "4X" as shown in Figure 10.20. Select *OK*.

26. Press [icon] and save the drawing.

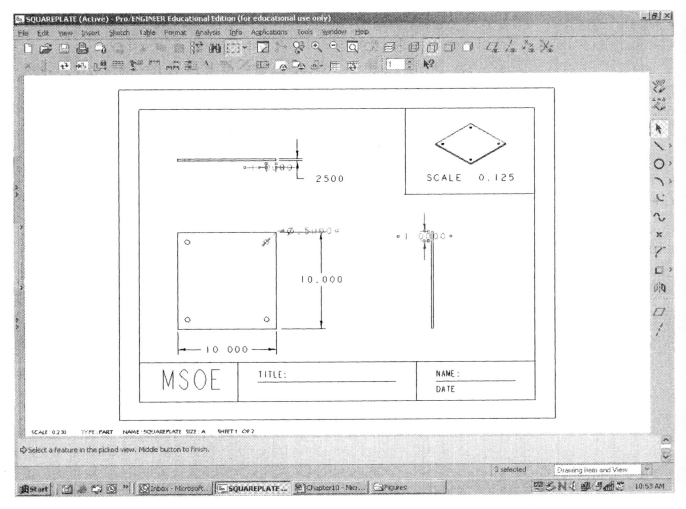

FIGURE 10.15 *Add the dimensions for the hole.*

10.3 AUXILIARY VIEWS AND PRO/E

An auxiliary view is constructed from an image plane that is not a frontal, horizontal, or profile plane. Auxiliary views are useful in visualizing an oblique plane; that is, a plane that does not appear true size and true in any orthogonal view. An example of an oblique plane is shown in Figure 10.21. In order to visualize the hole on surface A in true size and shape, a projection must be constructed on a plane parallel to surface A and then revolved.

An auxiliary view can be created with Pro/E by using the *Drawing* mode after a model has been created. The auxiliary view is then constructed from a frontal, horizontal, or profile view.

10.3.1 TUTORIAL 10.4: A DRAWING OF RODSUPPORT WITH AN AUXILIARY VIEW

This tutorial illustrates:

- Creating a drawing with orthogonal and auxiliary views

1. Select *File* and *New* or click on ▢

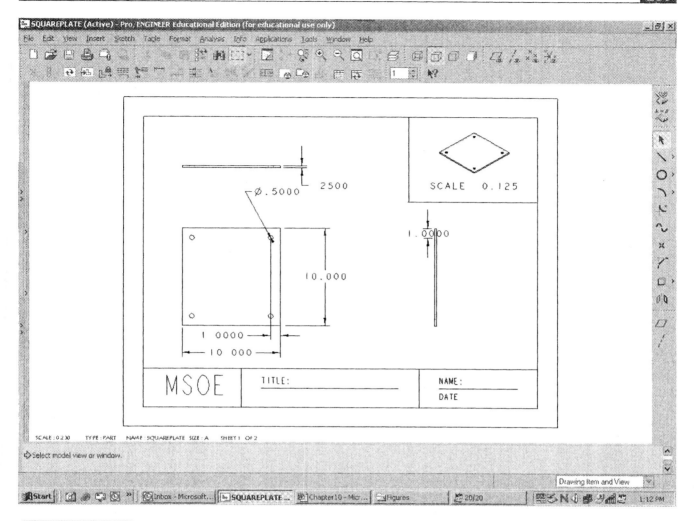

FIGURE 10.16 *Move the dimensions of the hole to the same position shown.*

2. Choose the drawing mode and enter the name: "RodSupport."
3. Click on the *OK* button.
4. Use *Browse* or enter the model name (RodSupport.prt) and then select the *Empty with Format* option.
5. Use *Browse* or enter "AFormat" in the cell provided.
6. Choose *OK*.
7. Change the display to isometric. Choose *Tools, Environment,* and change the default orientation to *Isometric*.
8. Select *OK* to save the change.
9. Add the first view. Choose ▨.
10. Accept the defaults in the *VIEW TYPE* menu by selecting *Done*.
11. Click somewhere in the lower left-hand portion of the screen. Pro/E will place an isometric view of the model.

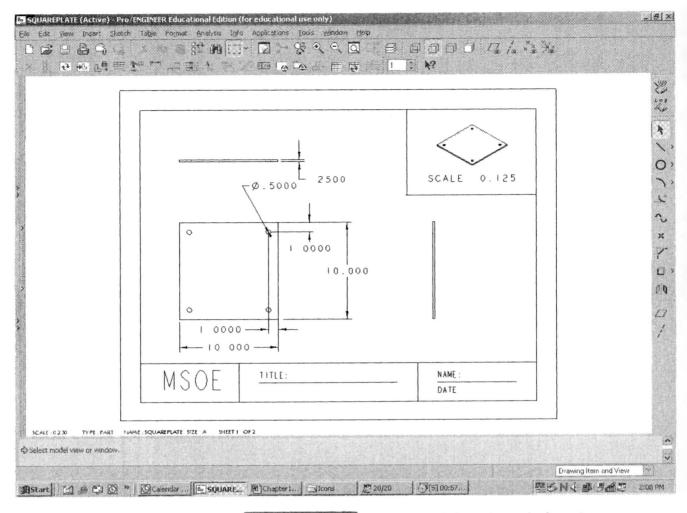

FIGURE 10.17 *Move the 1.000 dimension to the front view.*

12. Reorient the model so that it appears as in Figure 10.22. Select datum plane Front as *Reference 1* and Top as *Reference 2*. Choose *Done.*

13. Add a top view. Choose .

14. Accept the defaults in the *VIEW TYPE* menu by selecting *Done.*

15. Click somewhere above the first view.

16. Choose .

17. Accept the defaults in the *VIEW TYPE* menu by selecting *Done.*

18. Click somewhere to the right of the first view.

19. A right-hand side view will be added.

20. Choose .

21. Select *Auxiliary* and *Half View.* Then click on *Done.*

22. Locate the half view by clicking on the screen near the front view. Again, choose Edge A.

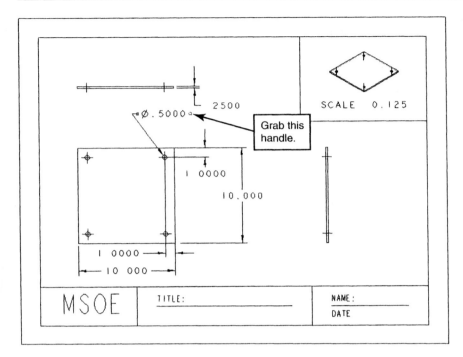

FIGURE 10.18 *Grab the handle to move the dimension text.*

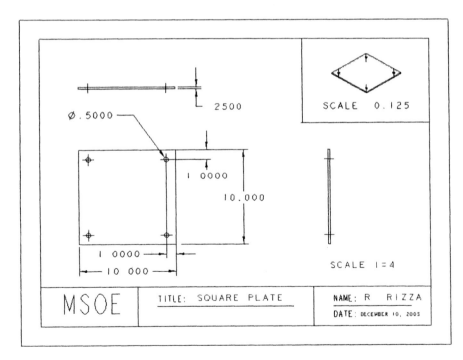

FIGURE 10.19 *With all the dimensions properly located, your drawing should appear as in this figure. The drawing is shown with the dimension and appropriate notes added.*

23. Pro/E will place the view with all the datum planes showing. Click on the DTM2 when asked to do so. Arrows will appear indicating the direction of the view to be saved. These arrows should point toward the single hole. If this is the case, choose *OK;* otherwise flip the arrows.

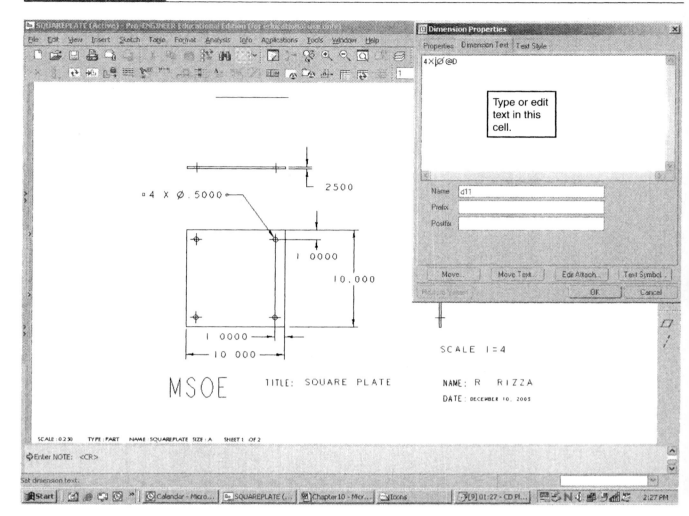

FIGURE 10.20 *Add the text "4X" to the diameter dimension.*

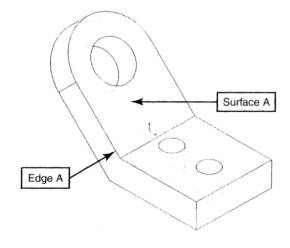

FIGURE 10.21 *An example of a part with an inclined oblique plane. Plane A will not appear true size or true shape on any orthogonal view.*

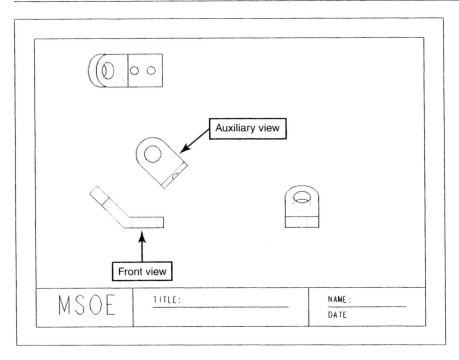

FIGURE 10.22 *A drawing containing the three primary orthogonal views and an auxiliary view of surface A. The drawing is incomplete since dimensions would normally be added.*

24. The partial auxiliary view will contain only the part of the model above the DTM2 plane. Your drawing should appear as in Figure 10.22.

25. Save the drawing.

10.4 SUMMARY

The drawing containing various multiviews of a model may be created by using the *Drawing* mode. The software creates drawing files in a manner analogous to the creation of part files.

A standard format can be added to the drawing, but it must be added during the creation of the drawing file. This is accomplished by using the *Empty with format* option. Format files may be created by using the *Format* mode. The geometry in the format is created by sketching entities.

After a drawing is created, views may be added to the drawing by using . The first view added to the drawing is a *General* view. Subsequent views may be *Projections, Auxiliary, Revolved,* and *Detailed* views.

Dimensions that were used to construct the model can be displayed in the drawing by using the *Show/Erase* option. Such dimensions often need to be moved in the drawing. This is accomplished by simply dragging the dimension with the mouse. Additional dimensions can be added to the drawing by using the *Insert* and *Dimension* options. The user needs to select the entities defining the dimension.

10.4.1 STEPS FOR CREATING A FORMAT

In general, a format may be created by using the following sequence of steps:

1. Choose *File* and *New* (or select ▢).
2. Select the *Format* option.

3. Enter a name for the format and select *OK*.

4. Choose the paper size and orientation. Select *OK*.

5. Draw the entities in the format using the *Sketch* option.

6. Save the format.

10.4.2 STEPS FOR CREATING AND ADDING VIEWS TO A DRAWING

The general procedure for creating a drawing is as follows:

1. Choose *File* and *New* (or select ⬚).

2. Select the *Drawing* option.

3. Enter a name for the drawing and select *OK*.

4. Choose the paper size and orientation.

5. Select *Empty with format* and enter the name of the format.

6. Choose *OK*.

Views may be added to the drawing with the *Add View* option by following this procedure:

1. Select ⬚.

2. Select the view type from the *VIEW TYPE* menu. Note that the first view placed in the drawing must be of type *General*.

3. Choose *Done*. Click on the screen.

4. Reorient the view if desired.

5. If desired, modify the scale of the view using *Edit* and *Value*.

6. Move the view by dragging the view.

10.4.3 STEPS FOR ADDING DRIVEN DIMENSIONS TO THE VIEWS

The driven dimensions (dimensions used in constructing the model) can be displayed on the drawing by using the *Show/Erase* option. The procedure is as follows:

1. Select *View* and *Show and Erase* or select ⬚.

2. Choose the *Show* button.

3. Depress the *Dimension* button.

4. Choose how to show the dimensions. The possible options are by *Feature, Part, View, Feat_View* (Feature and View), and *Part_View* (Part and View). We suggest displaying the dimensions by feature.

5. Choose which dimensions to keep.

6. Select *Close*.

Likewise, the display of the driven dimensions may be hidden. The procedure is:

1. Select *View* and *Show and Erase* or select ⬚.

2. Choose the *Erase* button.

3. Select the *Dimension* button.

4. Choose how to erase the dimensions. The available options are *Selected Items, Feature, Feaure and View, Part, Part and View,* and

View. We suggest using *Selected Items*. This will allow you to erase the dimensions one at a time.

5. Click on each desired dimension.
6. Select *Close*.

10.5 ADDITIONAL EXERCISES

For the exercise selected, create a fully dimensioned multiview drawing of the model.

10.1 Cement trowel (See Exercise 7.1.)
10.2 Side guide (See Exercise 7.2.)
10.3 Control handle (See Exercise 7.3.)
10.4 Bushing (See Exercise 7.4.)
10.5 Pipe clamp top (See Exercise 7.5.)
10.6 Pipe clamp bottom (See Exercise 7.6.)
10.7 Side support (See Exercise 7.7.)
10.8 Guide plate (See Exercise 7.8.)
10.9 Single bearing bracket (See Exercise 7.9.)
10.10 Double bearing bracket (See Exercise 7.10.)
10.11 Gasket (See Exercise 8.5.)
10.12 Wing plate (See Exercise 8.6.)
10.13 Skew plate (See Exercise 8.7.)
10.14 Ratchet (See Exercise 8.8.)
10.15 Index (See Exercise 8.9.)
10.16 Pinion gear (See Exercise 8.11.)
10.17 Rack (See Exercise 8.12.)
10.18 Post hinge (See Exercise 8.14.)
10.19 Create a drawing containing an auxiliary view of the slotted support (see Exercise 4.12). The drawing must contain at least the views shown in Figure 10.23.

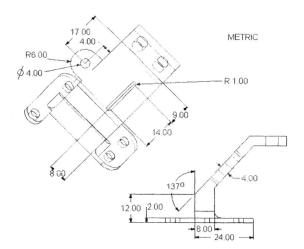

FIGURE 10.23

10.20 Create a drawing of the clamp base (see Exercise 4.13) containing the views shown in Figure 10.24.

FIGURE 10.24

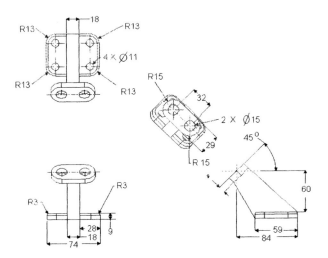

CHAPTER 11

CREATING A SECTION

INTRODUCTION AND OBJECTIVES

The drawing module may be used to create views containing either full or partial sections. Pro/E allows the user to create a section in the part, assembly, or drawing modes and then place the section in the drawing. In this chapter, creation of all sections will be taking place in the drawing mode. The sections will be constructed from previously created models.

In general, sections can be full, half, or partial sections, depending on how the cutting plane is passed through the part. In this chapter, we will consider how Pro/E can be used to create a full, offset, half, or revolved section.

The Pro/E software allows multiple pages (sheets) in a drawing. This option is helpful when several different drawings of a part are required.

The objectives of this chapter are to

1. Become familiar with the procedure required in creating either a full, offset, half, or revolved section
2. Use the *Sheets* option to add multiple pages to a drawing

11.1 CREATING AND PLACING A SECTION

As was seen in Chapter 10, when adding a view to a drawing, the user is given the option to create a section (see Figure 10.7, the *VIEW TYPE* menu). In doing so, the user must choose the *Section* option. After choosing this option, the user is presented the cross-section type menu (*XSEC TYPE*). This particular menu is reproduced in Figure 11.1.

The default type of section is *Full.* A *Full* section is created by passing a cutting plane entirely through the model. Half sections are used to show a section, wherein half of the model is cut with the cutting plane. The part of the model that is not cut is represented in a normal fashion. Examples of *Full* and *Half* sections are shown in Figure 11.2.

FIGURE 11.1 *The* XSEC
TYPE *menu is used to define the type
of section desired.*

*VERY IMPORTANT TO SEE
INTERIOR*

INTERNAL DRAWING W/

SECTION VIEWS

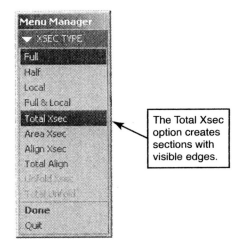

The Total Xsec
option creates
sections with
visible edges.

If the *Local section* option is chosen, a section of the model within a
defined boundary is created. The *Full & Local* option is used to create a full
section with local cross sections.

In creating sections, the preferred standard is to show not only the surface in contact with the cutting plane, but also the edges, which become
visible on the model after the cut has been made. In Pro/E, this can be
achieved by choosing the *Total Xsec* option. If the edges are not desired,

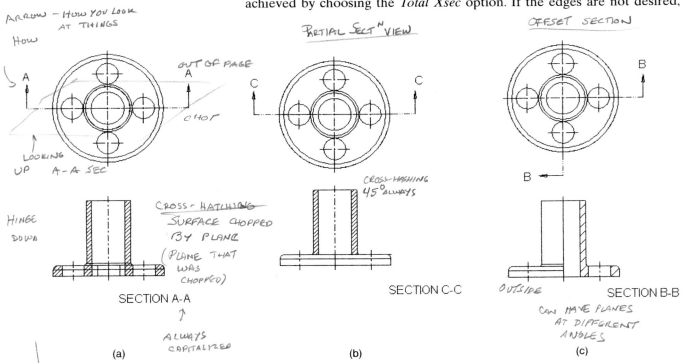

*ARROW — HOW YOU LOOK
AT THINGS*

How

OUT OF PAGE

PARTIAL SECT^N VIEW

OFFSET SECTION

*LOOKING
UP   A-A SEC*

*CROSS-HATCHING
45° ALWAYS*

*HINGE
DOWN*

*CROSS-HATCHING
SURFACE CHOPPED
BY PLANE
(PLANE THAT
WAS
CHOPPED)*

SECTION A-A

*ALWAYS
CAPITALIZED*

(a)

SECTION C-C

(b)

OUTSIDE

SECTION B-B

*CAN HAVE PLANES
AT DIFFERENT
ANGLES*

(c)

FIGURE 11.2 *Examples of* Full, Half, *and* Offset *sections. The model for
these sections is given next to the offset section, (c). In (a), a* Full *section is shown
of the surface containing the holes. The drawing above the section is the view
without the added section. In (b), a* Half *section of the upper part of the model is
given. The view without the section is shown directly above. Finally, in (c), an
offset section is shown.*

then the *Area Xsec* option may be selected. In this book, the *Total Xsec* option will always be used.

The *Total Align* option creates a total (*Total Xsec*) type section that is revolved and aligned. The *Total Align* option is useful when making offset sections. For an *Offset* section, the cutting plane is varied so that it goes through important features. The section produced is often difficult to visualize unless it is revolved.

The *Unfold Xsec* and *Total Unfold* options are also useful in offset cross sections, because they can be used to construct a projection of the offset. The projection is drawn flat. Both options are only available if the *Full* cross-section type is chosen.

An offset section is shown in Figure 11.2, along with the original model. The cutting plane was created by using the Sketcher to draw it. A sketching plane parallel to the two holes was used.

In defining the offset cutting plane, an open section is drawn through the desired features. After the contour of the plane has been sketched, it must be located with respect to the part and regenerated.

If the cross section was created in another mode, say the *Part* or *Assembly* mode, then the cross section can be retrieved. Otherwise, the cross section must be created using the *Create* option from the *XSEC ENTER* menu shown in Figure 11.3.

Upon creating a section, the section is given a name. After the section is displayed, the name of the section is displayed as well.

A section may be created using the options shown in Figure 11.4. Included in these options are the *Planar* and *Offset* options familiar to students of engineering graphics. In Pro/E, when creating a planar section, the user must define the cutting plane. This plane must be parallel to the screen. Therefore, the view from which the plane is defined must be properly chosen and oriented. Often default datum planes can be used for this purpose. If no suitable plane exists, then a datum plane may be created.

The user may modify the crosshatch in a section. Simply select the crosshatch, press your right mouse button and select *Properties*. As shown in Figure 11.5, the *Spacing, Angle,* and *Line Style* are among the features that can be changed. The standard for crosshatch spacing is 1/16″ to 1/8″ (1.5 mm to 3 mm) apart.

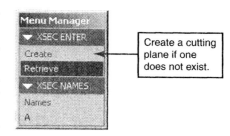

FIGURE 11.3 *If the cross section already exists, it can be retrieved by name; otherwise it must be created and named.*

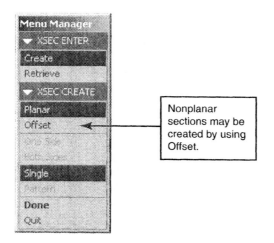

FIGURE 11.4 *The options for creating a cross section.*

FIGURE 11.5 *The options available to modify the cross-hatching.*

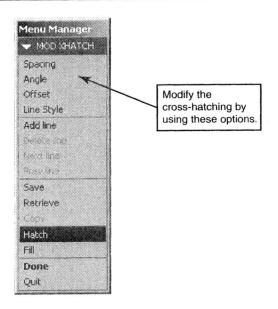

Modify the cross-hatching by using these options.

11.1.1 TUTORIAL 11.1: A FULL SECTION

This tutorial illustrates:

- Creating a full section
- Modifying the spacing of a crosshatch

1. Select *File* and *New* (or simply ☐). Choose *Drawing*.
2. Call the drawing called "PipeFlangeSection."
3. Select *OK* and enter the model name "PipeFlange."
4. Choose *Empty with format* and enter "AFormat."
5. *OK*.
6. Choose ▦ . Select *General, Full View, No Xsec, No Scale,* and *Done*.
7. Click on the top of the drawing, centered from left to right. Reorient the view. For *Reference 1*, select "front" and then click on datum plane Top. For *Reference 2*, select "right" and pick datum plane Right.
8. Choose *OK*.
9. Add a second view below the first. Select ▦ .
10. This view will be sectioned, so choose *Projection, Full View, Section,* and *Done*.
11. Select *Full, Total Xsec,* and *Done*.
12. Click on the screen below the first view.
13. We will need to create the section. In order to do so, choose *Create, Planar,* and *Single* defaults by selecting *Done*.
14. Enter the name A and hit the *return* key.
15. Click on datum plane Front.

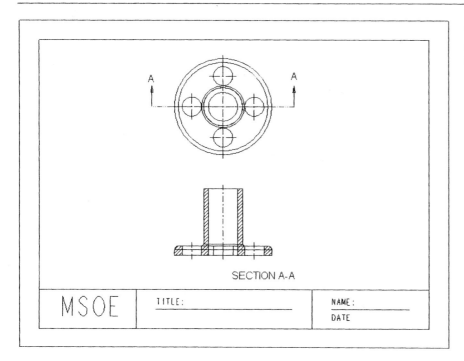

16. We want the arrows that define the location of the cutting plane to reside in the top view. Click on the first view (this is the top view).

17. You may need to move the views, so that the drawing looks uniform.

18. Change the spacing of the hatch lines to 1/16" (0.0625). First select the cross-hatching; then press your right mouse button and select *Properties.* Then choose *Spacing* and *Value.*

19. Enter 1/16 and hit the *return* key. Select *Done.*

20. Select . Choose *Show,* the *Axes* button, and *Show All.*

21. Choose *Yes, Accept All,* and *Close.*

22. The drawing is shown in Figure 11.6. Add dimensions if assigned. Save the drawing by using ▢.

11.1.2 TUTORIAL 11.2: AN OFFSET SECTION

This tutorial illustrates:

- Creating an offset section
- Adding a sheet to a drawing

1. Retrieve the drawing "PipeFlangeSection." Select ▨ and double-click on the file.

2. Select *Insert* and *Sheet.* Then choose *Add.* Notice that the software has loaded the format on the new page of the drawing.

3. Choose ▨. Select *General, Full View, No Xsec, No Scale,* and *Done.*

4. Click on the top of the drawing, centered from left to right. Reorient the view. For *Reference 1,* select "front" and then click

FIGURE 11.7 *The pipe flange model with the appropriate surfaces for the sketching plane and the orientation plane.*

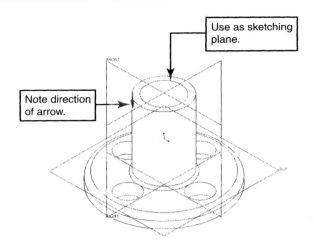

Use as sketching plane.

Note direction of arrow.

on datum plane Top. For *Reference 2,* select "right" and pick datum plane Right.

5. Choose *OK.*

6. Add a second view below the first. Select ▧.

7. This view will be sectioned, so choose *Projection, Full View, Section,* and *Done.*

8. Select *Full, Total Xsec,* and *Done.*

9. Click on the screen below the first view.

10. We will need to create the section. In order to do so, choose *Create, Offset,* and *One Side,* and *Done.*

11. Enter the name B and hit the *return* key.

12. We need a sketching plane. Select the surface shown in Figure 11.7.

13. Note the direction of the arrow in Figure 11.7. If yours is different, press *Flip;* otherwise, select *OK.*

14. Choose "right" from the menu and click on datum plane Right.

15. Use ◣ and sketch the two lines shown in Figure 11.8. Note that the endpoints of the lines are aligned with the edge of the circular protrusion.

16. Press ◡ and build the section.

17. Click on the top (the first) view to add the cutting plane arrows.

18. You may need to move the views, so that the drawing looks uniform.

19. Change the spacing of the hatch lines to 1/16" (0.0625). First select the cross-hatching, then press your right mouse button and select use *Properties.* Then choose *Spacing* and *Value.*

20. Enter 1/16 and hit the *return* key. Select *Done.*

21. Select ▦. Choose *Show,* the *Axes* button, and *Show All.*

22. Choose *Yes, Accept All,* and *Close.*

23. The drawing is shown in Figure 11.9. Add dimensions if assigned. Save the drawing by using ▢.

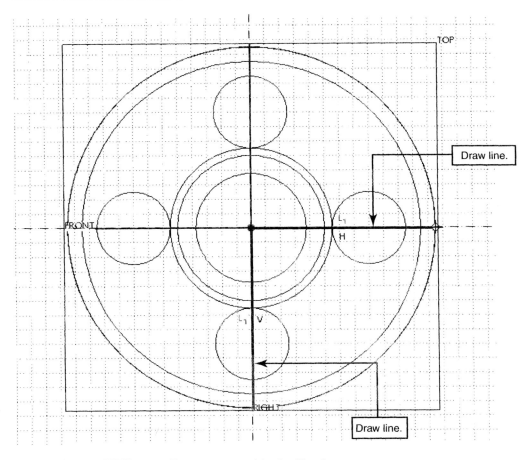

Draw line.

Draw line.

FIGURE 11.8 *The flange reoriented in the Sketcher.*

OPTIONAL

11.1.3 TUTORIAL 11.3: A HALF SECTION

This tutorial illustrates:

- Creating an half section
- Adding a sheet to a drawing

1. Retrieve the drawing "PipeFlangeSection." Select ⬚ and double-click on the file.

2. Select *Insert* and *Sheet*. Then choose *Add*. Notice that the software has loaded the format on the new page of the drawing.

3. Choose ⬚. Select *General, Full View, No Xsec, No Scale,* and *Done.*

4. Click on the top of the drawing, centered from left to right. Reorient the view. For *Reference 1*, select "front" and then click on datum plane Top. For *Reference 2*, select "right" and pick datum plane Right.

5. Choose *OK*.

6. Add a second view below the first. Select ⬚.

7. This view will be sectioned, so choose *Projection, Full View, Section,* and *Done.*

FIGURE 11.9 *The completed drawing of the offset section located in the front view of the model.*

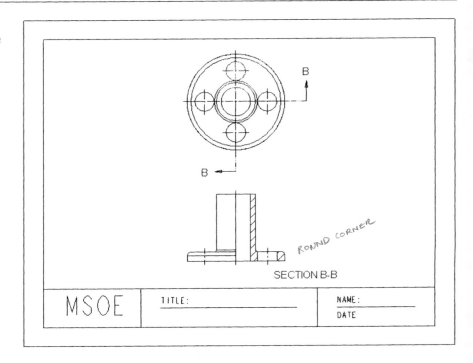

SECTION B-B

MSOE   TITLE: NAME:

DATE

8. Select *Half* and *Done*.
9. Click on the screen below the first view.
10. Select datum plane Top as the reference plane.
11. Note the direction of the arrows in Figure 11.10. Use *Flip* if the arrows on your screen do not correspond to the ones in the figure. Select *OK*.

FIGURE 11.10 *The figure shows the* Half *section of the angle bracket.*

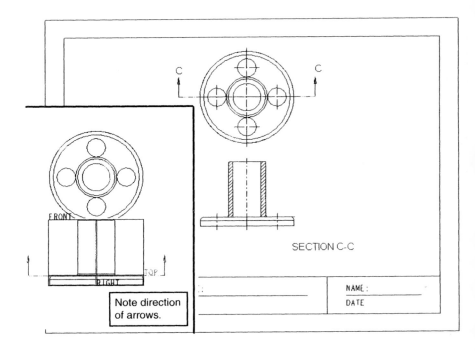

SECTION C-C

Note direction of arrows.

NAME:

DATE

12. Choose *Planar, Single,* and *Done.* Then, enter the name C and hit the *return* key.

13. Click on datum plane Front.

14. We want the arrows that define the location of the cutting plane to reside in the top view. Click on the first view (this is the top view).

15. You may need to move the views so that the drawing looks uniform.

16. Change the spacing of the hatch lines to 1/16" (0.0625). First select the cross-hatching, then press your right mouse button and select use *Properties.* Then choose *Spacing* and *Value.*

17. Enter 1/16 and hit the *return* key. Select *Done.*

18. Select ▨. Choose *Show,* the *Axes* button, and *Show All.*

19. Choose *Yes, Accept All,* and *Close.*

20. The drawing is shown in Figure 11.6. Add dimensions if assigned. Save the drawing by using ▨.

11.1.4 Tutorial 11.4: A Revolved Section of the Connecting Arm

This tutorial illustrates:

- Creating a revolved section

1. Select ▨. Choose *Drawing.*

2. Call the drawing called "ConnectingArmSection."

3. Select *OK* and enter the model name "ConnectingArm."

4. Choose *Empty with format* and enter "AFormat."

5. *OK.*

6. Choose ▨. Select *General, Full View, No Xsec, No Scale,* and *Done.*

7. Click on the top of the drawing, centered from left to right. Reorient the view. For *Reference 1,* select "front" and then click on datum plane Top. For *Reference 2,* select "right" and pick datum plane Right.

8. Choose *OK.*

9. Add a second view below the first. Select ▨.

10. Select *Projection, Full View, No XSec,* and *Done.*

11. Click below the first view. The views are shown in Figure 11.11.

12. Choose ▨. Select *Revolved, Full View, Section,* and *Done.*

13. Click on front view (the second view). This is the *Parent view.*

14. Click on *Create, Planar, Single,* and *Done.*

15. Enter the name "A" and hit *return.*

16. Select datum plane Right as the cutting plane.

17. An axis of symmetry may be defined. For our purposes, this is not necessary, so click your middle mouse button.

18. You may need to move the views so that the drawing looks uniform.

19. Change the spacing of the hatch lines to 1/16" (0.0625). First select the cross-hatching, then press your right mouse button and select use *Properties.* Then choose *Spacing* and *Value.*

FIGURE 11.11 *A drawing showing the first and second (top and front) views of the connecting arm.*

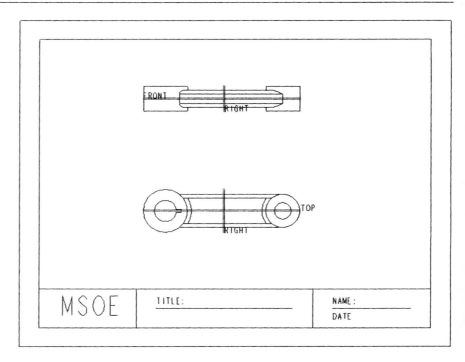

20. Enter 1/16 and hit the *return* key. Select *Done*.
21. Select 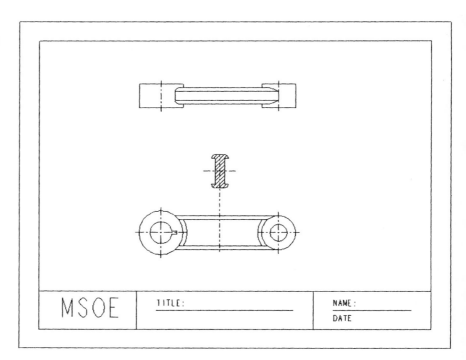. Choose *Show,* the *Axes* button, and *Show All*.
22. Choose *Yes, Accept All,* and *Close*.
23. The drawing is shown in Figure 11.12. Add dimensions if assigned. Save the drawing by using .

FIGURE 11.12 *A drawing of the connecting arm showing a revolved section of the rib.*

11.2 SUMMARY AND STEPS FOR CREATING A SECTION

A section is useful for illustrating internal features in a part. With Pro/E, the user can create several different types of sections. These sections may be full or partial sections. Standard practice is to create a section on one of the multiviews. Thus, at least two views must be added to create a drawing containing a section.

The steps for creating a section are identical for most types of sections (excluding *Revolved* sections). The procedure begins by adding the first view, which is not sectioned. Then a second view is added. This view is sectioned. The sequence of steps is as follows:

1. Add the first view. Select *Insert* and *Drawing View* or simply select ⊡.
2. Choose the view type. This may be a *Projection*. However, if the drawing contains no other view, then use the *General* option.
3. Select *No Xsec* and *Done*.
4. Locate the view on the drawing.
5. Add the second view, by choosing ⊡.
6. This view has to be a projection, so choose *Projection, Full View, Section,* and *Done*.
7. Locate the second view on the drawing.
8. Create or retrieve a section by selecting either *Create* or *Retrieve*.
9. If you have chosen to create the section, choose from *Planar* or *Offset*.
10. If the choice is *Planar,* choose the datum or construct the datum. If the choice is the *Offset* option, select the sketching plane and the reference plane. Then sketch the geometry of the edge of the cutting plane.
11. Select the first view as the view to locate the cutting plane line.

A revolved section requires an additional view. The procedure is the same through step 5. After this step, follow the sequence:

1. Choose *Projection, No Xsec,* and *Done*.
2. Locate the second view on the drawing.
3. The third view will be a section. Choose ⊡.
4. Select *Revolved, Full View, Section,* and *Done*.
5. Locate the third view with respect to the second.
6. *Create* or *Retrieve* the cutting plane.
7. For symmetric sections, you may wish to select an axis of symmetry. If wish to do so, select the appropriate axis. Otherwise, depress your middle mouse button to locate the section.

Additional pages (sheets) may be added to a drawing. This is advantageous if several sections of a part are desired. Add a sheet to a drawing by choosing *Insert* and *Sheet*. Use the *Next* and *Previous* options to cycle between the sheets. A sheet may be deleted from the drawing by using the *Edit, Remove,* and *Sheets*.

When adding a sheet to a drawing, Pro/E automatically inserts the format used on the first page onto any additional sheet. You may remove the format by selecting *Sheets, Format,* and *Remove.* A different format may be added to a selected sheet by using *File* and *Page Setup.* Click on the sheet and enter the name of the format or double-click to bring up a browse option.

11.3 ADDITIONAL EXERCISES

11.1 Create a drawing of the part bushing (see Exercise 4.3). Use Figure 11.13 as a reference.

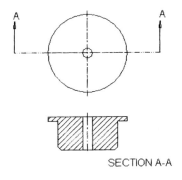

SECTION A-A

11.2 Construct a drawing having a section view of the bumper part (see Exercise 4.4). The section is shown in Figure 11.14.

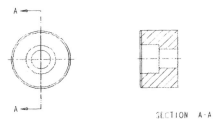

SECTION A-A

11.3 Use Figure 11.15 to construct a drawing containing the section view of the angle block (see Exercise 4.9).

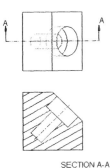

SECTION A-A

11.4 Create a drawing with the section view of the slotted hold down (see Exercise 5.3). Figure 11.16 shows the desired section view.

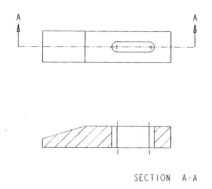

FIGURE 11.16

SECTION A-A

11.5 Construct a drawing with a section view of the coupler (see also Exercise 6.9). Use Figure 11.17 as a reference.

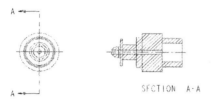

FIGURE 11.17

SECTION A-A

11.6 Figure 11.18 shows a revolved view of the rib in the lift lever (see Exercise 6.10). Construct a drawing with this view.

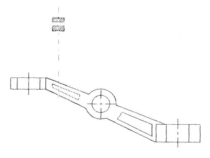

FIGURE 11.18

11.7 In Figure 11.19, a half, planar section of the double bearing bracket is shown. Generate a drawing with this view (see also Exercise 6.13).

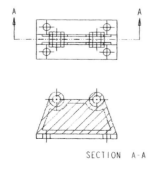

FIGURE 11.19

SECTION A-A

11.8 Generate a drawing with a full planar section of the double bearing bracket as shown in Figure 11.20. *Hint:* You will need to create a datum through the holes (see also Exercise 6.13).

FIGURE 11.20

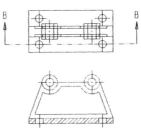

SECTION B-B

11.9 Create a drawing with a revolved section of the double bearing bracket (see Exercise 6.13). Use Figure 11.21.

FIGURE 11.21

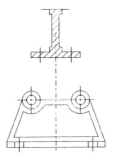

11.10 Use Figure 11.22 to construct a drawing with a revolved view of the single bearing bracket (see Exercise 6.12).

FIGURE 11.22

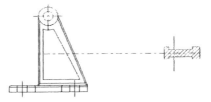

11.11 Use Figure 11.23 to construct an offset section view of the angle bracket with side reinforcements (see Exercise 6.11).

FIGURE 11.23

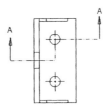

SECTION A-A

11.12 Construct the drawing shown in Figure 11.24 with the planar section
       of the angle bracket with side reinforcements (see Exercise 6.11).

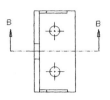

FIGURE 11.24

SECTION B-B

CHAPTER 12

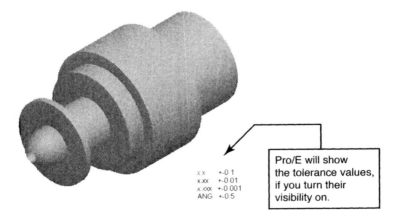

ADDING TOLERANCES TO A DRAWING

INTRODUCTION AND OBJECTIVES

In the manufacturing process, the tolerance of a feature is required. This chapter concerns specifying tolerances on drawings. Among the methods to be discussed in this chapter is Geometric Dimension and Tolerancing (GDT).

In Pro/E, tolerances should be prescribed when parts are created, since they determine how a feature will be constructed. If the tolerances are not prescribed by the user, then Pro/E uses default tolerance values. To see what the default values are, turn their visibility on by selecting *Tools* and *Environment*. Then choose the *±01 Dimensional Tolerances* option. After a model is loaded or created, Pro/E will display the default tolerance at the bottom of the drawing window as shown in Figure 12.1.

In order to change or set the bounds on any feature, select *Edit, Setup,* and *Dim Bound.* From the *DIM BOUNDS* menu reproduced in Figure 12.2, the user may set the dimension to its upper, lower, middle, or nominal value. The tolerance standard may be changed by choosing *Tol Setup* from the

FIGURE 12.1 *An example of a part with displayed tolerances.*

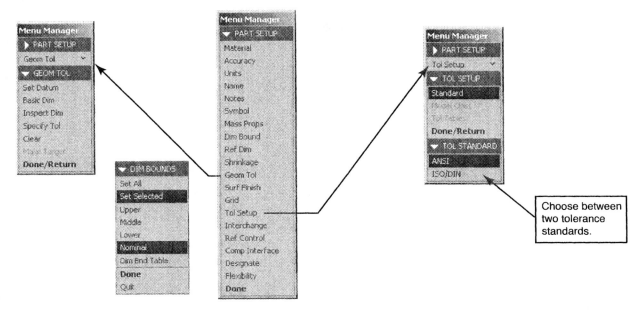

FIGURE 12.2 *Both traditional and geometric tolerances may be set for a model by using the options in the* PART SETUP *menu.*

DIM BOUND menu. In the *TOL SETUP* menu, also shown in Figure 12.2, the user may select between *ANSI* and *ISO/DIN* standards. The default setting is *ANSI*.

A geometric tolerance may be specified by using the *Geom Tol* option in the *PART SETUP* menu. By using the options in the *GEOM TOL* menu (Figure 12.2), the user may define the reference datums, the geometric tolerance, and set basic and inspection dimensions.

These menus are useful for specifying traditional and geometric tolerances for a model in the part mode. If a tolerance is defined for a model in the part mode, the dimensions will be displayed with the tolerance in the drawing mode. In the sections that follow, we will create the tolerance within the drawing mode.

The objectives of this chapter are to

1. Add a traditional or geometric tolerance to a model.
2. Add traditional or geometric tolerances to a multiview drawing.

12.1 ADDING TRADITIONAL TOLERANCES TO A DRAWING

Whether or not the user prescribed certain tolerances on the model when it was built is immaterial, because as was seen in the Introduction, the model was constructed with default tolerance values. Furthermore, these default values may be modified within the drawing mode, and Pro/E will update the changes in the model.

In order to add traditional tolerances to a dimension in the drawing mode, all that remains to do is to turn on the visibility of the tolerance values. This may be accomplished by loading the drawing configuration file and

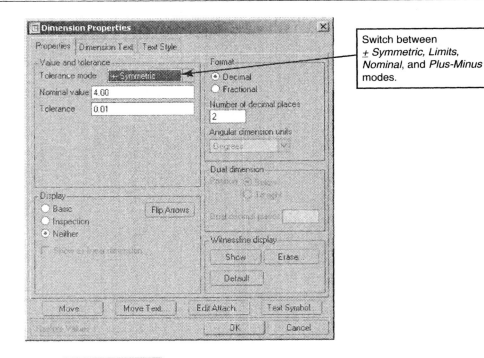

Switch between
± *Symmetric, Limits,*
Nominal, and *Plus-Minus*
modes.

FIGURE 12.3 *The* Dimension Properties *box. This dialog box contains options for displaying tolerances.*

changing the option *tol_display* from *No* to *Yes*. The sequence of steps needed to complete this task is:

1. Select *File, Properties,* and *Drawing Options.*
2. Scroll down until the option *tol_display* is found.
3. Change the setting from *No* to *Yes.*
4. Select *Add/Change* button and *OK.*
5. Repaint the screen by pressing ⃞.

The options *Edit* and *Properties* may be used to change the display of the tolerances on any or all the dimensions. Pick the desired dimensions with your mouse pointer and then choose *Edit* and *Properties.* This will load the *Dimension Properties* dialog box shown in Figure 12.3. The options contained in the box allow the user to change the tolerance among ± *Symmetric, Limits, Nominal,* and *Plus-Minus* modes. Depending on the mode desired, the user might modify the upper and lower limits or the tolerance values.

12.1.1 TUTORIAL 12.1: TRADITIONAL TOLERANCES FOR THE DRAWING "SQUAREPLATEWITHHOLES"

This tutorial illustrates:

- Adding traditional tolerances to a drawing

1. Retrieve the drawing "SquarePlatewithHoles" using 📂.
2. If the visibility of the tolerance values is turned off, turn it on using the procedure detailed in Section 12.1.

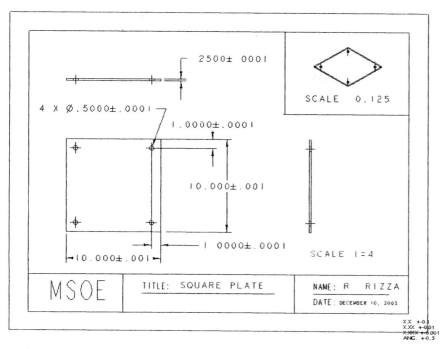

FIGURE 12.4 *The drawing of the square plate with added tolerance values.*

3. The software will display the tolerance values. The display should be in the ± *Symmetric* mode. If not, select the dimension(s), choose Edit and *Properties*, change to ± *Symmetric,* and then select *OK*.

4. You may need to move the views and/or dimensions in order to properly space all the elements of the drawing. Use Figure 12.4 as a reference and move the elements as necessary. Save the drawing unless told not to do so.

12.2   GEOMETRIC DIMENSIONING AND TOLERANCING

Geometric tolerances may be added to a part in the part mode or assembly mode. The user may also add the tolerances in the *Drawing* mode.

Unlike traditional tolerancing, Geometric Dimensioning and Tolerancing (GDT) is more than just adding bounds on a dimension. It is a way of defining features based on how they are to function in engineering practice. Besides requiring a control on a feature, GDT also describes how the feature is to be controlled.

Since *Reference Datums* are used in Geometric Tolerancing, before adding a GDT, the user should make sure that the proper datums have been defined. Often one or more of the default datum planes can be used as a reference datum. Then, all the user needs to do is to set the desired default datum plane as a reference plane.

If none of the existing planes can be used as a reference plane, then the appropriate datum plane must be created. The plane may be created in the part mode or in the drawing mode. To create the datum in the drawing mode, use the following series of options: *Insert, Model Datum,* and *Plane.*

An existing datum *must* be set as a reference datum before it can be used to define a geometric tolerance. In order to change an existing datum to a reference datum, select the datum with your mouse and then choose *Edit*

FIGURE 12.5 *The* Datum *dialog box. This box is used to create a datum.*

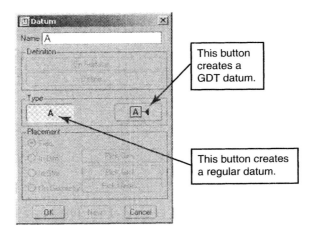

and *Properties.* The *Datum* dialog box shown in Figure 12.5 will appear; change the datum type to a reference datum and select *OK*.

Note that the datum reference symbol shown in Figure 12.5 is the ASME symbol. In Pro/E, the ANSI symbol is the default. Change from the default symbol to the ASME symbol by using the following sequence:

1. Select *File, Properties,* and *Drawing Options.*
2. Scroll down until the option g*tol_datums* is found.
3. Change the setting from *std_ansi* to *std_asme.*
4. Select *Add/Change* button and *OK.*
5. Repaint the screen by pressing ⬚.

After reference datums have been defined, dimensions may be added to the drawing and geometric tolerances added as needed. Geometric Tolerances may be created with the *Insert* and *Geometric Tolerance* options or by pressing ⬚.

The *Geometric Tolerance* box shown in Figure 12.6 may be used to define the geometric tolerance. The box contains several folders with options for

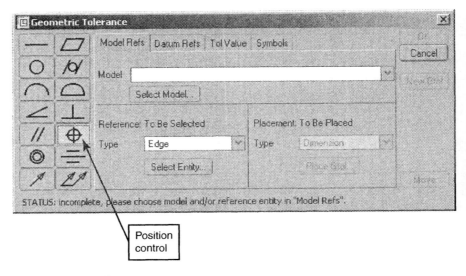

FIGURE 12.6 *The* Geometric Tolerance *dialog box.*

creating the geometric tolerance. It is possible that not all the options will be used for creating the GDT.

In creating the GDT, the user must select the model. The name of the default model is placed in the *Model* field by Pro/E. To change the *Model*, depress the *Select Model* button.

The *Reference* can be one of the following: *Edge, Axis, Surface, Feature, Datum,* and *Entity*. After one of these options has been chosen, the *Select Entity* button may be depressed and the actual edge, axis, etc., may be selected from the screen using the mouse pointer.

The geometric tolerance may be located by attaching the feature control box to a dimension (*Dimension* option). It may also be attached to another control box (*gtol* option), with a leader (*Leader, Tangent Ldr,* and *Normal Ldr* options), or as a free note (*Free Note* option).

The reference datums may be placed in the feature control box by choosing the *Datum Refs* folder. The primary, secondary, and tertiary datum references are chosen by selecting from a list of available datums.

The fourteen buttons along the left-hand side of the box are the available GDT control symbols. By placing your mouse pointer over the desired button and holding it there for a few seconds you can access their meaning.

Pro/E allows the user to define linear as well as curvilinear (rotational) reference datums. When creating a geometric tolerance on a rotational datum use the following approach:

1. Select *Insert, Model Datum,* and *Plane* or select ▱ .
2. Enter the name of the datum.
3. Choose the GDT datum type button.
4. Pick the *In Dim* option.
5. Select the *Pick Dim* button and pick the diameter dimension of the feature with your mouse.
6. Choose the *Define* button.
7. Pick the *Through* option and select the edge of the circular boundary.
8. Choose *Done*.
9. Select *OK*.

For a rotational reference datum the software will attach the datum symbol to the control by default. In order to attach the symbol to the dimension:

1. Select *File, Properties,* and *Drawing Options*.
2. Scroll down the list and find the option *asme_dtm_on_dia_dim_gtol*.
3. Change the setting for this option from *on_tol* to *on_dim*.
4. Select the *Add/Change* and *OK* buttons.

Notice that the number of decimal digits may also be set using the *Dimension Properties* box. Furthermore, the GDT specifications, *Basic Dimensions,* and *Inspection Dimensions* may also be set.

Basic Dimensions are dimensions that are theoretical values. Such dimensions do not have tolerance values associated with them. The GDT symbol for *Basic Dimensions* is a dimension enclosed in a rectangle. *Inspection Dimensions* are used for inspection. They are represented by a

FIGURE 12.7 *An example of a* Basic Dimension *and an* Inspection Dimension. *The* Basic Dimension *is indicated with the dimensional value enclosed in a rectangle. The* Inspection Dimension *is shown with the dimensional value enclosed in an oval.*

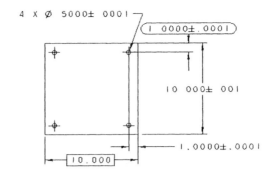

dimension enclosed in an oval. An example of a *Basic* and *Inspection Dimension* is given in Figure 12.7. Tolerances may be prescribed in an *Inspection Dimension.*

12.2.1 TUTORIAL 12.2: CREATING A DRAWING WITH GEOMETRIC TOLERANCES

This tutorial illustrates:

- Constructing a drawing with geometric tolerances
- Creating a geometric reference datum plane
- Adding a position control to a drawing

1. Select *File* and *New* or click on ⬜.
2. Choose the drawing mode. Enter the name: "SquarePlatewith-HolesGDT."
3. Click on the *OK* button.
4. Use *Browse* or enter the model name (SquarePlatewithHoles. prt) and then select the *Empty with Format* option.
5. Use *Browse* or enter "AFormat" in the cell provided.
6. Choose *OK*.
7. Use the procedure outline in Section 12.2 to change from *std_ansi* to *std_asme.*
8. Add the first view. Choose 🔲.
9. Accept the defaults by selecting *Done.*
10. Click somewhere in the lower left-hand portion of the screen. Pro/E will place an isometric view of the model.
11. Reorient the model so that it appears as in Figure 12.8. For *Reference 1* select datum plane Top as the "Front" and for *Reference 2*, choose datum plane Right as the "Right." Choose *Done.*
12. Add the additional views by using the *Projection* option.
13. The datum planes used in the construction of this part are not suitable for use as geometric reference datum planes. Therefore, select 🔲.
14. Now, in the name cell type in A. Then press 🔲.
15. Select *On Surface* and click on the edge shown in Figure 12.8. Choose *OK*.

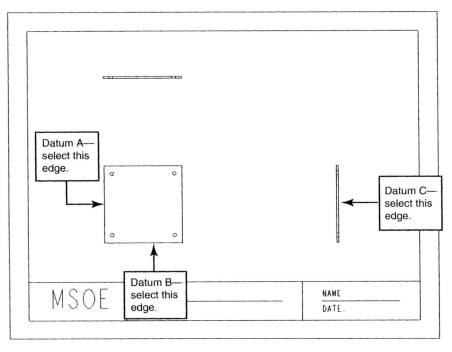

FIGURE 12.8 *Create a drawing of the square plate.*

16. Repeat steps 13 through 15 for the two remaining reference datum planes. Use the edges shown in Figure 12.9.

17. Use the pointer ▨ and drag the reference datum plane labels away from the views as shown in Figure 12.9.

18. Press ▨. Press the *Axis* button. Deselect the *Dimension* button. Press the *Show All* button. Choose *Yes, Accept All,* and *Close.*

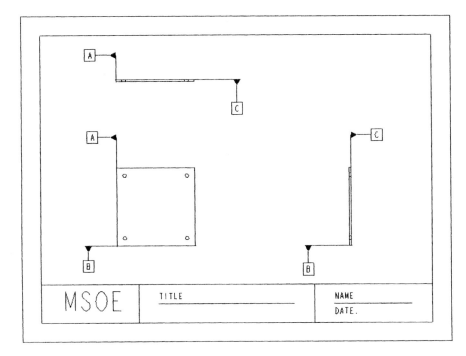

FIGURE 12.9 *The square plate with the GDT datum planes A, B, and C.*

FIGURE 12.10 *Before adding any controls, add the following dimensions to your drawing.*

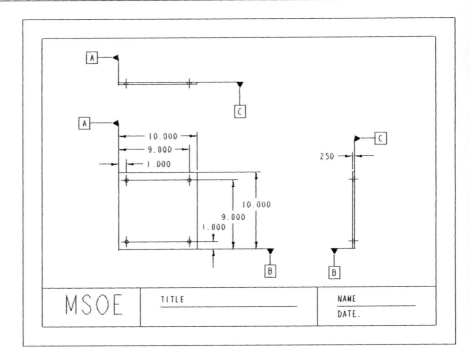

19. We will now create dimensions relative to the reference datum planes. This will involve clicking on the appropriate reference datum plane and then on the feature. For dimensions involving the center of the hole. You will need to select the crosshair mark indicating the hole's center. You may wish to zoom into the hole in order to do this.

20. Press ▨ or use *Insert, Dimension,* and *New References.*

21. Now, use Figure 12.10 as a guide to create the necessary dimensions. In the front view, click on datum reference A and the corresponding edge. Place the dimensional value by clicking on the screen with the middle mouse button.

22. Add the remaining dimensions. Press *Return* when finished. Move the dimensions and/or flip the arrows of the dimensions, as necessary.

23. Three decimal places are sufficient for the dimensions. If necessary change the number of decimal places by drawing your mouse across the drawing. Then, press and hold your right mouse button down and select *Properties.* Change the number to three decimal places.

24. Press ▨. Choose *Show* and click on the hole feature. Use *Sel to Keep* and keep only the diameter dimension.

25. Based on how this part is to be used, we need to place a position control on the holes. Choose ▨.

26. Make sure the position control is selected (see Figure 12.11). The *Reference* type is *Edge.* Press *Select Entity* and click on the circle (you may need to zoom in).

27. For the *Placement,* press *Place Gtol* and select the diameter dimension.

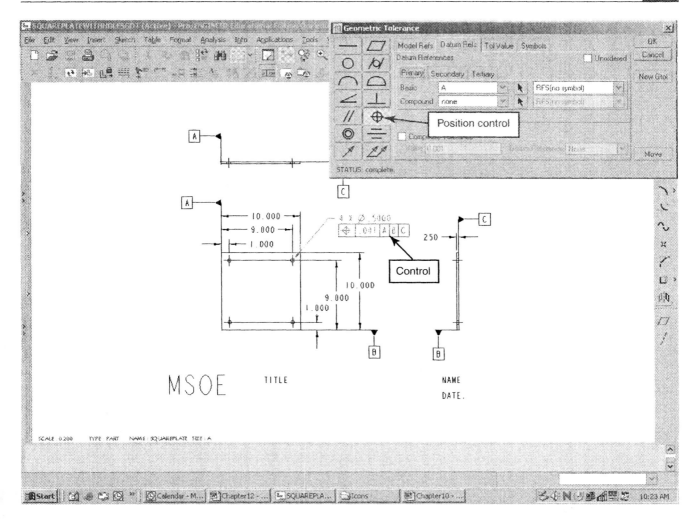

FIGURE 12.11 *The drawing with a position control added to hole in the front view.*

28. Now, select the *Datum Refs* tab. For the *Primary* tab, use the pull down and select: A.

29. For the *Secondary* tab, use the pull down and select: B.

30. For the *Tertiary* tab, use the pull down and select: C. Note the control on the drawing and in Figure 12.11.

31. Select the *Tol Value* tab. For the *Material Condition,* use the pull-down and select *MMC*.

32. Choose the *Symbol* tab. Check the diameter symbol. Choose *OK*.

33. You may answer *yes* to the query regarding basic dimensions. But, because we created the dimensions locating the holes ourselves, we will need to convert the dimensions to basic manually.

34. Select the two 9.000 and two 1.000 dimensions. Press and hold your right mouse button. Then, select *Properties*.

FI**G**U**R**E** 1**2**.**1**2** *The drawing with the complete feature control box added to the hole.*

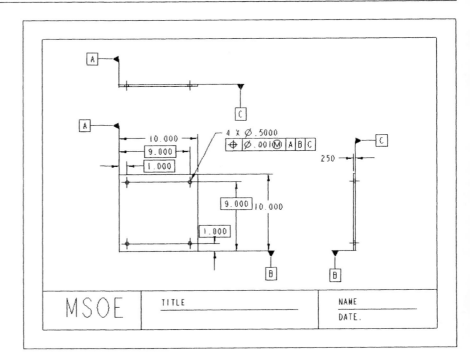

35. Click on the *Basic* option and select *OK*.
36. Reposition the dimensions and control as needed by dragging them.
37. Save the drawing. The drawing is shown in Figure 12.12.

12.3 SUMMARY AND STEPS FOR ADDING TOLERANCES TO A DRAWING

Because Pro/E assumes default tolerance values during the creation of a part, traditional tolerances may be placed in a drawing by simply turning on their visibility. This is accomplished by simply changing the option *tol_display* from *No* to *Yes*.

The user may change the default tolerance values in the *Part* mode or in the *Drawing* mode. In the *Part* mode, the tolerances may be changed by using *Edit* and *Set Up*. In the drawing mode, they may be modified by using the options *Edit* and *Properties* (a dimension must be selected). With the use of the *Dimension Properties* box, the user may select between ± *Symmetric, Limits, Nominal,* and *Plus-Minus* as the modes for displaying traditional tolerance values.

Geometric tolerances may be added to a part as well as to a drawing. The user must define GDT datums. In the drawing mode, this is done by selecting the datum and then using the *Edit* and *Properties* options. After a dimension has been displayed on a drawing, a feature control box may be added. The general procedure for adding a feature control box is as follows:

1. Select *Insert* and *Geometric Tolerance* or select ▨.
2. The *Geometric Tolerance* box will appear.

3. Enter the model or drawing name.

4. Select the *Reference* field—that is, feature, edge, etc., for which the geometric tolerance is to be defined.

5. Depress the *Select Entity* button and pick the appropriate entity.

6. Select the *Placement* option.

7. Depress the *Place Gtol* button and locate the geometric tolerance.

8. Choose the *Datum Refs* folder and select the datum references.

9. Select the *Tol Value* folder and enter the tolerance value. If desired, select a material condition.

10. Choose *OK*.

12.4   ADDITIONAL EXERCISES

For the exercises that follow, create multiview drawings with traditional or geometric tolerances as assigned by your instructor. Save your work unless instructed not to do so.

12.1   Cement trowel (See Exercise 10.1.)
12.2   Side guide (See Exercise 10.2.)
12.3   Control handle (See Exercise 10.3.)
12.4   Bushing (See Exercise 10.4.)
12.5   Pipe clamp top (See Exercise 10.5.)
12.6   Pipe clamp bottom (See Exercise 10.6.)
12.7   Side support (See Exercise 10.7.)
12.8   Guide plate (See Exercise 10.8.)
12.9   Single bearing bracket (See Exercise 10.9.)
12.10   Double bearing bracket (See Exercise 10.10.)
12.11   Gasket (See Exercise 10.11.)
12.12   Wing plate (See Exercise 10.12.)
12.13   Skew plate (See Exercise 10.13.)
12.14   Ratchet (See Exercise 10.14.)
12.15   Index (See Exercise 10.15.)
12.16   Pinion gear (See Exercise 10.16.)
12.17   Rack (See Exercise 10.17.)

CHAPTER 13

DATUM POINTS, AXES, CURVES, AND COORDINATE SYSTEMS

INTRODUCTION AND OBJECTIVES

In Chapter 2, we considered datum planes that can be used in construction of base features. These planes were either of the default type or user created. In addition to datum planes, Pro/E allows the user to define datum axes, points, curves, and coordinate systems (Figure 13.1).

Datum axes can be used in creating or placing a feature by using the axes as a reference. Datum points are used to specify target points for placement of features or to specify load application points in mesh generation. These points may be referenced by a datum coordinate system. Datum curves, which can be constructed through a series of datum points or from model geometry, are used to create features and for definition of sweep trajectory (sweeps are discussed in Chapter 15).

The objectives of this chapter are to

1. Construct a datum axis, datum coordinate system, datum points, and a datum curve.
2. Use these datum features to place additional features on a model.

13.1 CONSTRUCTING A DATUM AXIS

Because a datum axis may be used to reference a feature, the creation of an axis is similar to a datum plane. A datum axis is given a name by Pro/E of the form A_#, where # is the number of the datum axis. As with datum planes, datum axes are selected by choosing the name label of the axis.

The options for constructing a datum axis are:

1. *Through.* As with a datum plane, the axis goes through the selected reference.
2. *Normal.* A datum axis is created normal to a chosen plane or surface.

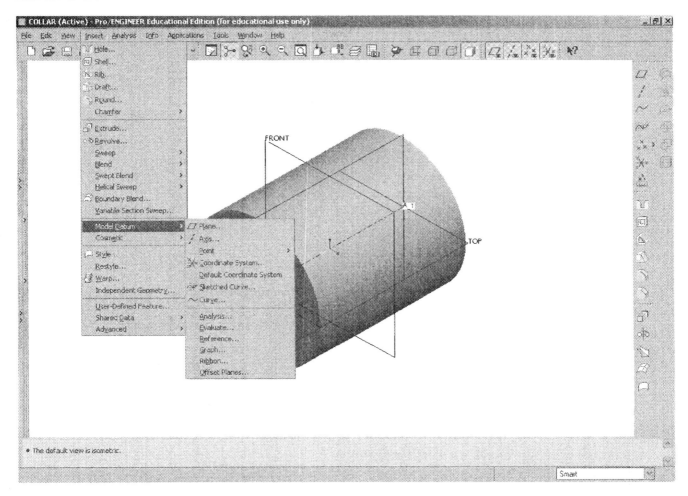

FIGURE 13.1 *In addition to datum planes, Pro/E allows the user to construct datum, axes, curves, points, and coordinate systems.*

3. *Tangent.* A datum axis is created that is tangent to the selected reference.

A part that could use the advantage of a datum axis is the collar shown in Figure 13.2. The hole on the lateral surface of the cylinder can be created in several ways. The approach, which we believe is the easiest and is used in Tutorial 13.1, is to construct a datum axis. Then, by using the coaxial option the user may place the hole.

13.1.1 TUTORIAL 13.1: A PART REQUIRING A DATUM AXIS

This tutorial illustrates:

- Creating a datum axis
- Using a datum axis to generate a feature

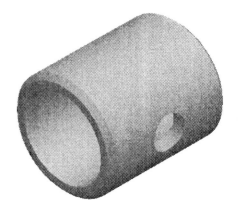

FIGURE 13.2 *The hole in the side of the collar is located by constructing a datum axis.*

FIGURE 13.3 *Sketch a circle to create the section.*

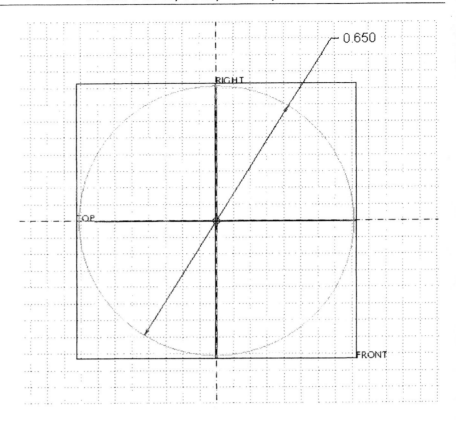

1. Select .
2. Make sure that the *Use default template* option is checked.
3. Enter the name "Collar" in the *Name* field.
4. Select *OK*.
5. First, construct the base feature, a cylinder. Select .
6. Choose the *Sketch* button in the dashboard.
7. With your mouse, choose datum plane Front as the sketching plane.
8. If necessary, use for the *Orientation* field and to "right." Then, click on datum plane Right.
9. Again, the arrow should point into the screen. Use *Flip* if it does not.
10. Select the *Sketch* button .
11. Sketch a circle as in Figure 13.3. Double-click on the dimensional value and modify to 0.650.
12. Press .
13. Select . Enter a depth of 0.75.
14. Press .
15. The datum axis to be constructed is located at the intersection of datum planes Front and Top. Select Insert, Model Datum, and Axis or simply select .

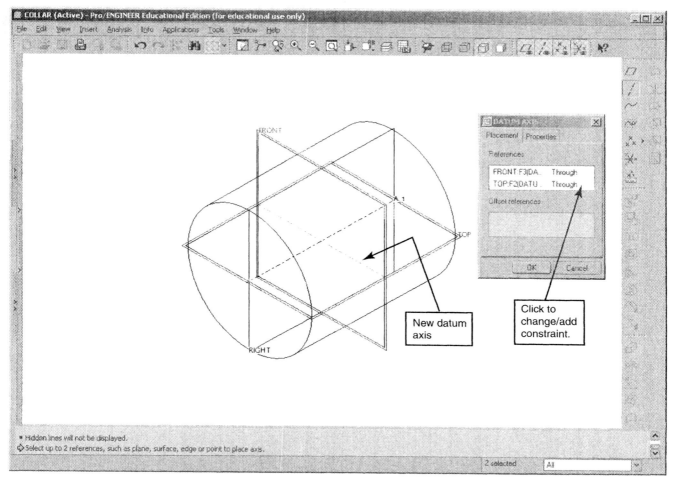

FIGURE 13.4 *The datum axis added by the user.*

16. Click on datum plane Front. Click in the Reference cell (see Figure 13.4). Use ⌄. Change to *Through*.

17. Click on datum plane Top. Click in the Reference cell and change to *Through*.

18. Press *OK*. Note the new axis, A_2.

19. In order to place the coaxial hole along this axis, we need a placement plane. This means a datum plane needs to be created normal to the axis. Select ⬚.

20. Now, click on datum plane Right. Make sure the Offset option is active. If not, click in the Reference cell (see Figure 13.5). Use ⌄. Change to *Offset*.

21. Enter 0.65/2=0.3250 for the offset. Choose *OK*. Note the new datum plane, DTM1.

22. Select ⬚.

23. Select *Straight, Simple,* and *Through all* options.

24. Enter a diameter of 0.200.

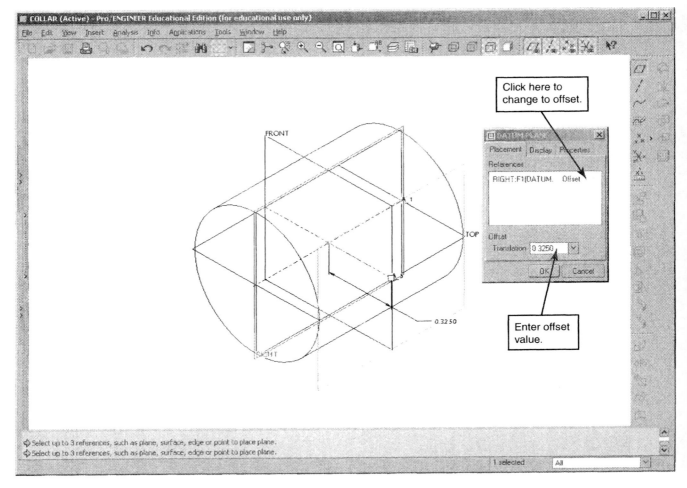

FIGURE 13.5 *The base feature for the collar, along with the additional datum plane DTM1.*

25. With your mouse click on the axis A_2.

26. Press the *Placement* option in the dashboard. Click in the Secondary references field.

27. Click on the new datum plane, DTM1.

28. Build the feature by pressing ☑.

29. Add a coaxial hole along axis A_1 by repeating steps 22 through 28. Select axis A_1 and use the surface shown in Figure 13.6 as the placement plane. The diameter of the hole is 0.500.

30. Finally, complete the model by chamfer both ends of the collar. Select *Insert* and *Chamfer* or simply ◻.

31. Select both edges of the collar (see Figure 13.6).

32. Use the 45 × D option and enter a value of 0.030.

33. Build the feature by pressing ☑.

34. Use ◻ and save the part.

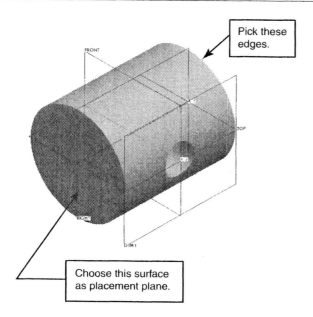

Pick these edges.

Choose this surface as placement plane.

FIGURE 13.6 *The model of the collar after creating the datum axis (axis A_2).*

13.2 DATUM COORDINATE SYSTEMS

Datum coordinate systems are used to:

1. Obtain mass properties of a model.
2. Assemble components.
3. Define constraints for finite element analysis.
4. Define the location of a feature such as a datum point.
5. Define a coordinate direction.

As with datum planes and axes, datum coordinate systems are defined by selecting references (see Figure 13.7). This is accomplished by selecting the datum plane, axis, or surface. Press the CTRL key to add additional references.

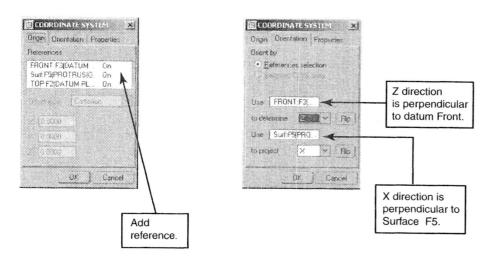

Add reference.

Z direction is perpendicular to datum Front.

X direction is perpendicular to Surface F5.

FIGURE 13.7 *Coordinate systems may be created using several options. The options are located in this OPTIONS menu.*

In addition, the orientation of the coordinate system needs to be defined (see Figure 13.7). The orientation may be defined by selecting direction references for two of the three coordinate axes.

The *Offset* option 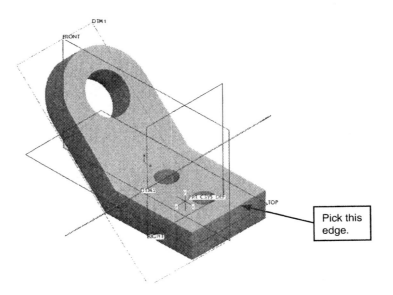 allows the user to construct a datum coordinate system offset from a *Cartesian, Cylindrical,* or *Spherical* system. In addition, the system may be created by entering the coordinates from a transformation data file. The file must have an extension ".trf." The format of the file is:

$$
\begin{array}{cccc}
X1 & X2 & X3 & T1 \\
Y1 & Y2 & Y3 & T2 \\
Z1 & Z2 & Z3 & T3
\end{array}
$$

The columns define vectors. The first column in the matrix determines the X-axis direction while the second column determines the Y direction. Any values can be used for the third column, as the Z direction is determined by using the right-hand rule. The last column contains the elements that define the translation of the new origin.

13.2.1 TUTORIAL 13.2: A COORDINATE SYSTEM FOR A ROD SUPPORT

This tutorial illustrates:

- Creating a datum coordinate system
- Defining the orientation of a datum coordinate system

1. We want to add a datum coordinate system in order to determine mass properties relative to a specific system. Retrieve the model "RodSupport." Select ⬚ and double-click on the file.
2. Select ✳.
3. We need three references. Select datum plane Front. Press your CTRL key and pick on datum plane Top. Then, press your CTRL key and select the edge in Figure 13.8.
4. Now, select the *Orientation* tab in the *Coordinate System* dialog box.

FIGURE 13.8 *Use the shown surface to locate the coordinate system.*

Note Z is perpendicular
to datum front and
X is perpendicular
to surface F5.

FIGURE 13.9 *Define the directions of the axes.*

FIGURE 13.10 *The coordinate system "CS0" has been created.*

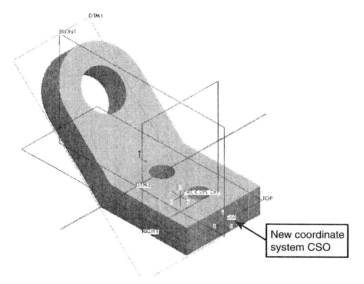

New coordinate
system CSO

5. Note Figure 13.9. In the first *Use* cell, datum plane Front should be listed. If it is not, select the cell and then click on datum plane Front.

6. Now, we want *Z* to be perpendicular to datum plane Front. Select the pull-down below the *Use* cell and choose *Z*.

7. Likewise, in the second *Use* cell, Surf: F5 should be listed. If it is not select the cell and then click on the edge shown in Figure 13.8.

8. We want the *X*-axis to be perpendicular to this surface. So, select the pull-down below the *Use* cell and choose *X*.

9. Note that the positive direction of each axis may be changed by pressing the corresponding *Flip* button.

10. Press *OK*.

11. Press 🖫 to save the model. The model with the coordinate system is shown in Figure 13.10.

13.3 DATUM POINTS

Datum points are displayed with an *X* and given a name in the form Pnt *N*, where *N* is the number of the point. An array containing multiple points can be created. This array is a single feature, which makes modifying or referencing the points easy.

Datum points are specified by three coordinates. The coordinates of the points may be *Cartesian* (x,y,z), *Cylindrical* (r,θ,z), or *Spherical* (r, θ,ϕ). A datum point may be created by using references and the Datum Point Tool ▨.

For multiple points it is best to create a point by entering its coordinates from a data file. In this case the *Offset CSys Datum Point* tool ▨ is used.

The file, which must have an extension ".pts," must be an ASCII file containing rows of numbers. If each row contains three numbers, Pro/E assumes that the three numbers are the coordinates of the point. If there are more than three numbers per row, then the software will read the second, third, and fourth numbers as the coordinates of the point. This allows the user to use the first number in each row as a point number.

Pro/E allows the user to treat the points as parametric or nonparametric features. Parametric is the default. For a nonparametric array, select the button: *Convert to Non Parametric Array* and no names will be given to the individual points in the point array. This definition can be changed by using the *Redefine* option, but the number of points contained in a datum array cannot.

13.3.1 TUTORIAL 13.3: DATUM POINTS FOR A FRAME

This tutorial illustrates:

- Creating a parametric datum point array by typing in the data points

1. Select ▢.
2. Make sure that the *Use default template* option is checked.
3. Enter the name "2DFrame" in the *Name* field.
4. Select *OK*.
5. Select ▨.
6. Click on the default coordinate system (PRT_CSYS_DEF).
7. Make sure the *Type* is set to *Cartesian*.
8. Click in the cell as shown in Figure 13.11. The first point will be added with the coordinates (0,0,0).

FIGURE 13.11 *Type in the coordinates of the datum points.*

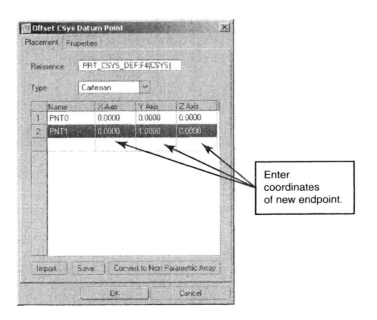

Enter coordinates of new endpoint.

TABLE 13.1 THE DATUM POINTS COORDINATE FOR THE 2DFRAME

Point no.	X	Y	Z
0	0.0	0.0	0.0
1	0.0	1.0	0.0
2	0.0	1.9	0.0
3	0.9	1.0	0.0
4	0.0	1.0	0.0

9. Additional points need to be added. The coordinates of these points are given in Table 13.1. Click below the first point and modify the coordinates by clicking on the cell containing the coordinate. Add the remaining points from the table using the same approach.

10. Note that the points may be saved. It is not necessary to save the points but, if desired, press the *Save* button and save the buttons.

11. Select *OK*.

12. The model is shown in Figure 13.12. Save the part by pressing 🔲.

13.3.2 TUTORIAL 13.4: DATUM POINTS FOR A TRUSS

This tutorial illustrates:

- Creating a parametric datum point array by reading in the data points
- Changing the units of a part to foot-pound-seconds

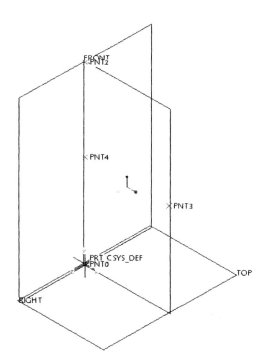

FIGURE 13.12 *The part "2Dframe" with the added datum points.*

FIGURE 13.13 *The truss may be created by defining points at the intersections of the structural members.*

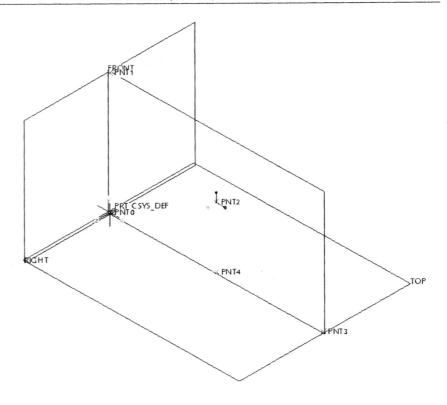

1. Select .
2. Make sure that the *Use default template* option is checked.
3. Enter the name "KingPostTruss" in the *Name* field.
4. Select *OK*.
5. Choose *Edit, Setup,* and *Units.* Then, choose the foot-pound-seconds system of units.
6. Press the *Set* button and *OK.* The select *Close* and *Done.*
7. Select .
8. Click on the default coordinate system (PRT_CSYS_DEF).
9. Make sure the *Type* is set to *Cartesian.*
10. Press the Import button. Find the file KingPostTruss.pts. This file is bundled with this book. Double-click on the file to open it.
11. Select OK. The software will add the points.
12. The model is shown in Figure 13.13. Save the part by pressing .

13.4 USING POINTS IN THE SKETCHER

The boundary of a sketch in the Sketcher may be generated by using a series of points and the spline option using the *Create Points* tool button . The points are added to the section and dimensioned relative to reference planes as was done in the previous tutorials or relative to a coordinate system. The coordinate system may created by using . This

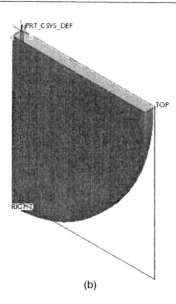

(a)   (b)

FIGURE 13.14 *The geometry of the butterfly valve plate is defined by a series of points.*

button is accessed by selecting the pull-down icon next to the *Create Points* tool button.

A spline may be used to generate a smooth curve through the points with the *Spline* tool button ⌇. The spline is placed by clicking with the left mouse button each point in succession.

The geometry of the butterfly valve plate shown in Figure 13.14(a) may be constructed by defining five points in the Sketcher. A smooth curve is then placed through the points. This will generate one quarter of the total part (Figure 13.14(b)). The rest of the part is constructed by using a copy mirror process.

13.4.1 TUTORIAL 13.5: DATUM POINTS IN THE SKETCHER

This tutorial illustrates:

- Feature construction using points and spline

1. Create a new part called "ButterFlyValve_Plate" by selecting ▢.
2. Select ▣.
3. Press ☑.
4. Pick datum Front as the sketching plane and datum plane Right as "right."
5. The arrow should point toward you. If it does not, use the *Flip* button.
6. Choose the *Sketch* button ▭.
7. This section has dimensions using three decimal places. Change the visibility of the number of decimal places by selecting *Sketch*, *Options*, and *Parameters*.

FIGURE 13.15 *Use this figure to dimension the location of the points in the Sketcher.*

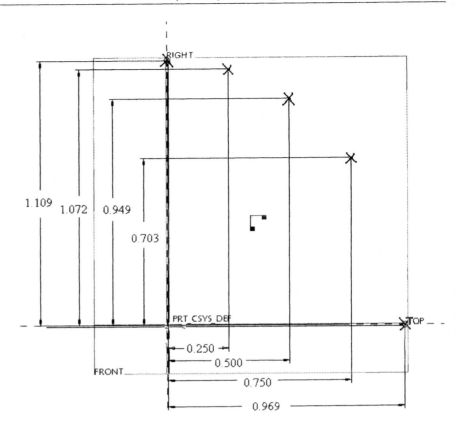

8. Enter 3 in the *Num Digits* field. Select the check mark button.

9. Press the *Create Points* button ⊠ in the toolbar.

10. By clicking on the screen, generate five points as shown in Figure 13.15.

11. Select the *Dimension* button and dimension the points as shown in Figure 13.15.

12. Select ▨ from the toolbar. Click on all the dimensions.

13. Uncheck the *Regenerate* option in the *Modify Dimensions* box.

14. In the cells, enter the appropriate dimensional values given in Figure 13.15.

15. Press the check mark button. This will regenerate the model with the new dimensions.

16. Select the *Spline* button ⌁. Click on each point.

17. This is a solid section and must be closed. Choose the *Line* tool and draw two lines as shown in Figure 13.16.

18. Press ✔ to build the section.

19. Choose *Symmetric* ▣ and enter a depth of 0.063.

20. Choose ✔ to build the model. The model at this point is shown in Figure 13.14(b).

21. Press ▢.

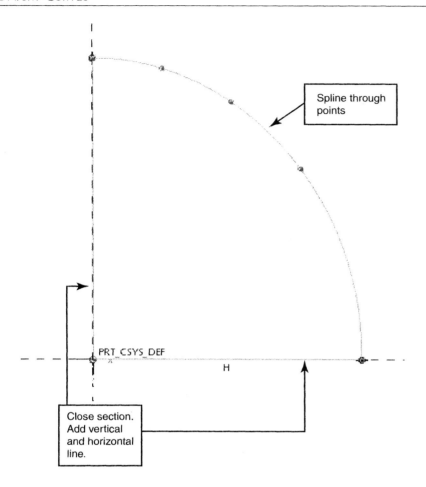

FIGURE 13.16 *Put a spline through the points.*

13.5 DATUM CURVES

Datum curves can be used to construct surfaces or as trajectories in sweeps (the *Sweep* option is examined in Chapter 15). The *Datum Curve* tool allows the user to generate a datum curve by using predefined points (*Thru Points* option), points read in from a file (*From File* option), from a cross section (*Use Xsec* option), or from an equation (*From Equation* option). These options may be found in the *CRV* (Curve) *OPTIONS* menu. This menu is reproduced in Figure 13.17.

Datum curves can be sketched using the *Sketch* option . In this case, a sketching and reference orientation plane must be chosen. Then, the Sketcher is used to sketch the curve geometry.

Datum curves constructed with the *Thru Points* option may be constructed with the additional options given in the *CONNECT TYPE* menu. This menu is reproduced in Figure 13.18. The *Spline* option allows the user to place a three-dimensional spline through the points. For a curve that twists and bends through a three-dimensional shape, a radius may be prescribed along the bends. This may be done by using the *Single Rad* or *Multiple Rad* options. In cases where the same radius is used

FIGURE 13.17 *The options in the CRV (Curve) OPTIONS menu are used to construct different types of curves.*

FIGURE 13.18 *The* CONNECT TYPE *menu.*

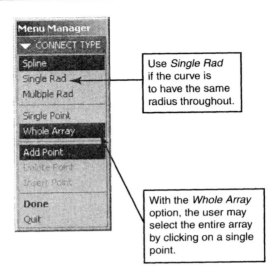

Use *Single Rad* if the curve is to have the same radius throughout.

With the *Whole Array* option, the user may select the entire array by clicking on a single point.

throughout the curve, use the option *Single Rad;* otherwise use the *Multiple Rad* option.

The *Whole Array* option allows selection of datum points that were created as a single array. This saves time, since each point does not have to be chosen. The option selects the points in sequential order.

Additional points may be added to a curve using the *Add Point* or *Insert Point* options. The *Add Point* option is used to add existing points to a curve, while the *Insert Point* option is used to add a new point between pre-chosen endpoints. In order to remove a point from a curve, use the *Delete Point* option.

For frames constructed by bending a single piece of stock into a pre-scribed shape, a single datum curve is sufficient to construct the model of the frame. This will be the case for the 2DFrame in Tutorial 13.6 (Figure 13.19). However, many parts are composed of several structural elements. Datum curves may be generated for such parts by using datum points and the *Spline* option.

Consider the king post truss shown in Figure 13.20. The elements of the truss are composed of straight 2×4 dimensional lumber (actually 1.5×3.5). These elements may be created in Pro/E in three steps. The first step is to use datum points to define the ends of the elements (we did this in Tutorial 13.4). The second step is completed by creating a datum curve for each element using the *Spline* option. Finally, by sweeping a rectangular section along each datum curve, the truss is completed. Notice that because of the symmetry in the model, we may create half of the truss and use the *Copy Mirror* option to complete the model.

Another example of the use of the *Spline* option is in the construction of an aircraft wing. The geometry of the section is generated using datum points and a spline curve through the datum points. Then a surface quilt (see Chapter 18) may be used to generate the surface of the wing. The airfoil under consideration is an NACA 4412 section to be used for a model in a wind tunnel. Data for this and other airfoil sections are available in

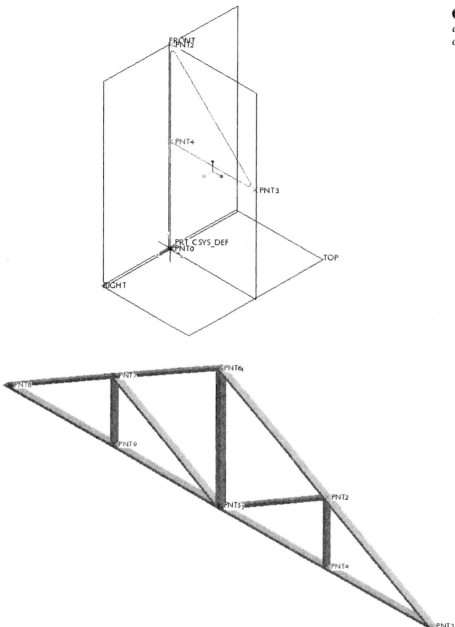

FIGURE 13.20 *The model of the king post truss.*

Abbott and Doenhoff (1959)[1]. The total number of datum points is 35, which describe an airfoil 3" in length. There are quite a few points in the list, so we have bundled the file "Cs0.pts" that contains these datum points.

[1]See the Bibliography for this reference.

13.5.1 TUTORIAL 13.6: ADDING A DATUM CURVE TO THE TWO-DIMENSIONAL FRAME

This tutorial illustrates:

- Adding a datum curve with the *Thru Points* option
- Using the *Single Rad* option

1. Retrieve the model "2DFrame." Select ⬜ and double-click on the file.
2. Press ⬜.
3. From the *CRV OPTIONS* menu (Figure 13.17), select *Thru Points* and *Done*.
4. We will use a single radius of 0.03, so select *Single Rad* from the *CONNECT TYPE* menu.
5. Select "Pnt0" and enter the bend radius of 0.03. Hit *return*.
6. Select *Done* and *OK*.
7. The model is shown in Figure 13.18. Press ⬜ and save the model.

13.5.2 TUTORIAL 13.7: ADDING A DATUM CURVE TO THE KING POST TRUSS

This tutorial illustrates:

- Adding a datum curve with the *Thru Points* option
- Using the *Spline* option

1. Retrieve the model "KingPostTruss." Select ⬜ and double-click on the file.
2. Press ⬜.
3. From the *CRV OPTIONS* menu (Figure 13.17), select *Thru Points* and *Done*.
4. Choose *Spline, Single Point,* and *Add Point*.
5. Pick points "Pnt1," "Pnt2," and "Pnt3."
6. Select *Done* and *OK*.
7. The software will create the datum curve shown in Figure 13.21.
8. Repeat steps 2 through 6 but replace step 5 with "Pnt0" and "Pnt1."
9. Repeat the steps a second time replacing step 5 with: "Pnt0," "Pnt4," and "Pnt3."
10. Follow the procedure one more time replacing step 5 with: "Pnt4" and "Pnt2."
11. For the last time, replace step 5 with "Pnt2" and "Pnt0."
12. The model, with all the curves added, is shown in Figure 13.22. Save the part; it will be completed in Chapter 15.

13.5.3 TUTORIAL 13.8: DATUM CURVES FOR AN AIRCRAFT WING

This tutorial illustrates:

- Adding a datum curve with the *Thru Points* option
- Adding a spline through points defining curvilinear geometry

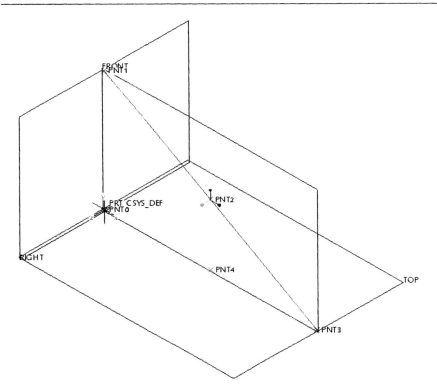

FIGURE 13.21 *The first curve is added to the model by passing through the points "Pnt1," "Pnt2," and "Pnt3."*

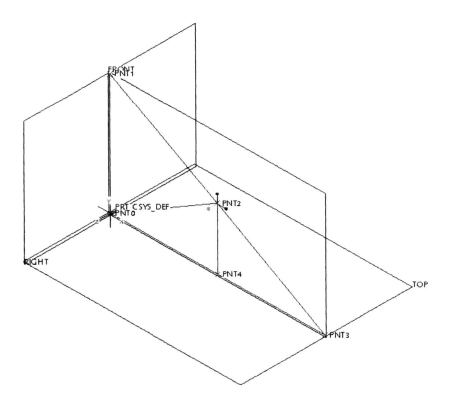

FIGURE 13.22 *Using the appropriate procedure, the rest of the curves may be added.*

1. Retrieve the model "Wing." This part is bundled with the text.
2. Select ⬚ and double-click on the file.
3. Select ⬚.
4. Click on the default coordinate system (CS0).
5. Make sure the *Type* is set to *Cartesian*.
6. Press the *Import* button. Find the file CS0.pts. This file is bundled with this book. Double-click on the file to open it.
7. Select *OK*. The software will add the points.
8. Press ⬚.
9. From the *CRV OPTIONS* menu, select *Thru Points* and *Done*.
10. Choose *Spline, Single Point,* and *Add Point*.
11. The part, with the added datum points, is shown in Figure 13.23. In the part, points 1 through 18 run along the upper part of the airfoil from the leading to the trailing edge. Points 19 through 35 run along the lower part of the airfoil from trailing to leading edge.
12. At this point, you will most likely need to zoom in to the model. Place the curve through each of the points by selecting each point individually (see Figure 13.24). Start with the first point (Pnt 0) and finish with the last point (Pnt 17). Choose *Done* and *OK*.
13. Repeat the process (steps 8 through 10) for the lower half. This time start with point 34 and end with point 17.
14. Save the part. The part will be completed in Chapter 15. Note the part with the added datum curves in Figure 13.25.

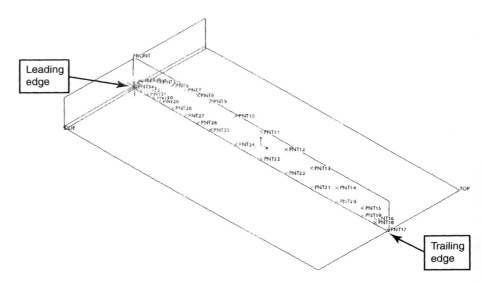

FIGURE 13.23 *The airfoil section begins to take shape after loading the datum points.*

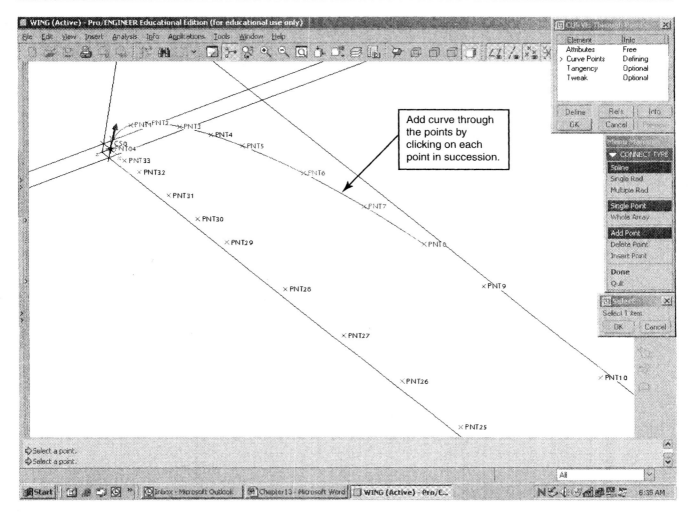

FIGURE 13.24 *Place a datum curve through the points.*

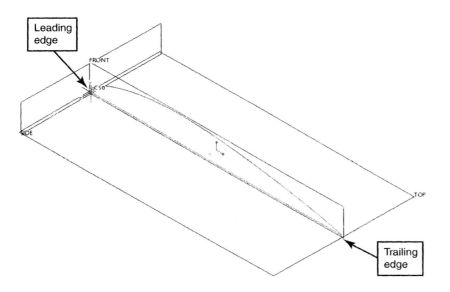

FIGURE 13.25 *The figure shows the airfoil with the datum curves. The display of the datum points has been turned off.*

13.6 SUMMARY

In addition to datum planes, the user may construct datum coordinate systems, points, axes, and curves.

The coordinate system may be used as a reference for datum points. In addition, it may be used as a coordinate system for calculating material properties such as moments of inertia or in exporting the model, such as a reference for an IGES transfer.

Datum points may be created as a parametric or nonparametric array. If the datum points are to be updated on an individual basis in the future, then they should be created as a parametric array.

Datum curves may be placed through datum points. These curves may be used as trajectories for sweeps. The use of datum points and datum curves allows the user to construct structural components such as frames and trusses.

Datum axes are often helpful in locating additional features in a model. The datum axis may be constructed by using selecting references in the same way as one would create a datum plane.

13.6.1 STEPS FOR CREATING A DATUM AXIS

The options for the actual placement of a datum axis are found in the *DATUM AXIS* menu (Figure 13.1). The general steps for placing an axis are as follows:

1. Select *Insert, Model Datum,* and *Axis* or press ▨.
2. Choose the appropriate references.
3. Select *OK*.

13.6.2 STEPS FOR CONSTRUCTING A DATUM COORDINATE SYSTEM

In general, the procedure for constructing a datum coordinate system is:

1. Select ▨ to create an offset datum plane. Select ▨ to create a datum coordinate system using references.
2. If using ▨, select the appropriate references. Then choose the Orentation tab and set the orientation of the coordinate axes.
3. In the case of ▨, chose the appropriate offset coordinate system or read in the location of the new coordinate system.

13.6.3 STEPS FOR ADDING DATUM POINTS TO A MODEL

The general steps for adding datum points are:

1. Select ▨ to create a datum point defined by references. To create a datum point or series of datum points by typing in or reading in the coordinates of the points use ▨.
2. In the case of ▨, select the appropriate references. For ▨, select the coordinate system, then either type in the coordinates of the points or select the *Import* button to read them in.
3. Select *OK*.

13.6.4 STEPS FOR CREATING A DATUM CURVE

A datum curve may be placed through a series of points or through other features. In general, the sequence of steps for creating a datum curve is:

1. Select *Insert, Datum,* and *Curve,* or press .
2. From the *CRV OPTIONS,* select *Thru Points* and *Done.*
3. If the curve has a radius, select *Single Rad* or *Multiple Rad* from the *CONNECT TYPE* menu. Enter the value of the radius or radii as appropriate.
4. If the curve is to go through several points, select *Spline, Single Point, Add Point.* Use *Whole Array* as opposed to *Single Point* to choose all the points with one click.
5. Select the points.
6. Choose *Done* and *OK.*

13.7 ADDITIONAL EXERCISES

For the exercises that follow, construct the trajectories using datum coordinates, points, and curves. Use a start part if desired. The trajectories will be used to construct sweeps in the additional exercises to Chapter 15.

13.1 Construct a datum curve through the points shown in Figure 13.26. The trajectory may be used to create a model of the bell crank.
13.2 Highway sign. Use the *Single Rad* option (0.75 radius) and the given geometry in Figure 13.27 to construct the datum curve through the points. Note the units of this model are in feet.

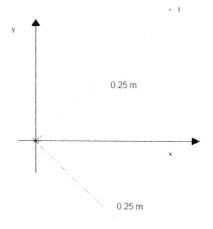

FIGURE 13.26

FIGURE 13.27

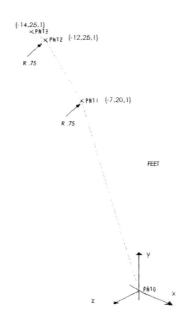

13.3 The geometry of a flat roof truss is given by:

$$y = A \cos n\pi x$$

In this equation, x and y are horizontal and vertical points determining the centerline of the truss in feet. A is amplitude describing the maximum vertical height of the truss. The parameter n describes the number of peaks in the interval. Generate a series of points for a 2-foot span of the truss similar to the one shown in Figure 13.28. Save these points in a text file. Construct a datum curve through these points.

FIGURE 13.28

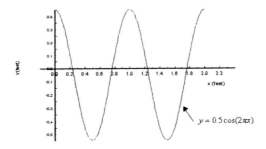

Use A and n from Table 13.2 as assigned by your instructor.

TABLE **13.2**   VALUES OF A AND n FOR EXERCISE 13.3

	A	n
A	0.5	1
B	2.0	1
C	0.5	2
D	2.0	2

13.4 Use the *Through Edge* option and a datum axis to place the hole in the rest shown in Figure 13.29.

FIGURE 13.29

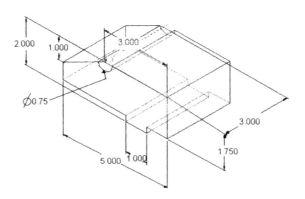

13.5 Construct the model of the key collar illustrated in Figure 13.30. Use a *Datum Axis* to locate the hole.

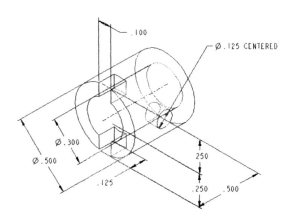

FIGURE 13.30

13.6 Complete the model of the key bushing given in Figure 13.31. Use a *Datum Axis* to locate the hole.

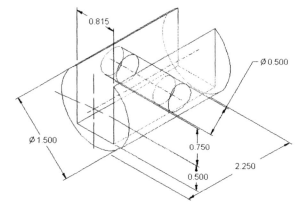

FIGURE 13.31

13.7 Complete the model of the external lug by using a *Datum Axis* to locate the hole. Consult Figure 13.32 for the geometry and location of the hole.

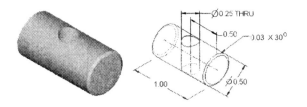

FIGURE 13.32

13.8 Using the geometry in Figure 13.33, create a model of the Fink truss (half section).

FIGURE 13.33

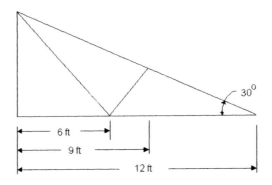

13.9 Create a model of the Howe roof truss (half section) shown in Figure 13.34.

FIGURE 13.34

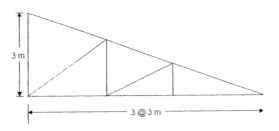

13.10 For the bridge truss (half section), construct a part with datum curves using the geometry shown in Figure 13.35.

FIGURE 13.35

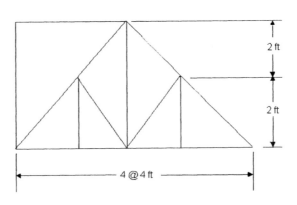

CHAPTER ▉14

THE REVOLVE OPTION

INTRODUCTION AND OBJECTIVES

The *Revolve* option is used to construct a feature whose section may be revolved around a centerline. Thus, adding a revolved section to a base feature requires a sketch and a centerline. Actually, we have already constructed a revolved section. For example, both the neck and flanges in Chapters 5 and 6, respectively, are sections that were created by revolving a sketch about a given centerline.

Revolved sections may be solid or thin in a manner analogous to solid and thin protrusions. A cut may also be revolved. In this chapter, we examine three models. The first model is constructed using a sold protrusion, whereas the second is created using the thin protrusion option. The third model is a thin revolved parabolic section.

The third part in this chapter serves to illustrate two points. The first is the construction of a conic section, and the second is the creation of a more advanced thin revolved section.

The objectives of this chapter are to

1. Create solid and thin features using the *Revolve* option.
2. Construct a conic section and revolve the section.

▉14.1▉ REVOLVED SECTIONS

For a revolved section, Pro/E allows the user to choose the angle of rotation. This can be done by entering the angle of revolution into the appropriate cell in the dashboard (see Figure 14.1). There are four preset revolution angles; these values are 90°, 180°, 270°, and 360°.

A revolved section can be constructed either as a solid or as a thin. In the case of a thin section, the user must define the thickness.

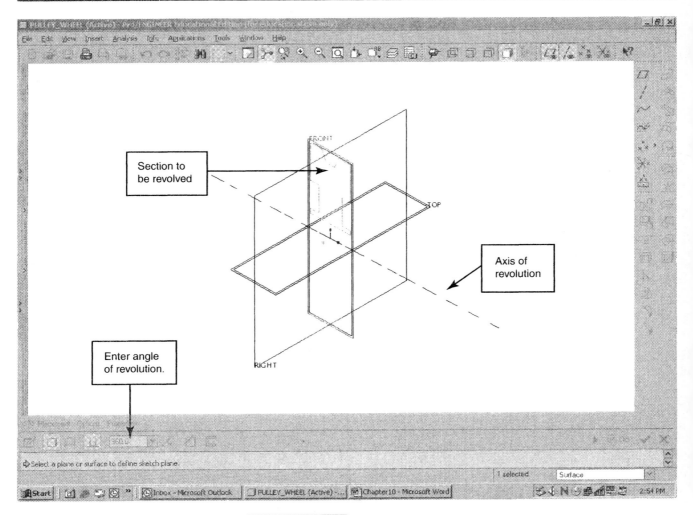

For the *Revolve* option, the section can be sketched on only one side of the centerline. If you sketch on both sides, then Pro/E will issue an error message.

14.1.1 TUTORIAL 14.1: A PULLEY WHEEL

This tutorial illustrates:

- Drawing a solid cross section and revolving the section to create a feature

1. Consider the cross section of the pulley wheel in Figure 14.2.
2. Select *File* and *New* (or simply 🗋).
3. Make sure that the *Use default template* option is checked.
4. Enter the name "PulleyWheel" in the *Name* field.
5. Select *OK*.
6. Select *Insert* and *Revolve* (or simply ⊛).

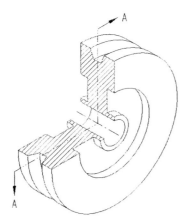

FIGURE 14.2 *A pulley wheel with an offset cut section. This section shows the interior geometry of the wheel, which can be used to generate the model.*

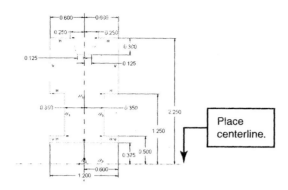

FIGURE 14.3 *The geometry required for constructing the revolved section of the pulley wheel.*

Place centerline.

7. Choose the *Sketch* button [icon] in the dashboard.

8. With your mouse, choose datum plane Front as the sketching plane.

9. If necessary, use [icon] for the *Orientation* field and to "right." Then, click on datum plane Right.

10. Again, the arrow should point into the screen. Use *Flip* if it does not.

11. Select the *Sketch* button [icon].

12. Use the *Line* tool and sketch the geometry shown in Figure 14.3. Dimension using [icon] as necessary and modify the dimensional values to those shown in the figure.

13. Change the *Line* tool to centerline and draw a centerline so that it is located along datum plane Top (see Figure 14.3).

14. Build the section by pressing [icon].

15. Sketch the geometry shown in Figure 14.3.

16. Dimension as shown in the figure. The locations of the dimensions have been chosen so that the section will regenerate symmetrically about the edge of datum Right.

17. Build the feature.

18. The software will reorient the model as shown in Figure 14.1. Enter 360.

19. Build the feature by pressing [icon].

20. Save the part. The completed model is shown in Figure 14.4.

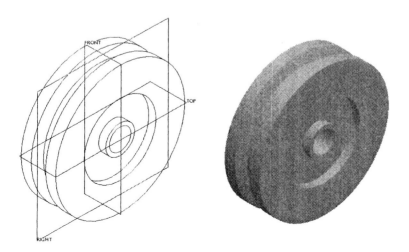

FIGURE 14.4 *A no hidden line and shaded model of the pulley wheel generated by revolving the section shown in Figure 14.3.*

FIGURE 14.5 *The section needed to create the body of the mug.*

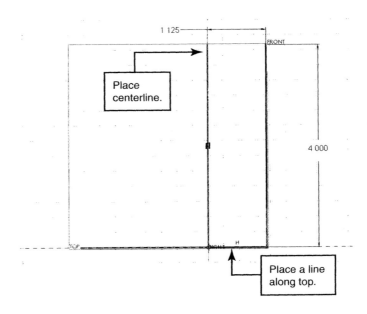

Place centerline.

FRONT

1 125

4 000

Place a line along top.

14.1.2 Tutorial 14.2: Creating a Mug Using a Revolved Thin

This tutorial illustrates:

- Drawing a thin cross section and revolving the section to create a feature

1. Select *File* and *New* (or simply 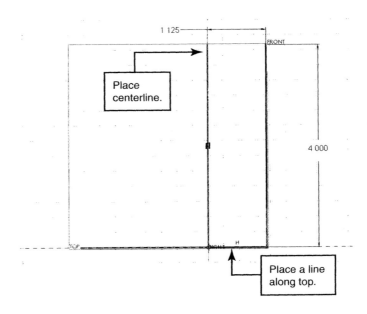).
2. Make sure that the *Use default template* option is checked.
3. Enter the name "CoffeeMug" in the *Name* field.
4. Select *OK*.
5. Select *Insert* and *Revolve* (or simply).
6. Select the Thin button . Enter a thickness of 0.125.
7. Choose the *Sketch* button in the dashboard.
8. With your mouse, choose datum plane Front as the sketching plane.
9. If necessary, use for the *Orientation* field and to "right." Then, click on datum plane Right.
10. Again, the arrow should point into the screen. Use *Flip* if it does not.
11. Select the *Sketch* button .
12. Use the Line tool and sketch the two lines shown in Figure 14.5. Use and dimension the section. Use and change the dimensional values of the dimensions to those given in the figure.
13. Press and build the section.
14. The thickness is to be filled inward. Press to change the direction of the thickness.
15. Enter 360 for the degree of revolution.
16. Build the feature by selecting .
17. Press and save the part (see Figure 14.6). We will add a handle to the mug in the next chapter.

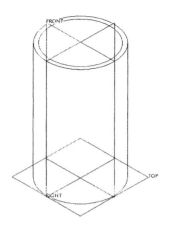

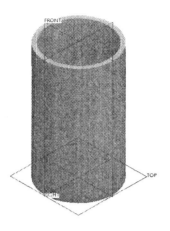

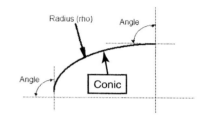

FIGURE 14.6 *After revolving the section given in Figure 14.5, the mug begins to take shape.*

14.2 CONIC SECTIONS

A conic section can be created in the Sketcher by using the conic button. In Pro/E, a conic is constructed by selecting the ends of the conic spline and a shoulder. The tangency angles of the endpoints must be prescribed. This can be achieved by defining the angle with an angular dimension, an adjacent entity, or a centerline. This is called the "rho" method. It is shown in Figure 14.7.

FIGURE 14.7 *A conic is constructed by defining the endpoint angles and the rho value.*

In order to dimension a conic using the "rho" method, the value of rho, the radius, must be given. This is done by selecting the conic with the left button and placing the value with the middle button. Initially, Pro/E sets rho equal to 0.5. This value can be modified in the following manner:

1. $0.05 < rho < 0.5$ for an ellipse [a closed ellipse has a value (SQRT (2) − 1); in this case enter the formula]
2. rho = 0.5, for a parabola
3. $0.5 < rho < 0.95$ for a hyperbola

In order to dimension the angle complete the following steps:

1. Select the conic, then the endpoint where the tangency is to be defined.
2. Pick the entity, centerline, or other conic to which the angle is to be defined.
3. Place the dimension by clicking on the screen, at the desired location, with your middle mouse button.

We are going to construct a conic section in the next tutorial. The part, a parabolic reflector, makes use of the *Revolve* option.

14.2.1 TUTORIAL: 14.3: A PARABOLIC REFLECTOR

This tutorial illustrates:

• Drawing a thin conic cross section and revolving the section to create a feature

1. Select *File* and *New* (or simply 🗋).
2. Make sure that the *Use default template* option is checked.

FIGURE 14.8 *First, place these lines so that the endpoints of the conic may be more readily drawn.*

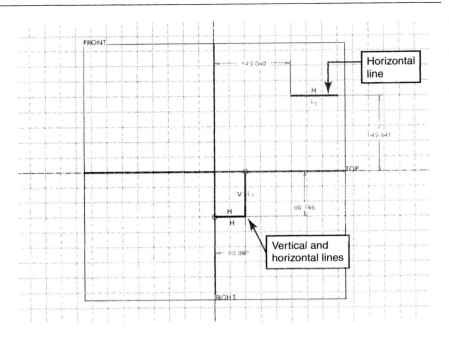

3. Enter the name "EmergencyLightReflector" in the *Name* field.

4. Select *OK*.

5. Select *Insert* and *Revolve* (or simply 🔲).

6. Select the Thin button 🔲. Enter a thickness of 0.100″.

7. Choose the *Sketch* button 🔲 in the dashboard.

8. With your mouse, choose datum plane Front as the sketching plane.

9. If necessary, use 🔽 for the *Orientation* field and to "right." Then, click on datum plane Right.

10. Again, the arrow should point into the screen. Use *Flip* if it does not.

11. Select the *Sketch* button 🔲.

12. Place a centerline along the edge of datum plane Right.

13. Begin constructing the sketch by drawing the three lines shown in Figure 14.8. We draw these lines first in order to more easily place the conic.

14. Select the *Arc* button 🔲 and change the type to conic 🔲.

15. Select the tip of the vertical line as the start point for the conic. This start point, as well as the end of the conic, is shown in Figure 14.9. After locating the start and endpoints of the line, "tweak" the conic; that is, add curvature to the line by dragging your mouse sideways. It does not matter how much curvature you add to the conic, as we will dimension the sketch shortly.

16. In order to place the angular dimensions; first select the conic, then the endpoint of the conic, and finally the corresponding line. Place the dimension by clicking the middle button on your mouse. Use Figure 14.10 as a guide.

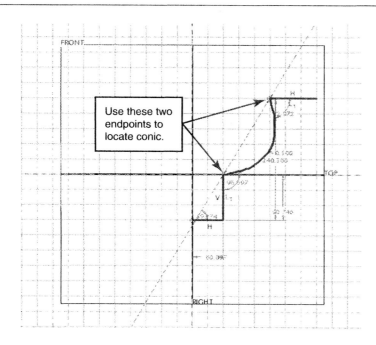

Use these two endpoints to locate conic.

FIGURE 14.9 *Sketch the conic using the lines from Figure 14.8.*

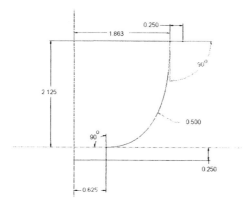

FIGURE 14.10 *The dimensions for the conic section.*

17. Add the remaining dimensions shown in Figure 14.10 in the normal manner. Note that since the conic section is a parabola, "rho" has a value of 0.5.

18. Delete the dotted line shown in Figure 14.9. If you do not do this then the section will not be entirely on one side of the centerline and the section will fail to regenerate.

19. Select ☑ to build the section.

20. Enter 360 to revolve the section 360°.

21. Choose ☑ and build the feature. The model is shown in Figure 14.11.

22. Let us complete this part by adding a cut. Choose 🔲 and select ⬜.

23. Choose a depth of *Through all* 🔳.

24. Select the *Sketch* button ▭.

25. Choose the surface shown in Figure 14.12 as the sketching plane.

FIGURE 14.11 *The reflector after revolving the section in Figure 14.10 through 360°.*

FIGURE 14.12 *The parabolic reflector rotated to show the base.*

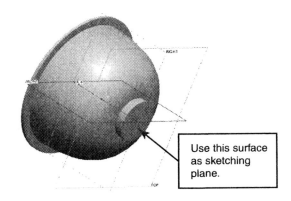

Use this surface as sketching plane.

26. If necessary, use 🔽 for the *Orientation* field and to "top." Then, click on datum plane Front.
27. Use ⭕ and sketch the circle shown in Figure 14.13.
28. Use 🖊 and sketch the lines in Figure 14.13. Draw the lines so that they extend over the circle. We will break the lines using the *Divide* tool next.
29. Select the *Divide* tool ⭐ and click on the intersections of the lines and circle. Select the unwanted pieces of the line and circle and press the *delete* key to remove them.
30. Dimension the sketch using the values shown in Figure 14.14, where we have "zoomed in" to show the detail and dimensions of the cut.
31. Press ⭣ and modify the dimensional values.
32. Build the section by pressing ✔.
33. Remove the inside of the section. Use 🔀 to change the direction of the cut.
34. Press ✔ and build the feature.
35. Save the completed model, shown in Figure 14.15.

FIGURE 14.13 *The cutout for the bulb holder.*

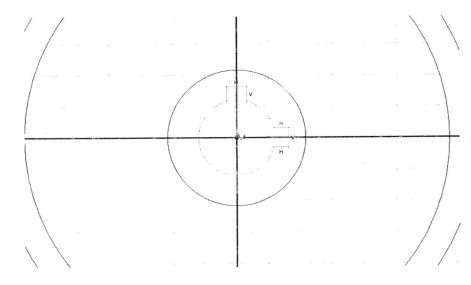

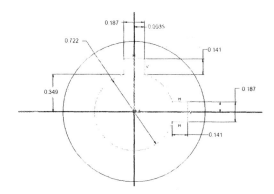

FIGURE 14.14 *The dimensions needed for regenerating the cutout.*

14.3 SUMMARY AND STEPS FOR USING THE *REVOLVE* OPTION

Some models have features that may be constructed by revolving a section about an axis of revolution. In Pro/E, the option for carrying out such a task is called the *Revolve* option. In using this option, the user must define a sketching plane. Once a sketching plane is defined and the model reoriented in the Sketcher, the cross section is sketched. A centerline must be placed by the user. This centerline acts as the axis of revolution. The user must define the angular dimension through which the section is revolved. This is accomplished by entering any value from 0° to 360°.

The general sequence of steps for constructing a revolved section is:

1. Select *Insert*.
2. Select the type of feature, for example, *Protrusion*.
3. Select *Revolve*.
4. Choose the sketching and reference orientation plane.
5. In the Sketcher, place a centerline and sketch the cross section.
6. Dimension the section. Modify the values as necessary.
7. Build the section by pressing ✔.
8. Enter the degree of rotation.
9. Build the feature.

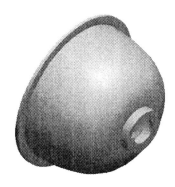

FIGURE 14.15 *The parabolic reflector after adding the cutout for the lightbulb.*

14.4 ADDITIONAL EXERCISES

14.1 Use the section shown in Figure 14.16 to create the model of the roller.

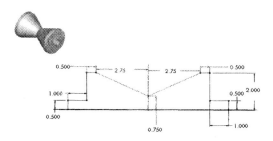

FIGURE 14.16

14.2 Use the section shown in Figure 14.17 to construct the model of the baseball bat.

FIGURE 14.17

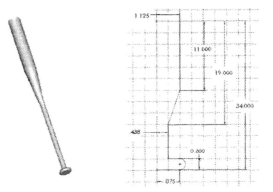

14.3 Use the geometry in Figure 14.18 to create the model of the door-knob.

FIGURE 14.18

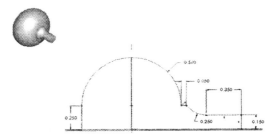

14.4 Use the *Revolve* option and the given sketch in Figure 14.19 to create the model of the handle ball. The hole has a diameter of 0.20″ and is 0.25″ deep.

FIGURE 14.19

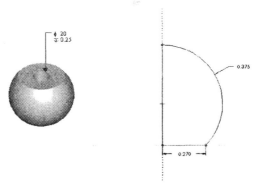

14.5 Construct the steady rest base, shown in Figure 14.20, using the dimensions provided.

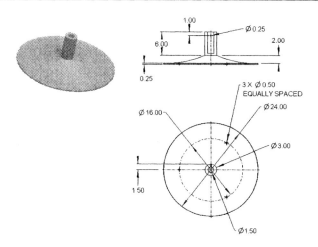

FIGURE 14.20

14.6 The ANSI standard Venturi nozzle shown in Figure 14.21 has an elliptic inlet (0.400 rho). Generate the geometry and revolve the section. The part is a thin with a thickness of 0.1".

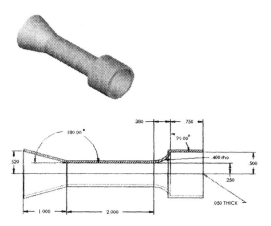

FIGURE 14.21

14.7 Use a 360° revolution and the given section in Figure 14.22 to create the base feature for the fixed bearing cup. Add the three equally spaced holes and the round as shown. This is a metric part.

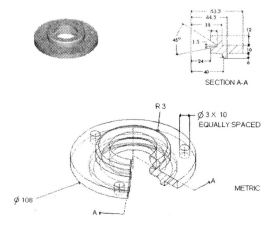

FIGURE 14.22

14.8 **Round Adapter.** For this metric part, use the *Revolve* option to construct the base feature. Add the slots and holes as required. The part is shown in Figure 14.23.

FIGURE 14.23

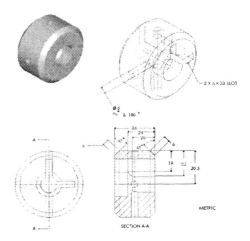

14.9 **Tapered Collar.** Construct the tapered portion of the collar using the section shown in Figure 14.24. Add the rest of the features as necessary.

FIGURE 14.24

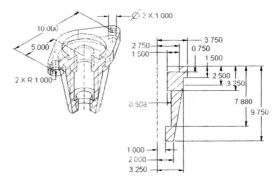

14.10 Construct the machine wheel, using Figure 14.25 as a reference.

FIGURE 14.25

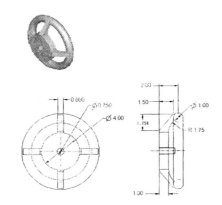

14.11 Construct base of the ladle using the *Revolve* option. Use Figure 14.26.

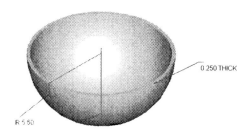

FIGURE 14.26

0.250 THICK

R 5.50

14.12 Create the model: "Flood Light Cover." Use the *Revolve* option and Figure 14.27.

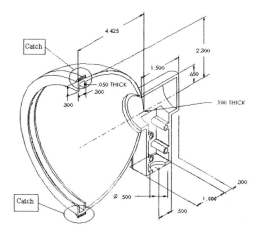

FIGURE 14.27

14.13 Add the "catch" and the four bosses to the model of the floodlight cover as shown in Figure 14.28.

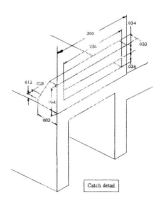
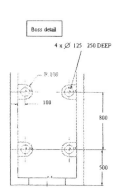

FIGURE 14.28

14.14 Create a mirror copy of the floodlight cover as shown in Figure 14.29. Add the counter bore holes and the latch. The geometry of the latch is shown in Figure 14.30.

FIGURE 14.29

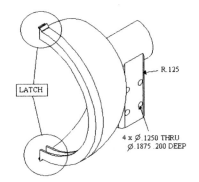

FIGURE 14.30

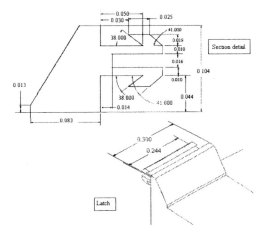

CHAPTER 15

FEATURE CREATION WITH SWEEP

INTRODUCTION AND OBJECTIVES

Many parts have features that are defined along a trajectory. Such features can be created in Pro/E by using the *Sweep* option. In creating a sweep, the user must define the sweep trajectory, as well as the section.

In Pro/E, a sweep can be a protrusion (material added to a model), or a cut (material removed from the model). Three-dimensional sweeps can be created by using a spline as the trajectory of the sweep. A *Helical Sweep* can be used to construct springs and parts with similar geometry. A helical sweep is produced by moving a section along a helical trajectory.

The objectives of this chapter are to

1. Develop the ability to construct a sweep trajectory or select a trajectory from suitable features on a model.
2. Attain an aptitude for constructing springs and threads using a *Helical Sweep*.
3. Use datum curves as a sweep trajectory.

15.1 THE *SWEEP* OPTION

Sweeps can be used to create features that have a cross section defined along a trajectory, provided the trajectory and the cross section of the protrusion can be easily defined.

The trajectory of the sweep can be defined by either selecting existing edges and/or curves, or by drawing the trajectory (see Figure 15.1). If the trajectory is sketched, then a sketching plane is required, wherein the path or trajectory of the sweep is drawn. The Sketcher is used to create the trajectory. After the trajectory is created, the section of the sweep is constructed using the Sketcher.

FIGURE 15.1 *The* SWEEP TRAJ *menu used to define the trajectory of the sweep.*

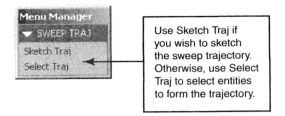

Use Sketch Traj if you wish to sketch the sweep trajectory. Otherwise, use Select Traj to select entities to form the trajectory.

FIGURE 15.2 *Coffee cup shows a sweep constructed with the* Free Ends *option.*

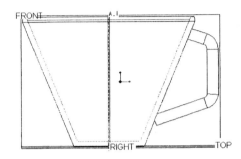

Pro/E chooses the point at which you start your trajectory as the start point for the sweep. In some cases, the start point is required at the end of the trajectory. By using the options *Sketch, Feature Tools,* and *Start Point* the user may change the start point for the sweep.

On some models, because of the geometry of the model and the sweep, the ends of the sweep may not blend into the model. This occurs because the end trajectory is not perpendicular to the model. Consider the coffee cup shown in Figure 15.2. Notice how the end of the handle does not merge with the cup surface.

If the user wants the sweep to blend into the model, then the *Merge Ends* option must be chosen. If the *Free Ends* option is chosen, the ends are not merged with the surface. The *Merge Ends* and *Free Ends* options are selected from the *ATTRIBUTES* menu shown in Figure 15.3.

Sweeps may fail if the section and trajectory are improperly defined. Often the problem is that during the sweep process, the section intersects itself. This will occur if the section is too large with respect to the sweep. The problem can be fixed by reducing the size of the section or changing the trajectory.

FIGURE 15.3 *The Pro/E* ATTRIBUTES *menu.*

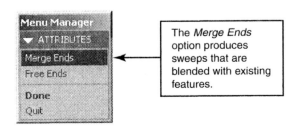

The *Merge Ends* option produces sweeps that are blended with existing features.

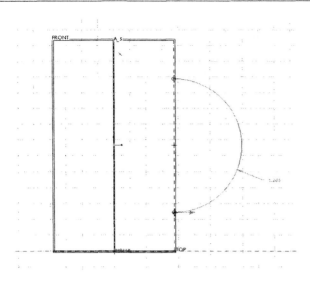

FIGURE 15.4 *The circle used to create the handle of the mug.*

15.1.1 TUTORIAL 15.1: A SIMPLE SWEEP

This tutorial illustrates:

- Creating a simple swept feature

1. Retrieve the model "CoffeeMug." Select [image] and double-click on the file.
2. Choose *Insert, Sweep,* and *Protrusion.*
3. We will sketch the trajectory of the sweep so choose *Sketch Traj* from the *SWEEP TRAJ* menu (Figure 15.1).
4. Click on datum plane Front to choose that default datum plane as the sketching plane. The arrow should point into the screen. Use *Flip* to change its direction.
5. If necessary, use [image] for the *Orientation* field and to "top." Then, click on datum plane Top.
6. In order to sketch the trajectory, use the arc button [image] to draw an arc as shown in Figure 15.4.
7. Dimension the arc as shown in Figure 15.5 using the given values.
8. Press [image] to build the geometry.
9. After the trajectory has been defined, Pro/E will reorient the model and ask the user to sketch the section to be swept. This is shown in Figure 15.6. The intersection of the dotted line and the sketching plane, which in this case is datum plane Front, forms the starting point for the sweep. Sketch a circle with the center near the starting point.
10. Dimension the diameter of the circle to 0.60.
11. Press [image] to build the geometry. Choose *Merge Ends* from the *ATTRIBUTES* menu when prompted to do so.
12. Now, complete the mug by adding rounds along the inner *and* outer edge of the top of the mug. Choose [image]. Click on the two edges shown in Figure 15.7.

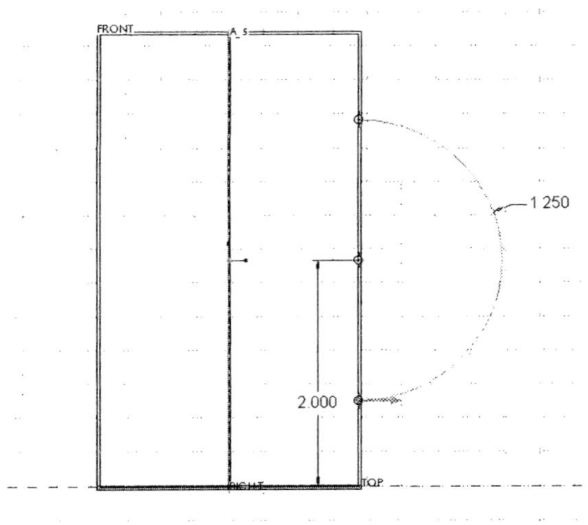

FIGURE 15.5 *The dimensions required for creating the handle trajectory.*

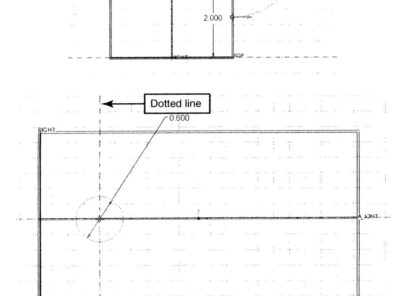

FIGURE 15.6 *Pro/E has reoriented the model to allow the user to sketch the section to be swept. The intersection of the dotted line and the sketching plane (in this case datum Front) forms the starting point.*

13. Use 0.1 for the radius of the rounds. Press ☒. Build the feature.

14. The final version of the mug should appear as in Figure 15.7. Save the part.

FIGURE 15.7 *The final version of the mug with the handle created, using a sweep.*

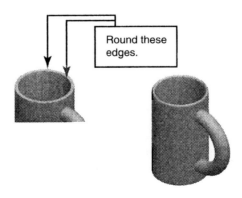

15.1.2 TUTORIAL 15.2: A GASKET

This tutorial illustrates:

- Creating a swept feature with more complicated geometry
- Changing units to metric

1. In order to construct the modal in Figure 15.8, select *File* and *New* (or simply ).
2. Make sure that the *Use default template* option is checked.
3. Enter the name "Gasket" in the *Name* field.
4. Select *OK*.
5. Choose *Edit, Setup,* and *Units.* Then, choose the *millimeter-newton-seconds* system of units.
6. Press the *Set* button and *OK.* The select *Close* and *Done.*
7. Select *Insert, Sweep,* and *Protrusion.*
8. We will generate one quarter of the total model as shown in Figure 15.9.
9. The trajectory of the sweep is shown in Figure 15.10 and will be sketched. Select the *Sketch Traj* option and sketch the trajectory using the given geometry.
10. With your mouse, choose datum plane Top as the sketching plane. The arrow should point down. Use *Flip* to change its direction.
11. If necessary, use ▦ for the *Orientation* field and to "bottom." Then, click on datum plane Front.
12. Note the location of the start point in Figure 15.10, which is indicated by the heavy black arrow. Make sure that the start point is located at the same point on your model. If not, use the options *Sketch, Feature Tools,* and *Start Point* to place the start point.
13. Dimension the trajectory to the values given in Figure 15.10.
14. Press ✔ to build the trajectory.
15. Select ◣ and sketch the cross section of the sweep. Use Figure 15.11.
16. Use ▥ and dimension the sketch.
17. Press ✔ to build the section.
18. Select *OK.* Pro/E will generate the sweep as shown in Figure 15.12.
19. Select ▣.

FIGURE 15.8 *A shaded model of the gasket. Note the symmetry of the model. Because of the symmetry, only one quarter of the model need be constructed. The rest of the model is completed using the* Mirror Geom *option.*

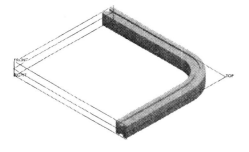

FIGURE 15.9 *One-quarter of the gasket produced using the* Sweep *option. The alignment tab and hole were added using a protrusion. Note the cross section of the sweep.*

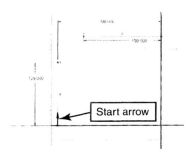

FIGURE 15.10 *The trajectory for the sweep is sketched using two lines and an arc. Note the location of the start arrow.*

FIGURE 15.11 *Sketch the geometry of the cross section using this figure. When dimensioning, be sure to dimension the section to the DTM2 plane and to the trajectory (the dotted line).*

20. With your mouse, choose datum plane Top as the sketching plane. The arrow should point down. Use *Flip* to change its direction.

21. If necessary, use [image] for the *Orientation* field and to "bottom." Then, click on datum plane Front.

22. Use the Line and Arc tools to generate the geometry shown in Figure 15.13. Note that the section must be closed.

23. Dimension and modify to the values given in Figure 15.13.

24. Press [image] to build the section.

25. Choose [image]. Enter a depth of 20 mm. The part after the extrusion is shown in Figure 15.14.

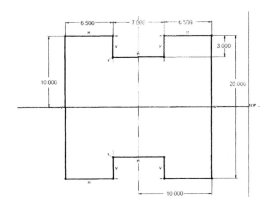

FIGURE 15.12 *The gasket after sweeping the section in Figure 15.11.*

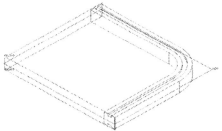

FIGURE 15.13 *The model in the Sketcher with the geometry of the alignment tab.*

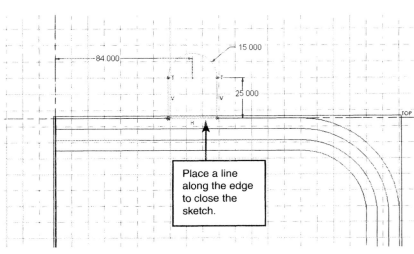

Place a line along the edge to close the sketch.

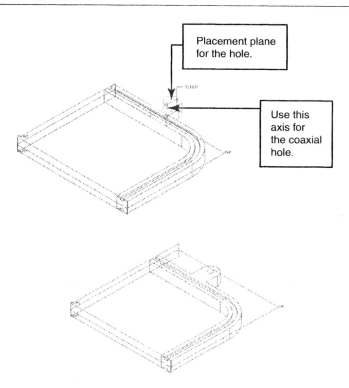

FIGURE 15.14 *The gasket after adding the alignment tab. Surface rounds need to be placed between the sweep surface and the tab.*

Placement plane for the hole.

Use this axis for the coaxial hole.

FIGURE 15.15 *The gasket after adding the hole on the tab.*

26. Select ▢.

27. Select *Straight, Simple,* and *Through all* options.

28. Enter a diameter of 10.000.

29. With your mouse click on the axis shown in Figure 15.14.

30. Press the *Placement* option in the dashboard. Click in the Secondary references field and click on the surface shown in Figure 15.14.

31. Build the feature by pressing ☑. The hole is shown in Figure 15.15.

32. Select ▢.

33. Select one of the two surfaces shown in Figure 15.16. Press your CTRL key and select the second surface.

34. Enter a radius of 10.

35. Build the feature by pressing ☑.

36. Repeat steps 33 through 36 for the other surface to surface round. The model with both rounds added is shown in Figure 15.17.

37. Complete the model using *Copy* and *Mirror.* Select *Edit, Feature operations,* and *Copy.*

38. Choose *Mirror, Select,* and *Dependent.* Select *All Feat.*

39. Choose *Done.* Click on datum plane Front, the mirror plane.

40. The software will add the feature as illustrated in Figure 15.18. Select *Done.*

41. Repeat steps 38 through 41, but click on datum plane Right.

42. The completed model is shown in Figure 15.19. Press ▢.

FIGURE 15.16 *The model rotated to show the surface to select when placing the rounds.*

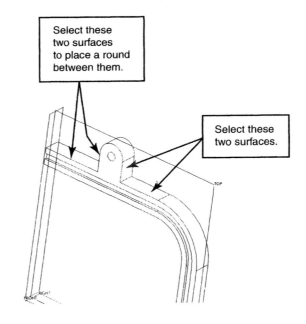

Select these two surfaces to place a round between them.

Select these two surfaces.

FIGURE 15.17 *The model of the gasket after the surface to surface rounds have been placed.*

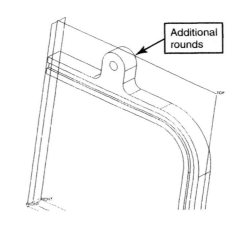

Additional rounds

FIGURE 15.18 *The gasket, after using Front to mirror the features. Half of the model is complete. A mirror using Right will complete the model.*

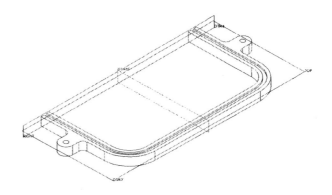

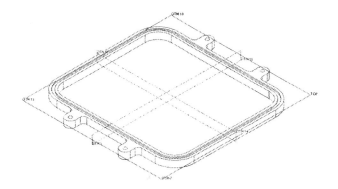

15.2 HELICAL SWEEPS

A *helical sweep* is constructed by moving a defined cross section along a helical path. Helical sweeps have practical applications in the construction of springs and thread surfaces. A *Helical Sweep* is a type of protrusion as shown in Figure 15.20.

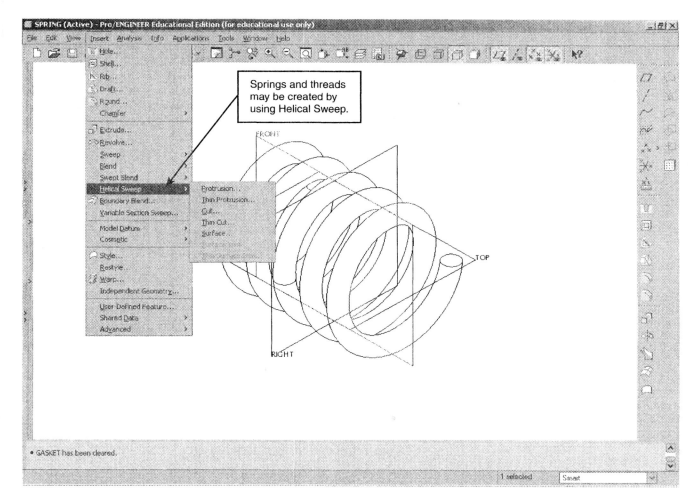

FIGURE 15.20 *The helical sweep may be a protrusion, cut, or surface.*

FIGURE 15.21 *The available attributes for creating a helical sweep.*

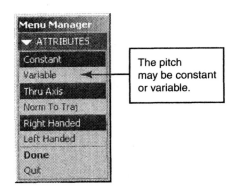

The distance between the coils of a spring or between the teeth of a thread is called the pitch. In constructing a helical sweep, the value of the pitch must be provided by the user. The other information required in construction of these types of features is entered by selecting from the options in the *ATTRIBUTES* menu shown in Figure 15.21.

From the menu, it is evident that either right-handed or left-handed sweeps may be constructed. Furthermore, a sweep with a *Constant* or *Variable* pitch may be created. The options *Thru Axis* or *Normal To Traj* refer to the orientation of the sweep cross section.

If a sweep is constructed with the *Thru Axis* option (the default), the cross section is defined so that it lies in a plane that contains the axis of revolution. The *Normal To Traj* option allows the user to create a sweep wherein the cross section is normal to the trajectory. Using the *Normal To Traj* option gives the cross section of the sweep a twist relative to the axis of revolution.

As we will see in Chapter 17, Pro/E has a built-in option for creating screw threads. However, this option is a cosmetic option; that is, Pro/E creates a surface that represents the threads, but does not actually draw them. One advantage is that model regeneration time is minimized. However, if the actual threads are desired, they can be created using a helical sweep.

In Tutorial 15.4, ANSI standard 0.25 pitch square threads will be placed on a screw. Although it takes time and effort to generate the threads, the thread feature is more visible. Furthermore, a stress analysis may now be performed on the screw as the thread feature is included in the model.

15.2.1 TUTORIAL 15.3: A SPRING CONSTRUCTED WITH A HELICAL SWEEP

Let us construct a spring using the *Helical Sweep* option. Use the following procedure to create this part.

This tutorial illustrates:

• Creating a spring using a *Helical Sweep*

1. Select ▢.
2. Make sure that the *Use default template* option is checked.
3. Enter the name "Spring" in the *Name* field.

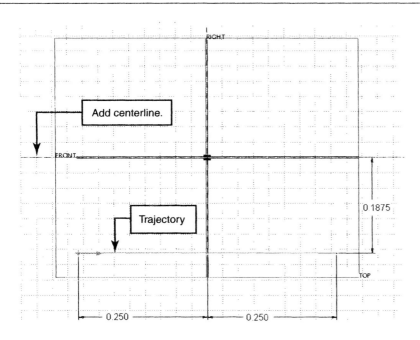

4. Select *OK*.

5. Select *Insert, Helical Sweep,* and *Protrusion.*

6. We will make a right-handed spring with a constant pitch, so accept the defaults from the *ATTRIBUTES* menu (Figure 15.21) by clicking on the *Done* option.

7. Pro/E will inquire about the sketching plane. Select datum plane Top. The arrow should point down.

8. Then choose datum plane Front as *Top* to orient the part in the Sketcher.

9. In the Sketcher, place a centerline along the edge of datum plane Front. Use 🔳.

10. Sketch a line parallel to the centerline and dimension as shown in Figure 15.22. The dimension will result in a spring with an inner bore of 0.19″ and a length of ½″.

11. Notice the solid black arrow in Figure 15.22, which defines the start point of the sweep.

12. Press 🔳 to build the trajectory.

13. Enter a value of 0.1 for the pitch.

14. Use the *Circle* tool and sketch the cross section of the spring shown in Figure 15.23.

15. Double-click on the diameter dimension and change the value to 0.06.

16. Press 🔳 to build the cross section.

17. Select *OK*.

18. Pro/E will generate the model. The spring is shown in Figure 15.24.

FIGURE 15.23 *In this sketch of the spring cross section the center of the circle is aligned to the start point.*

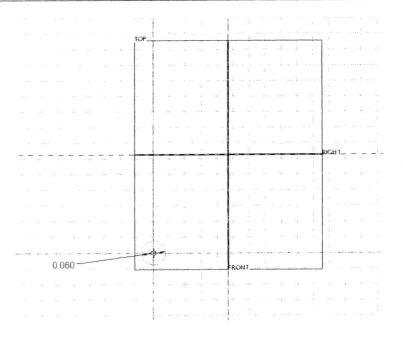

FIGURE 15.24 *Shaded model of the spring.*

15.2.2 TUTORIAL 15.4: CONSTRUCTING SCREW THREADS USING A HELICAL SWEEP

This tutorial illustrates:

- Construction of screw threads using a *Helical Sweep*

1. Retrieve the model "AdjustmentScrew." Select 📂 and double-click on the file.

2. After retrieving this part, select *Insert, Helical Sweep,* and *Protrusion.*

3. The thread is right-handed with a constant pitch. The surface to be threaded is given in Figure 15.25. Use the *Thru Axis* option.

4. Use datum plane Top as the sketching plane and datum plane Front as the *Top.* The arrow should point down.

5. Figure 15.26 shows the screw in the sketcher. Place a centerline along the edge of datum Right using 🔲.

FIGURE 15.25 *The adjustment screw with the surface to be threaded.*

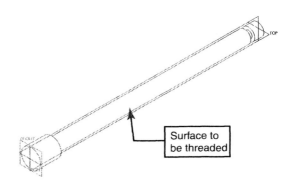

Surface to be threaded

6. Use the *Line* tool and sketch a line as shown in the figure. The starting point of the line is indicated by the big black arrow. The direction of the sweep is in the direction of the arrowhead.

7. Use 🔲 and dimension as shown in Figure 15.26. Modify the values using 📝.

8. Press ✅ to build the trajectory.

9. Enter a value of 0.25 for the pitch.

10. Sketch the cross section by using 🔲. Use Figure 15.27. The size of the thread is 0.125″ by 0.125″. In order to ensure that the tooth geometry intersects the adjustment screw during the sweep, the width of the thread is drawn slightly larger, say to a value of 0.130.

11. Press ✅ to build the feature.

12. Select *OK*.

13. The model with the sweep is shown in Figure 15.28.

14. In Figure 15.29, we have reproduced a top view of the screw that shows the spacing of the threads more clearly. Notice how the threads end at the appropriate points.

15. Press 🔲 and save the part.

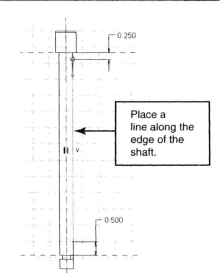

FIGURE 15.26 *The adjustment screw in the Sketcher showing the geometry for the trajectory of the sweep.*

FIGURE 15.27 *An enlarged sketch of the thread section. Note that the rectangle is drawn slightly into the adjustment screw. This will ensure that the helical sweep intersects the part.*

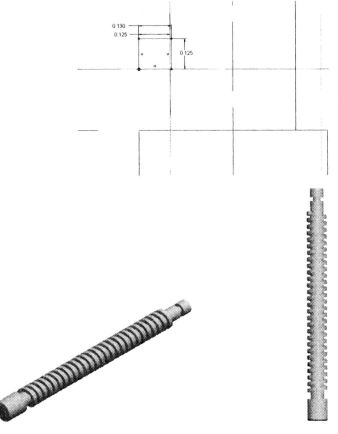

FIGURE 15.28 *A shaded model of the adjustment screw with added square threads.*

FIGURE 15.29 *The model has been rotated to readily illustrate the threads.*

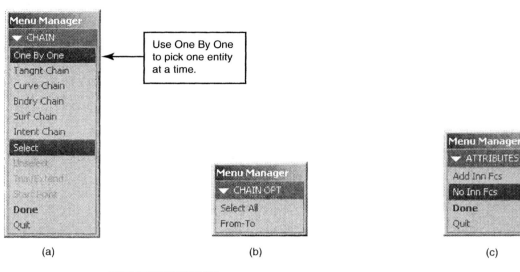

(a)   (b)   (c)

FIGURE 15.30 *The CHAIN and CHAIN OPT (options) menus.*

15.3 SWEEPS ALONG A DATUM CURVE

A sweep may be constructed along a datum curve by using the *Select Traj* option (Figure 15.1) and picking the datum curve as the sweep trajectory with the appropriate options from the *CHAIN* menu, shown in Figure 15.30(a). The options from this menu are:

1. *One By One.* Create a chain of curves and edges by selecting each of the features one at a time.

2. *Tangnt Chain.* The user selects an edge. The software selects all the edges tangent to the chosen edge. This completes the trajectory.

3. *Curve Chain.* Define the chain by selecting a curve. The *CHAIN OPT* menu shown in Figure 15.30(b) allows the user to select the endpoints of the chain. The *Select All* option selects all the curves that are connected to the curve selected by the user. The *From-To* option provides the user with the ability to create a composite chain by selecting vertices or ends of the features.

Bndry Chain and *Surf Chain* are used to construct a chain from the edges of a quilt (see Chapter 18 for a discussion of a *Quilt*). *Unselect* may be used to disable a choice.

In situations where the chain forms a closed loop, Pro/E can add top or bottom faces to the section as it sweeps along the trajectory. This is the *Add Inn Fcs* option shown in Figure 15.30(c). However, when using this option, the section must be an open section. For closed sections use the default option: *No Inn Fcs*.

15.3.1 TUTORIAL 15.5: COMPLETING THE TWO-DIMENSIONAL FRAME

This tutorial illustrates:

- A feature swept along a single curve

1. Retrieve the model "2DFrame." Select 📂 and double-click on the file.

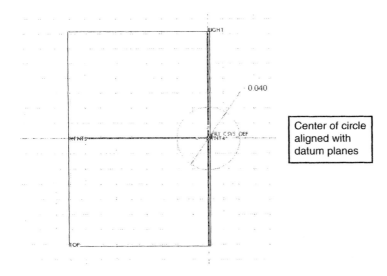

FIGURE 15.31 *The cross section of the two-dimensional frame.*

Center of circle aligned with datum planes

2. After retrieving this part, select *Insert, Helical Sweep,* and *Protrusion.*

3. From the *SWEEP TRAJ* menu (Figure 15.1), choose *Select Traj.*

4. Since we are going to select a datum curve, choose the *Curve Chain* option from the menu.

5. With your mouse, select the datum curve. Choose the *Select All* and *Done.* This will ensure that the entire curve is chosen for the trajectory.

6. A red arrow will appear. This arrow defines the upward direction for the sweep direction. Make sure it points towards you.

7. Sketch the circular cross section of the frame as shown in Figure 15.31.

8. Double-click on the diameter dimension and change the value to 0.04.

9. Press ☑ to build the feature.

10. Pro/E will sweep the cross section along the datum curve. The final part is shown in Figure 15.32. Save the part.

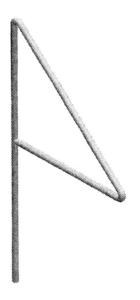

FIGURE 15.32 *The two-dimensional frame after sweeping the circular cross section along the datum curve.*

15.3.2 TUTORIAL 15.6: COMPLETING THE KING POST TRUSS

This tutorial illustrates:

- Sweeps along multiple curves
- Adding a feature to a swept feature

1. Retrieve the model "KingPostTruss." Select 🖼 and double-click on the file.

2. We now proceed to construct the outer loop made up of the three datum curves shown in Figure 15.33. Select *Insert, Sweep,* and *Protrusion.*

3. Choose *Select Traj* and *One By One.*

4. Pick the datum curves shown in Figure 15.33 to form the chain. Press CTRL to pick each additional curve. Select *Done.*

FIGURE 15.33 *The king post truss with the two loops.*

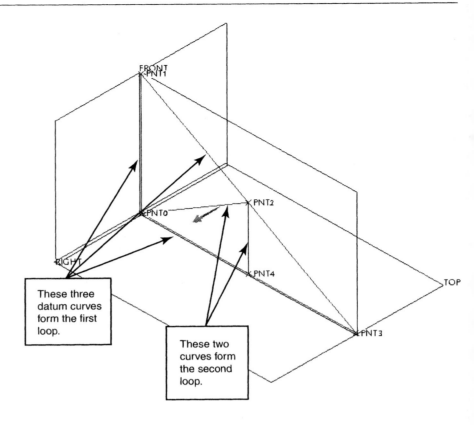

These three datum curves form the first loop.

These two curves form the second loop.

5. Since the section is closed, you must select *No Inn Fcs* and *Done*.

6. Notice the arrow in Figure 15.33. Make sure that the arrow on your screen is pointing in the same direction. If not, use the *Flip* option to change the direction of the arrow.

7. Use ▱ and sketch the rectangular geometry shown in Figure 15.34.

8. Use ▱ and dimension as shown.

FIGURE 15.34 *The cross section of the sweep is a rectangular section.*

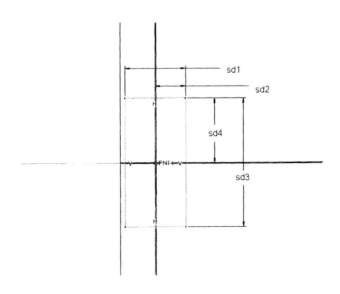

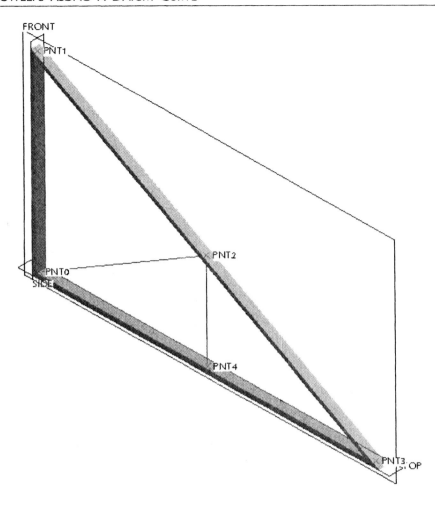

FIGURE 15.35 *The model of the king post truss after the first sweep.*

9. Select *Tools and Relations*. Enter the following relations (note the parameter names may be different):

$$sd2 = sd1/2$$
$$sd4 = sd3/2$$

Choose *OK*.

10. Modify the dimensions so that

$$sd1 = 1.5/12$$
$$sd3 = 3.5/12$$

We divide by 12 because the units of the model are in feet.

11. Press ✔ to build the feature.

12. Then select *OK*. The software will sweep the section.

13. Add the second sweep. Repeat steps 2 through 11. But use the second set of curves shown in Figure 15.35. Use the *Merge Ends* option.

14. Remove the extra triangular protrusion by cutting the feature. Select ⬜.

FIGURE 15.36 *The truss after completing the second sweep. Because of the geometry of this sweep, a small triangular feature protrudes beyond the first sweep.*

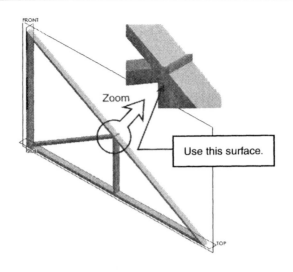

15. In the dashboard, choose the *Cut* option ⬜.
16. Use the surface shown in Figure 15.36 as the sketching plane.
17. To orient the model in the Sketcher, choose the *Top* option and select the datum plane called "Top." Make sure the orientation arrow points into the screen.
18. Use the *Line* tool and sketch three collinear lines as shown in Figure 15.37.
19. Build the section by pressing ✔.
20. Choose *Through all* for the depth of the cut. The effect of the cut on the model is shown in Figure 15.38.
21. Build the feature. Press ✔.
22. Select *Edit, Feature operations,* and *Copy.*

FIGURE 15.37 *The section needed for the cut. This is a "zoomed in" view of the intersection.*

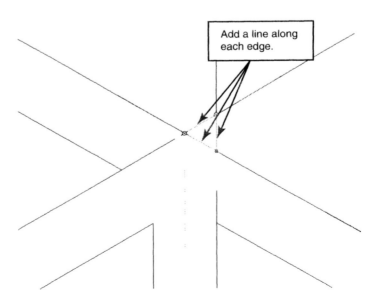

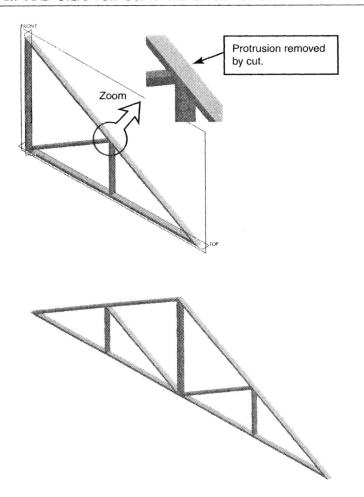

FIGURE 15.38 *After the use of the cut, the intersection under consideration takes on the appropriate appearance.*

FIGURE 15.39 *The king post truss after the use of the* Mirror *option. Other than the placement of nail plates at the joints, the model is complete.*

Protrusion removed by cut.

Zoom

23. Choose *Mirror, Select,* and *Dependent.* Then choose *All Feat.*

24. Choose *Done.* Click on the surface shown in Figure 15.38 as the mirror plane.

25. The software will add the feature. Select *Done.* See Figure 15.39.

15.4 SUMMARY AND STEPS FOR USING THE *SWEEP* OPTION

The *Sweep* option may be used to construct features that have a constant cross section defined along a trajectory. In creating a sweep, the user must define the trajectory as well as the cross section. The trajectory may be sketched using the Sketcher (*Sketch Traj* option) or constructed as a chain or entities that already exist on the model (*Select Traj*).

Datum curves may be used as the trajectory for a sweep by selecting the datum curve. Sweeps along datum curves are useful in creating models of structures such as trusses. A chain containing several curves may be created by using the *Curve Chain* option from the *CHAIN* menu (Figure 15.30(a)). In order to select individual datum curves, use the *One By One* option.

In general, a sweep is constructed by following these steps:

1. Select *Insert, Sweep,* and *Protrusion* (or *Cut*).
2. From the *SWEEP TRAJ* menu (Figure 15.1) choose *Select Traj* or *Sketch Traj.*
3. If you choose the *Sketch Traj* option, choose a sketching plane and an orientation plane. Then, sketch the trajectory and dimension. If you are selecting the trajectory, pick the appropriate entities to form the trajectory. Then choose *Done.*
4. Sketch the cross section of the sweep. Dimension to the appropriate values.
5. Choose *OK.*

A *Helical Sweep* is constructed by using a helical trajectory. Such sweeps are used to create springs and threads on fasteners.

The basic approach that may be used to create this type of sweep is:

1. Select *Insert, Helical Sweep,* and *Protrusion* (or *Cut*).
2. From the *ATTRIBUTES* menu (Figure 15.21), select the desired options—for example, a right-handed helix versus a left-handed helix.
3. Choose *Done.*
4. Pick a sketching and a reference orientation plane.
5. In the Sketcher, sketch and dimension the profile of the surface of revolution. Dimension the profile to the axis of revolution.
6. Build the section.
7. Enter the value of the pitch.
8. In the Sketcher, draw and dimension the cross section. Dimension to the appropriate values.
9. Build the cross section.

15.5 ADDITIONAL EXERCISES

15.1 Construct the handle shown in Figure 15.40 using the *Sweep* option. Note that this is a metric part.

FIGURE 15.40

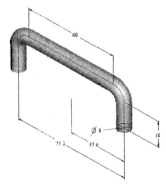

15.2 Use the *Sweep* option to construct the spring clip shown in Figure 15.41.

FIGURE 15.41

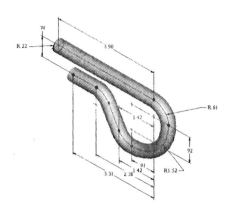

15.3 Using the *Sweep* option, create the metric model of the paper clip. Use Figure 15.42 as a reference.

FIGURE 15.42

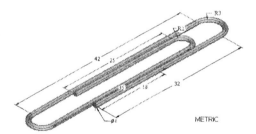

METRIC

15.4 Complete the model of the Fink truss using the *Sweep* option. Use 2 × 4 (1.5 × 3.5 actual) dimensional lumber for the elements of the truss. The base model was created in Exercise 13.8. The completed model is shown in Figure 15.43.

FIGURE 15.43

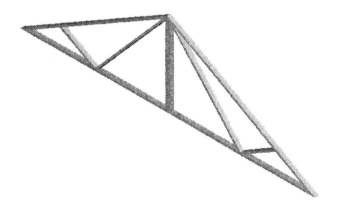

15.5 Complete the bridge truss, whose base feature was originally constructed in Exercise 13.10. Use 8 × 8 (7.5 × 7.5 actual) dimensional lumber for the elements of the truss. The completed model is shown in Figure 15.44.

FIGURE 15.44

15.6 Complete the highway sign frame using the datum points and curves generated in Exercise 13.2. Use Figure 15.45. The cross section of the sweep is a 4″ diameter circle. *Hint:* Use the *Sweep* option to create the protrusion through the frame. Then use *Copy Mirror* with datum Front as the mirror plane. Add the cross members by sketching on datum Front and extruding.

FIGURE 15.45

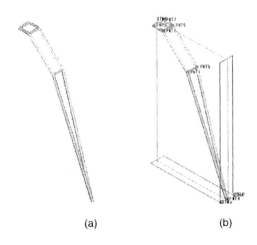

(a)   (b)

15.7 Create a model of the hook spanner using the geometry shown in Figure 15.46.

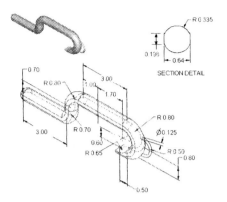

FIGURE 15.46

15.8 Construct a model of the shaft hanger using the appropriate Pro/E options and the geometry in Figure 15.47.

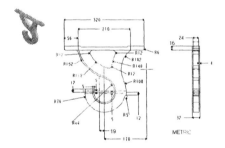

FIGURE 15.47

15.9 Create a model of the hand rail support. For the cross section of the sweep, use the given section in Figure 15.48.

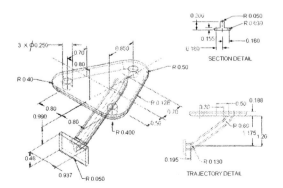

FIGURE 15.48

15.10 Construct the model of the dash pot lifter. Use the appropriate Pro/E options. The geometry of the part is given in Figure 15.49.

FIGURE 15.49

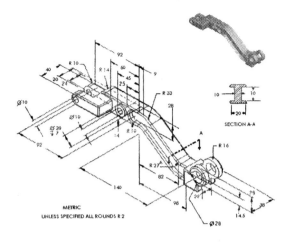

15.11 Add a handle to the ladle constructed in Exercise 13.11. Use the *Sweep* option and Figure 15.50.

FIGURE 15.50

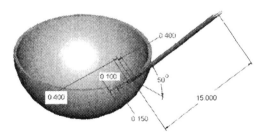

15.12 Add a grip to the ladle from Exercise 15.11. Use the geometry in Figure 15.51.

FIGURE 15.51

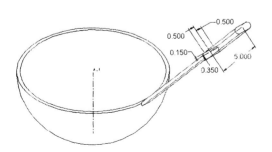

15.13 Construct the model of the tapered spring using Figure 15.52.

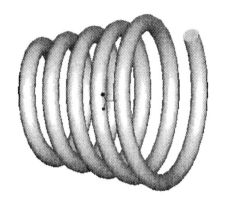

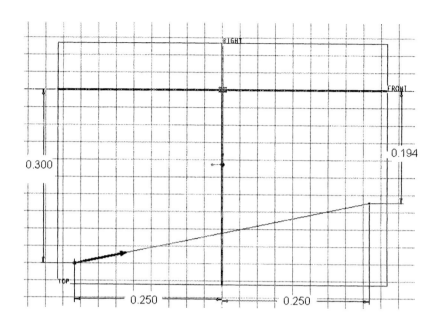

FIGURE 15.52

CHAPTER 16

BLENDS

INTRODUCTION AND OBJECTIVES

A *Blend* is a feature constructed of more than one section joined together by transitional surfaces. A blend may be a protrusion (solid or thin), cut (solid or thin), or surface. There are three types of blends (see Figure 16.1(a)): *Parallel, Rotational,* and *General.*

In a *Parallel* blend, all the sections lie on parallel planes. *Rotational* blends are created if the sections are rotated. Pro/E allows a maximum of 120° rotation. *General* blends allow both rotation and translation of the sections.

A blend along a trajectory may be created by using the *Swept Blend* option and its associated options (see Figure 16.1(b)). This option is sort of a combination of the *Sweep* and *Blend* options, wherein the user sketches the trajectory and sections. This option is useful for constructing features that would normally be created using the *Sweep* option, but whose cross section is not constant.

FIGURE 16.1 *The* BLEND OPTS *menu.*

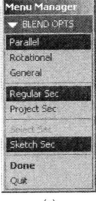

(a)

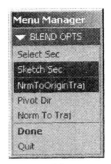

(b)

Pro/E assumes that all sections in a blend have the same number of entities. This presents a problem when, say, the user wishes to blend two sections that do not contain the same number of entities. An example is the blend between a circular and a rectangular section. Because the software considers a circle as one entity and the rectangle as four, the blend cannot be done unless the circle is divided into four elements. Of course, the circle must be properly divided.

The objectives of this chapter are to

1. Attain a proficiency in constructing blends
2. Create a model using the *Swept Blend* option
3. Attain experience in properly dividing a section so that it may be blended with another section containing a different number of entities

16.1 THE *BLEND* OPTION

In creating a blend, the user is asked to sketch, dimension, and regenerate each subsection before proceeding to the next one. In order to sketch the next subsection, the previous subsection must be grayed after successful regeneration. Selecting *Sketch, Feature Tools,* and *Toggle Section* in the Sketcher allows the user to toggle between sections.

After clicking on the *Toggle Section* option, the previous subsection is grayed and the Sketcher may be used to sketch the next subsection. The new subsection is drawn, dimensioned, and built. The process is repeated as many times as the number of subsections. Only after the final subsection has been successfully generated is the feature constructed by selecting ✔.

Pro/E blends the subsections by connecting the vertices of each subsection to the next. Thus, all the subsections must have the same number of entities. The connection between the corresponding vertices may be with straight lines (*Straight* option) or smooth curves (*Smooth* option). These options may be accessed via the *ATTRIBUTES* menu (see Figure 16.2). The type of connection, whether smooth or straight, depends on the geometry and design intent of the part.

In Tutorial 16.1, an HVAC takeoff will be constructed using a blend. The takeoff under consideration has two subsections, the first being a 6″ × 6″ square, and the second, a 6″ diameter circle. Since the number of entities for each subsection is not the same, we will need to divide the circle into four equal-length arcs. This will be accomplished by first sketching and regenerating the square. Then, diagonal lines will be drawn connecting the corners of the square and dividing the circle into the equal-length arcs.

It is also important to keep track of the start point for each section. If the start point is not the same for all the subsections, then a part with a twisted blend is obtained.

16.1.1 TUTORIAL 16.1: HVAC TAKEOFF

This tutorial illustrates:

- Creating a blend between two subsections
- Toggling between subsections
- Dividing a geometric entity so that it can be used in a blend

FIGURE 16.2 *The* ATTRIBUTES *menu showing the* blend *options* Straight *and* Smooth.

1. Select 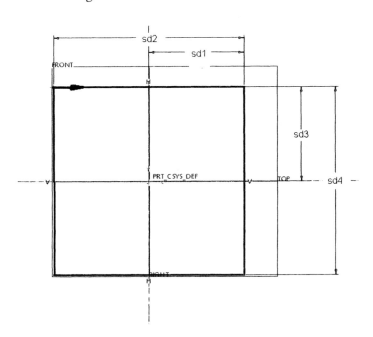.

1. Select [icon].
2. Make sure that the *Use default template* option is checked.
3. Enter the name "HeatTakeoff" in the *Name* field.
4. Select *OK*.
5. Select *Insert* and *Blend* and *Thin Protrusion*.
6. Select *Parallel* and *Regular Sec,* since the subsections will be parallel to one another.
7. After selecting *Done,* click on *Straight.*
8. Choose datum plane Front as the sketching plane.
9. The arrow should point toward you. Use *Flip* if it does not.
10. Select "Top" from the menu and click on datum plane Top.
11. Use [icon] and draw the square as dimensioned and as shown in Figure 16.3.
12. If necessary, use [icon] and dimension as shown in the figure.
13. Now, select the *Tool* and *Relations.*
14. Enter (note that your parameter names may be different):

$$sd2 = sd1/2$$
$$sd3 = sd4/2$$

Select *OK.*

15. Dimension the values so that sd1 = sd4 = 6.
16. Click on *Sketch, Feature Tools,* and *Toggle Section.* This will grey the square.
17. Now, draw the circle so that the center of the circle is aligned to the edges of datum planes Top and Right. See Figure 16.4.
18. Use [icon] and draw diagonal lines through the corners of the square as shown in Figure 16.4.

FIGURE 16.3 *The first subsection with added dimensions.*

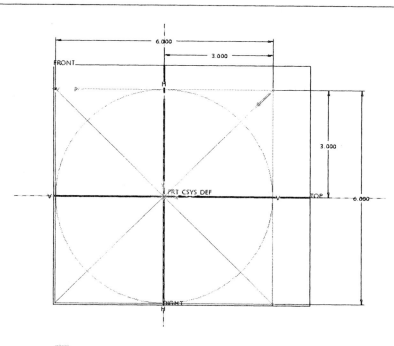

19. Select 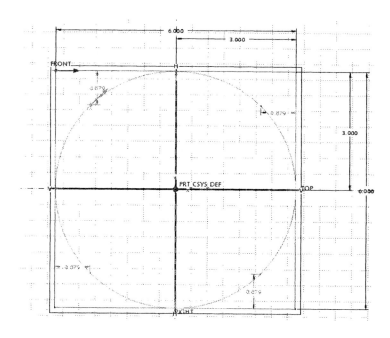, the *divide* button, and click at the four points where the lines intersect the circle. After obtaining these points, delete the lines.

20. Change the start point to the upper left-hand quadrant (see Figure 16.5) by clicking on the desired point and then selecting *Sketch, Feature Tools,* and *Start Point.*

21. Build the section by selecting ✔.

22. Fill the thickness of the part *inward.* Enter a value of 0.05″.

FIGURE 16.5 *In the circular subsection, notice the proper placement of the start point.*

FIGURE 16.6 *After constructing with a* Blend, *this shaded solid model of the HVAC takeoff may be obtained.*

23. Enter a depth of 12.0″.
24. Select *OK*.
25. The takeoff should appear as in Figure 16.6. Save the part.

16.2 SWEPT BLENDS

A *Swept Blend* is a blend along a defined trajectory. Multiple sections may be placed along the trajectory and their geometry defined. The trajectory may be closed or open. In the case of an open trajectory, the section *must* be sketched at the endpoints of the trajectory. For closed trajectories, the sections must be drawn at the start point and one other point along the trajectory. As with a regular sweep, the trajectory may be sketched or selected from predefined datums, curves, or edges.

After defining the trajectory, Pro/E will place points on the sketch where the section is to be sketched. The section can be rotated about the *z*-axis of the given coordinate system. The value of the rotation must be between +120° and −120°. The section may be placed normal to the trajectory (*Nrm to Spine* option) or normal to trajectory as viewed from a pivot point (*Pivot Dir* option). These options may be chosen from the *BLEND OPTS* menu shown in Figure16.1(b).

16.2.1 TUTORIAL 16.2: A CENTRIFUGAL BLOWER

This tutorial illustrates:

- Creating a swept blend

1. Select .
2. Make sure that the *Use default template* option is checked.
3. Enter the name "Blower" in the *Name* field.
4. Select *OK*.
5. First create the base part, which is a revolved section (see Figure 16.7). Choose .
6. Choose . Enter a thickness of 0.05.
7. Select and enter 360° of revolution.
8. Press .

FIGURE 16.7 *The blower.*

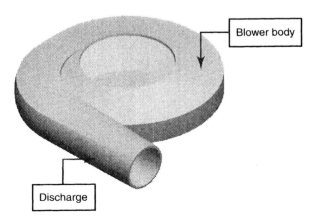

Blower body

Discharge

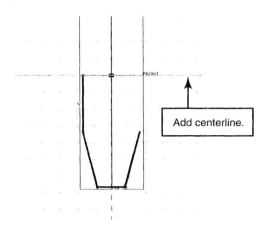

FIGURE 16.8 *The geometry necessary for constructing the body of the blower.*

Add centerline.

9. Select datum plane Right as the sketching plane.
10. The arrow should point into the screen.
11. Select "top" as the orientation and then click on datum plane Front.
12. Press ▦.
13. Use ⊡ and sketch a centerline along datum plane Front.
14. Use ▧ and sketch the geometry given in Figure 16.8.
15. Dimension the geometry shown in Figure 16.8. Use ▱; select all the dimensions and update the dimensional values to the ones given in Figure 16.9.
16. Press ✔ and build the section.
17. Press ✔ and build the part.
18. Select *Insert* and *Swept Blend* and *Thin Protrusion*.
19. Choose the *Nrm to Spine* option. Select *Sketch Traj*.
20. Use datum plane Top as the sketching plane. The arrow should point down. Choose "top" from the menu and select datum plane Front.
21. Now sketch vertical and horizontal lines as shown in Figure 16.10. Add an arc between the two lines.

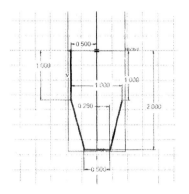

FIGURE 16.9 *Dimension the geometry as shown.*

FIGURE 16.10 *The blower in the Sketcher with the geometry needed to define the trajectory of the swept blend nozzle.*

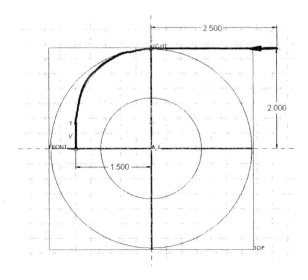

FIGURE 16.11 *The coordinate system on the blower is locating the first point of the trajectory.*

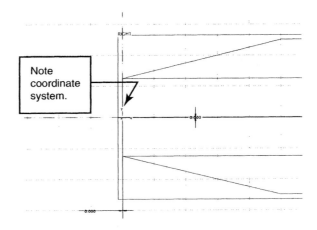

22. Dimension as shown in the figure.
23. Notice the position of the start point in Figure 16.10. If necessary, use *Sketch, Feature Tools,* and *Start Point.*
24. Press ☑ to build the section.
25. Use the *Next* option (if necessary, use the *Previous* option) and scroll through the points on the trajectory until Pro/E highlights the endpoint where the start arrow is located. Choose *Accept* when it has done so.
26. Notice that Pro/E has placed an *x-y-z* coordinate system at the point. Enter a value of zero for the rotation angle.
27. Pro/E will reorient the sketch as shown in Figure 16.11. Notice the coordinate system indicating the point.
28. Sketch a circle with the center of this circle aligned to the origin of the coordinate system, as illustrated in Figure 16.12.
29. Sketch two lines roughly at ±45° to the horizontal.
30. Use the *Divide* button 🖎 and divide the circle into the four arcs as shown in Figure 16.13.
31. Press 🖎 and delete the lines, as they are no longer needed.
32. Use 🖎 and dimension the geometry as shown in Figure 16.13.

FIGURE 16.12 *The circle bisected by two lines. These lines will be used to divide the circle into four equal arcs.*

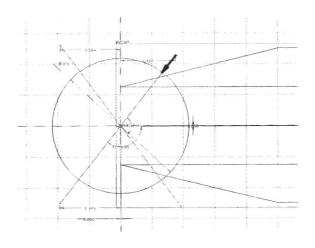

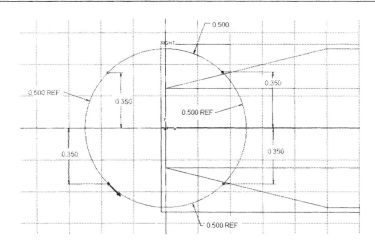

FIGURE 16.13 *After dividing the circle into four equal arcs, the section can be sized using the shown dimensions.*

33. Use ![] and change the dimensional values. Note that the vertical dimension of the points is $0.5 \times \sin 45 = 0.350$.

34. Make sure the start point is in the same quadrant as the one for the next section. If necessary, use *Sketch, Feature Tools,* and *Start Point* to move the start point.

35. Build the section by pressing ![].

36. Grow the thin section inward.

37. For the next section enter a rotation of 0.00.

38. Again, the software will reorient the model and place a coordinate system as shown in Figure 16.14 indicating the second point.

39. Use ![] and sketch three lines, two along the silhouette of the part body. This section is given in Figure 16.15.

40. Dimension and modify the value as necessary.

41. In Figure 16.15, notice the location of the start point for the second section. Again, move the start point, if necessary.

42. Press ![]. Build the section.

43. Grow the part inward. Enter a value of 0.05 for the thickness.

44. Select *OK*.

45. Pro/E will generate the feature. A shaded model of the blower is given in Figure 16.7. Save the part.

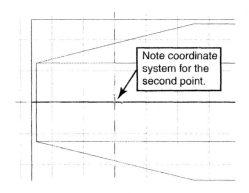

Note coordinate system for the second point.

FIGURE 16.14 *The blower with the second point locating the second section.*

FIGURE 16.15 *The geometry of the second section includes a vertical line.*

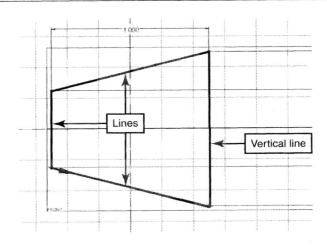

16.3 SUMMARY AND STEPS FOR CREATING BLENDS

A blend is a feature constructed by joining more than one section using smooth or linear transitional surfaces. The blend is created by sketching and dimensioning or by selecting each section.

The software blends the individual sections by connecting the vertices of each subsection. Therefore, in order to avoid a "twisted blend," the start arrow for each subsection must be located at the same vertex.

The general sequence of steps for creating a *Blend* is:

1. Select *Insert, Blend,* and Protrusion (or *Cut*).
2. Choose one of these options: *Parallel, Rotational,* or *General.* In addition, select from *Regular Sec* or *Project Sec.*
3. Select *Done.*
4. Select either *Straight* or *Smooth.* Then select *Done.*
5. Choose a sketching and an orientation plane.
6. Sketch and dimension the first subsection.
7. If necessary, locate the start point by using *Sketch, Feature Tools,* and *Start Point.*
8. Select *Sketch, Feature Tools,* and *Toggle Section.* This will gray the first subsection.
9. Repeat steps 6 through 8 for the additional subsections.
10. Enter the depth of the blend.

A blend may be created along a trajectory. This is called a *Swept Blend.* As with a regular blend, multiple sections may be defined. However, in this case, the sections are defined for discrete points along the trajectory. In general, the following steps may used to create a *Swept Blend:*

1. Select *Insert, Swept Blend,* and *Protrusion* (or *Cut*).
2. Select *Sketch Sec* in order to sketch the sections or *Select Sec* to pick the sections.

3. The trajectory needs to be defined. This may be accomplished by sketching (*Sketch Traj*) or by selecting the appropriate entities (*Select Traj*).

4. Depending on your response in step 3, either sketch or pick the trajectory. If you have chosen to sketch the trajectory, then the software will need a sketching plane and an orientation plane.

5. Sketch the subsection for each desired location. For an open trajectory, the section *must* be sketched at the endpoints of the trajectory.

16.4 ADDITIONAL EXERCISES

16.1 Construct the model of the Stop using the *Blend* option and Figure 16.16.

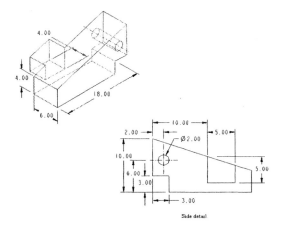

FIGURE 16.16

16.2 Create the model of the Holder using the *Blend* option. Use Figure 16.17 as a reference.

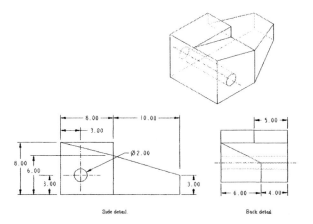

FIGURE 16.17

16.3 Using the *Blend* option and Figure 16.18 construct the model of the support.

FIGURE 16.18

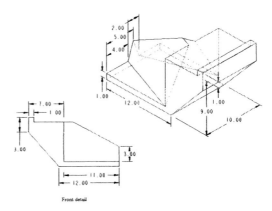

Front detail

16.4 Add a spout to the ladle from Exercise 15.12. Use the *Swept Blend* option and Figure 16.19.

FIGURE 16.19

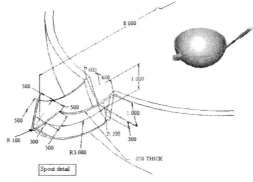

Spout detail

16.5 Create the faucet. Use the *Swept Blend* option. In Figure 16.20, note that there are four sections along the sweep trajectory.

FIGURE 16.20

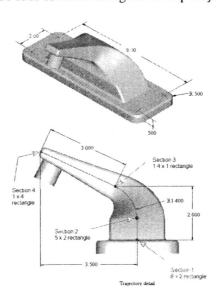

Trajectory detail

16.6 Add rounds and an aerator to the model of the faucet as shown in Figure 16.21. All rounds are 0.250.

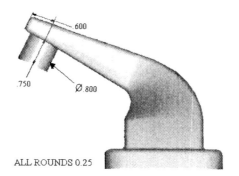

.600

.750

Ø.800

ALL ROUNDS 0.25

FIGURE 16.21

COSMETIC FEATURES

INTRODUCTION AND OBJECTIVES

The *Cosmetic* option is used to place special features, including text for company logos and serial numbers, projected sections, and threads. The various cosmetic options are accessed via the menu shown in Figure 17.1.

The cosmetic *Groove* is constructed by projecting a sketch onto a surface. The cosmetic *Groove* is useful in the manufacturing process *Groove,* where the groove is used to define the tool path.

The option *Sketch* is used to place logos and serial numbers. The *Thread* option is used to create a surface on the part that represents the threads. Pro/E does not actually draw the individual threads. If you want to draw the individual threads, see the section on helical sweeps in Chapter 15.

The objectives of this chapter are to

1. Place cosmetic threads on a surface
2. Create cosmetic text on a part

17.1 PRO/E COSMETIC OPTIONS

Using the *Cosmetic* feature, threads may be placed by defining the thread. In this case, the surface to be threaded (*Thread Surf*), the position to start threading (*Start Thread*), the depth to which the feature is threaded (*Depth*), and the major diameter of the thread (*Major Diam*) must be defined.

Pro/E will create both internal and external threads depending on the surface on which the thread is located. If the thread is placed on a hole, then an internal thread is created. If the thread is placed on a cylinder, then an external thread is created.

Pro/E sizes all threads 10 percent larger or smaller than the governing surface size. For example, if the thread is placed on a hole, then the thread

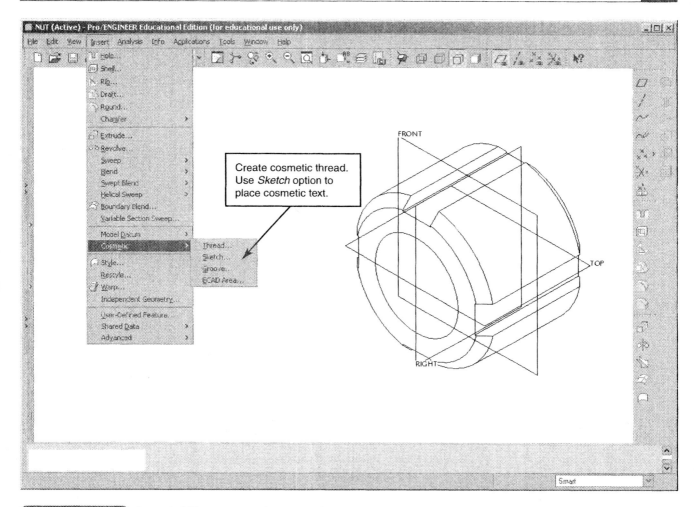

Several different types of cosmetic features may be created.

is sized 10 percent larger than the size of the hole. If the thread is placed on a shaft, then the thread is created 10 percent smaller than the diameter of the shaft.

The parameters of the thread may be accessed and modified by the *FEAT PARAM* menu shown in Figure 17.2. When the *FEAT PARAM* menu appears, select *Mod Params*. Pro/E will load the Pro/Table editor. The editor will list the parameters of the thread. Change the parameters of the thread as desired.

Another interesting use of the cosmetic option is to place a logo or text on a part. The text is often a serial or part number. Because the Sketcher is used to construct the text or logo, the cosmetic placed on the part can be quite complex and elaborate.

The text box, shown in Figure 17.3, contains the options for creating the text. The text is entered into the field provided. Several different types of fonts are available. These include fonts from PTC and True Type fonts as shown in Figure 17.4. The text, like any section created in the Sketcher, must

The FEAT PARAM menu may be used to access the parameters defining the thread.

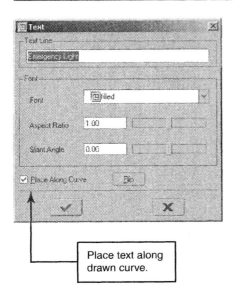

Place text along drawn curve.

FIGURE 17.3 *The* Text *box is used to define properties for the text.*

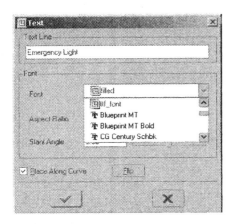

FIGURE 17.4 *Several different font types are available for cosmetic text; these include those by PTC as well as True Type fonts.*

be properly located with respect to the part. This is done by dimensioning the base of the sketch.

17.1.1 TUTORIAL 17.1: COSMETIC THREADS FOR THE NUT

This tutorial illustrates:

- Adding a cosmetic thread to a part

1. Retrieve the model "Nut." Select 📷 and double-click on the file.
2. Then select *Insert, Cosmetic,* and *Thread.*
3. Using your mouse, pick on the cylindrical surface, which forms the surface of the hole through the nut as shown in Figure 17.5.
4. Next, you will need to pick the starting surface. Use the figure for reference.
5. Pro/E will display an arrow indicating the direction of the thread creation. The direction of the arrow should be the same as in Figure 17.5. Flip the arrow if it is not in the same direction as the one shown in the figure. Choose *OK.*
6. We want the threads to extend along the entire surface of the hole, so pick the "back" of the nut as shown in Figure 17.5.
7. For the thread diameter, Pro/E will calculate 10 percent of the size of the hole and present the result as the default value. This will give a value of 0.33". In order to make the thread a standard size; enter the value .375 (3/8").
8. The ANSI standard pitch for a 3/8" major diameter UNC thread is 16 threads per inch. We can change the pitch of the thread by selecting the *Mod Params* option from the *FEAT PARAM* menu shown in Figure 17.2.
9. Pro/E will load the Pro/Table editor as in Figure 17.6. Click on the value for the pitch and enter 16.
10. Also, make sure that the metric option is set to *False.* Then select *File* and *Exit.*
11. From the *FEAT PARAM* menu, choose *Done/Return.*
12. Then select *OK.* Pro/E will load the threads but will not draw the individual teeth.
13. The threads will be indicated by an additional surface. This additional surface is illustrated in Figure 17.7. In this case, the surface is coincident with the surface forming the hole. Save the part.

17.1.2 TUTORIAL 17.2: COSMETIC TEXT FOR THE COVER PLATE

This tutorial illustrates:

- Adding cosmetic test to a part

1. Select ▢.
2. Make sure that the *Use default template* option is checked.
3. Enter the name "CoverPlate" in the *Name* field.
4. Select *OK.*

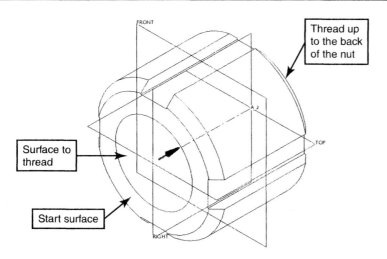

FRONT

Thread up
to the back
of the nut

Surface to
thread

Start surface

TOP

RIGHT

FIGURE 17.5 *The arrow is
indicating the direction for the creation
of the threads on the nut. The base of
the arrow is attached to the starting
surface.*

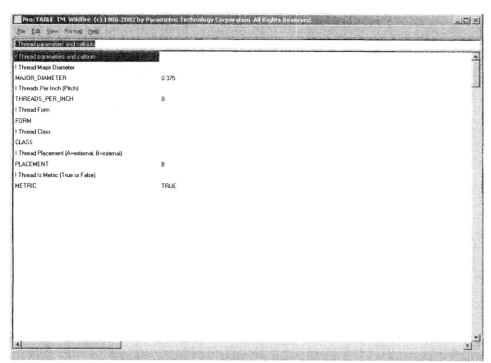

FIGURE 17.6 *The Pro/Table editor showing the parameters for the thread
on the nut. Change the* Threads per Inch *to read 16. Also, make sure that the*
Metric *option is set to False.*

5. Select ▣.

6. Choose the Sketch button ▣ in the dashboard.

7. With your mouse, choose datum plane Top as the sketching plane.

8. If necessary, use ▣ for the *Orientation* field and to "bottom."
   Then, click on datum plane Right.

9. The arrow should point down as in Figure 17.8. Use *Flip* if it does
   not.

FIGURE 17.7 *The nut after adding the cosmetic thread. Pro/E does not draw the individual thread teeth, rather, it inserts a surface that represents the threads. Compare this fugure with Figure 17.5.*

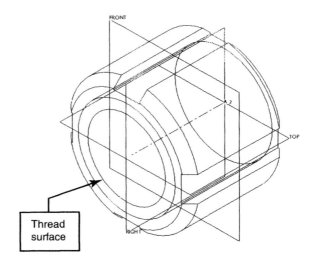

Thread surface

10. Select the *Sketch* button ⬚.
11. Use ⬚ and sketch the geometry in Figure 17.8.
12. Dimension the sketch as necessary. Then, use ⬚ and change to the values given in the figure.
13. Press ✓ to build the section.
14. Choose ⬚ and enter a depth of 0.150.
15. Press ✓ and build the feature.

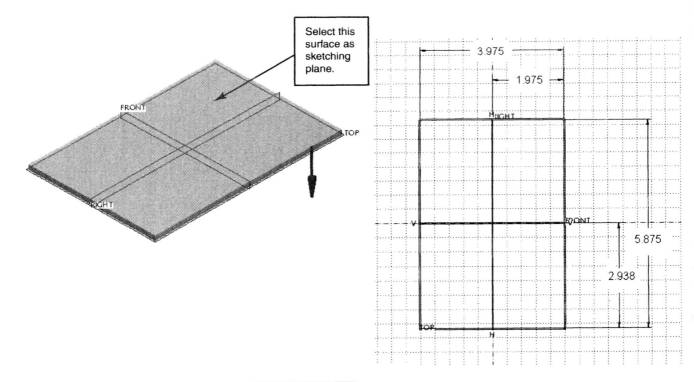

Select this surface as sketching plane.

FIGURE 17.8 *The model "CoverPlate" has the dimensions shown.*

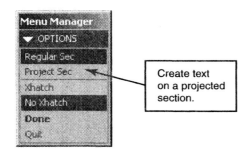

FIGURE 17.0 *Regular or projectedf text may be created with the* **OPTIONS** *menu.*

16. Select *Insert, Cosmetic,* and *Sketch.*
17. For this model, we will place the cosmetic text on the surface itself, so select *Regular Sec* (see Figure 17.9). Click on *Done.*
18. Select the top of the plate as shown in Figure 17.8.
19. Make sure that the arrow points in the direction shown in Figure 17.8.
20. Use datum plane Right as the "bottom" to reorient the model in the Sketcher.
21. Choose the *Text* button ▲.
22. Draw a horizontal line in the lower right-hand corner of the plate by dragging your mouse. This line doesn't have to be very long.
23. Enter the text "Emergency Light" in the *Text* box.
24. Check the option *Place Along Curve* and then select the line drawn in step 22. This will place the text in a horizontal fashion.
25. Click on the *Close* button in the *Text* dialog box (See Figure 17.3).
26. Change the value of the dimensions to the values given in Figure 17.10.
27. Build the feature by selecting ✔.
28. Pro/E will add the cosmetic text to the model. Save the part.

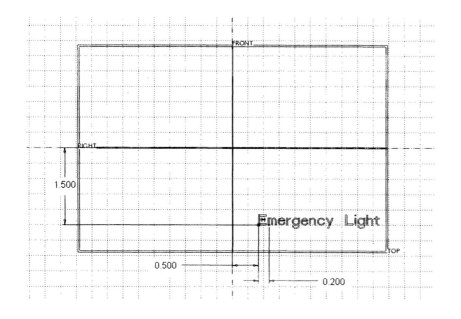

FIGURE 17.10 *Dimension the base of the text to the datum planes as shown.*

17.2 SUMMARY AND STEPS FOR ADDING COSMETIC FEATURES

The *Cosmetic* option may be used to place logos and text, such as serial numbers and threads, on a part. When Pro/E creates a *Cosmetic Thread*, it constructs a surface that represents the threads. A cosmetic *Sketch* is added to a part by using the Sketcher.

The general steps for adding a cosmetic sketch to the part are as follows:

1. Select *Insert, Cosmetic,* and *Sketch*.
2. Choose the sketching surface and an orientation plane.
3. Sketch the geometry. If you are placing text draw a line to indicate the position and size of the text.
4. Dimension the geometry to the part.
5. Build the section.

In general, cosmetic threads may be added to a surface by using the following procedure:

1. Choose *Insert, Cosmetic,* and *Thread*.
2. Select the surface to be threaded.
3. Choose the feature from which the threading is to begin.
4. Select the direction of the thread.
5. Indicate the depth of the thread.
6. Enter the parameters of the thread including its diameter.
7. Choose *Done/Return*.

17.3 ADDITIONAL EXERCISES

17.1 Add the cosmetic thread to the handle (Exercise 15.1). Use the geometry in Figure 17.11.

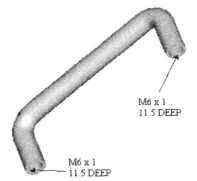

M6 x 1
11.5 DEEP

M6 x 1
11.5 DEEP

17.2 For the model of the coupler from Exercise 6.9, add the cosmetic thread as show in Figure 17.12.

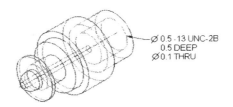

Ø 0.5 -13 UNC-2B
0.5 DEEP
Ø 0.1 THRU

17.3 Add cosmetic threads to the holes in the model of the pipe clamp bottom (Exercise 8.2). Use Figure 17.13 for the geometry of the threads.

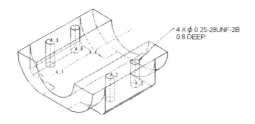

FIGURE 17.13

17.4 For the model of the offset bracket from Exercise 4.6, consider Figure 17.14 and add the cosmetic thread.

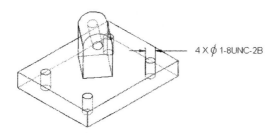

FIGURE 17.14

17.5 Add the threads to the holes on the side support model (Exercise 7.7). Use Figure 17.15 for the geometry of the threads.

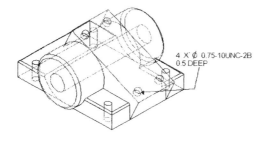

FIGURE 17.15

17.6 Add the cosmetic thread to the hole in the angle block (Exercise 4.9). Information on the thread is given in Figure 17.16.

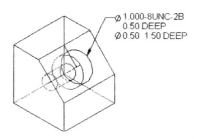

FIGURE 17.16

17.7 Add the text "RoboArm" to the model of the robot arm housing (Exercise 5.9). The location of the text on the model is shown in Figure 17.17.

FIGURE 17.17

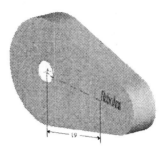

17.8 For the model of the electric motor cover, add the text "EMotors." Use Figure 17.18 for the location of the text. Also see Exercise 5.10.

FIGURE 17.18

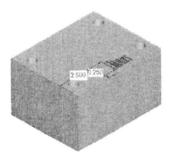

CHAPTER █18█

SURFACES, DRAFTS, LIPS, AND EARS

INTRODUCTION AND OBJECTIVES

The *Draft* option is used to add a draft angle on an existing feature. The angle must be between −30° and +30°. A *Lip* is a small protrusion often used to interlock one part with another. An *Ear* is a protrusion extending from a surface that is bent at its base. Both the *Lip* and *Ear* options are *Advanced* features, as shown in Figure 18.1.

A *surface* quilt is a nonsolid feature; that is, it has no thickness. A surface is created using the same fundamental options such as *Extrude, Revolve, Sweep,* and so on as shown in Figure 18.2. Surface quilts may be manipulated. A surface is useful when the design intent is to analyze the surface, not the interior of an object. For example, in aerodynamic studies, the surface geometry of a wing is analyzed. Therefore, only a model of the surface is required to calculate the lift and drag forces on the wing.

The objectives of this chapter are to

1. Use the *Draft, Lip,* and *Ear* options to place specialty protrusions on a part
2. Create a surface (quilt)

18.1 ADDING A DRAFT TO A BASE FEATURE

Drafts are placed on features that are cast so that they can be removed from molds. Bosses are often given a small draft, say of 5°—such as on the emergency light cover in Chapter 5.

A draft is constructed by moving a surface or surfaces through the given angle about a neutral plane, line, or curve. The only surfaces that can be drafted are surfaces of a cylinder or plane. They are called *draft surfaces.* The neutral plane, line, or curve must be indicated by the user; these are called *draft hinges.* The *pull* or *draft direction* is the direction used to measure

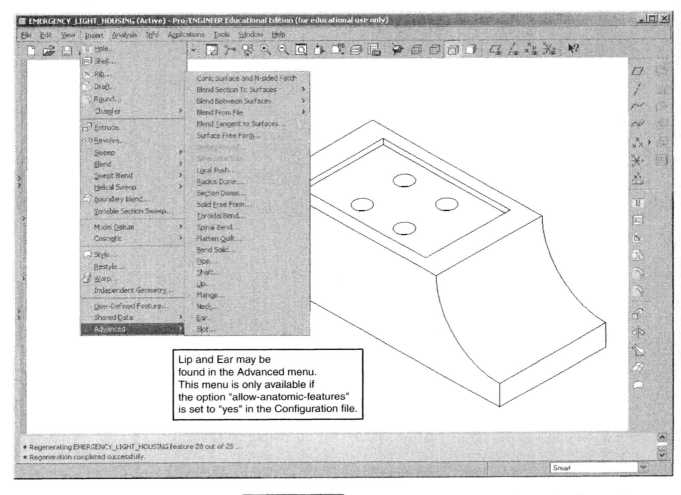

Lip and Ear may be
found in the Advanced menu.
This menu is only available if
the option "allow-anatomic-features"
is set to "yes" in the Configuration file.

FIGURE 18.1 *The ADVANCED menu contains options for creating specialty features.*

the draft angle. The draft direction is normal to the reference plane chosen by the user. The draft angle is measured from this normal.

If the draft surface is split, the draft angle may be different for each surface. The draft surface may be split by either the *draft hinge* or a different curve on the *draft surface*.

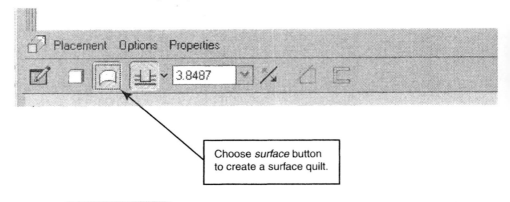

Choose *surface* button
to create a surface quilt.

FIGURE 18.2 *By using the* Surface *tool, the user may create a quilt.*

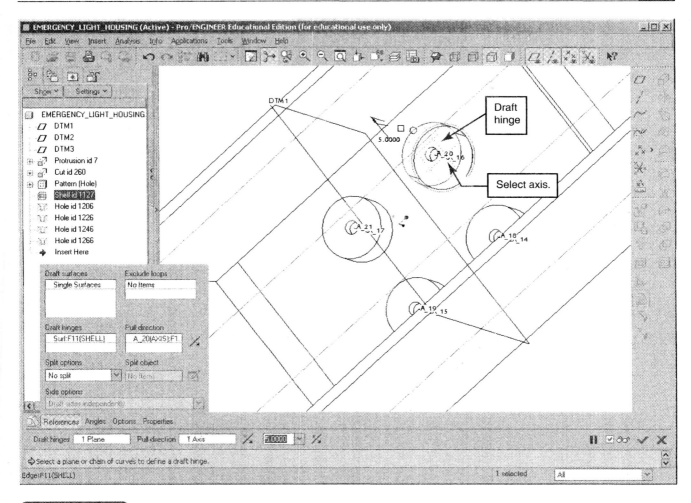

FIGURE 18.3 *The desired surfaces on one of the bosses. Note that the surface of the cylinder is actually two surfaces. Both of these surfaces must be selected.*

Pro/E divides a cylindrical surface into two separate surfaces. If a draft is to be performed on the entire cylindrical surface, then both surfaces must be chosen by the user.

Fillets pose a special problem in constructing a draft, since a draft cannot be performed on the surface edge. Thus, if a feature is to have both a draft and a fillet, the draft must be performed first, and then the fillet can be added.

Press the CTRL key to select additional draft surfaces. Note the draft options in the dashboard (Figure 18.3).

18.1.1 TUTORIAL 18.1: A DRAFT FOR THE EMERGENCY LIGHT COVER

This tutorial illustrates:

- Adding a draft to a cylindrical feature

1. Retrieve the model "EmergencyLightBase." Select ![icon] and double-click on the file.
2. Choose *Insert* and *Draft* or simply select ![icon].

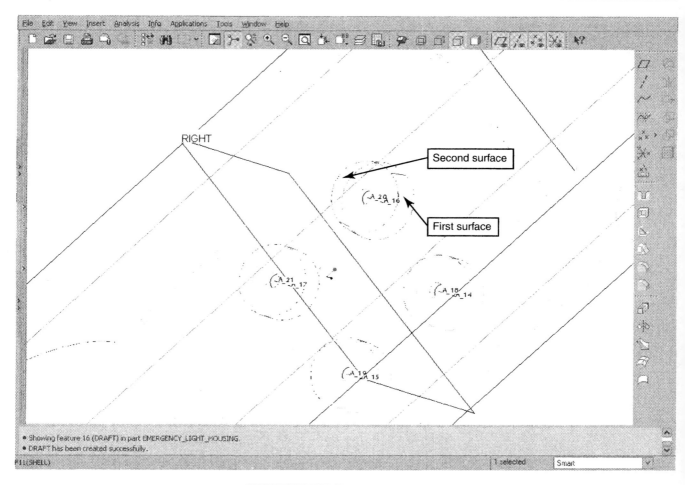

FIGURE 18.4 *The cover with the boss shown in Figure 18.3 tapered with a* Draft.

3. Click on one of the cylindrical surface shown in Figure 18.3. Press the CTRL key and select the other surface.
4. Click on the *Draft Hinges* cell. Pick the surface shown in Figure 18.3.
5. Click on the *Pull Direction* cell. Click on the axis running longitudinally through the cylinder, as in Figure 18.3.
6. Enter an angle of −5.000 in the cell.
7. Build the feature by selecting ✓. Note Figure 18.4.
8. Repeat steps 2 through 7 for each of the three remaining bosses. See Figure 18.5.
9. Save the part.

18.2 EARS

An *ear* is a small protrusion that is bent at its base. In Pro/E, the amount by which the protrusion is bent is determined by the user. Ears are used to align parts.

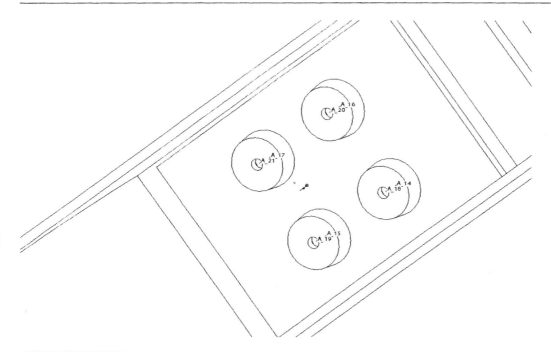

The cover with tapered bosses.

The section of an ear is Sketched using the Sketcher after a sketching plane has been identified. The length of the ear may be one of two types. For a *variable* ear, the length of the sketch includes the bent part of the ear. For a *tab* ear, the length of the sketched section includes the *projection* of the bent part of the ear.

In creating an ear, the software requires that the sketching plane be normal to the surface which the ear is attached. Furthermore, the sketch must be an open section with the ends of the section aligned to the attached surface. While the section of the ear may have any geometry, the part of the geometry that is attached to the surface must have edges that are parallel to each other and long enough to form the bend.

18.2.1 TUTORIAL 18.2: EARS FOR THE COVER PLATE

This tutorial illustrates:

- Adding an ear feature to a part
- Patterning an ear

1. Retrieve the model "CoverPlate." Select and double-click on the file.
2. Then select *Insert, Advanced,* and *Ear.*
3. The ear will be bent 90° to the attachment surface, so choose *90 deg tab* and *Done.*
4. Select the sketching plane shown in Figure 18.6. Notice the direction of the arrow in the figure. Use *Flip* to change the direction. Press *OK* when ready.

FIGURE 18.6 *Use the given surfaces as the sketching and orientation planes.*

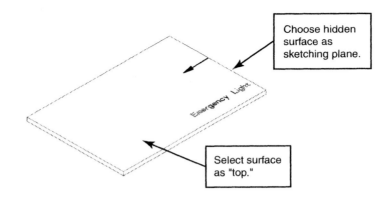

Choose hidden surface as sketching plane.

Select surface as "top."

5. Orient the model in the Sketcher by choosing the surface with the cosmetic text as the "top."

6. In the sketcher, draw the open section shown in Figure 18.7.

7. Use ⊟ to dimension as shown in Figure 18.7. Also, use ⊟ to modify the dimensional values to the ones given in the figure.

8. Build the feature by selecting ✓.

9. Enter 0.050 for the depth of the ear. The bend radius is also 0.050. The software will create the ear.

10. A second ear may be added to the part using the same procedure or the *Pattern* option. Select the ear from the screen, or after opening the model tree, click on the feature.

11. Choose *Edit* and *Pattern* (or use the right mouse button in the model tree).

12. Press *Options*. Select *General*.

13. Press *Dimensions*. Click on the 0.400 dimension as the first direction.

14. Click on the *Increment* cell and enter: 3.95-0.80-0.25. Note the value 3.95 is the depth of the plate. The value 0.80 will give an ear

FIGURE 18.7 *Draw the open section in the Sketcher.*

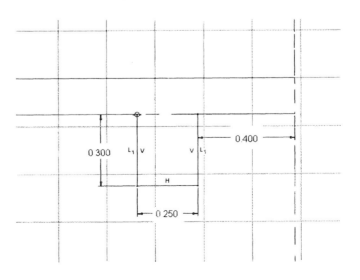

equidistant from either side of the plate. The value 0.25 is included in the formula in order to take into account the depth of the ear.

15. Enter 2 for the total number of instances.
16. Press ☑. The plate with both ears is shown in Figure 18.8.
17. Because of the ears, two small slots must be added to the model of the emergency light cover. These slots will not be added here. In order for the cover to lock in position, a latch must be added to the plate. Select ☑.
18. Choose ☑. Enter a thickness of 0.050.
19. Choose the *Sketch* button ☑ in the dashboard.
20. With your mouse, choose datum plane Right as the sketching plane.
21. If necessary, use ☑ for the *Orientation* field and to "top." Then, click on datum plane Top.
22. Again, the arrow should point into the screen. Use *Flip* if it does not.
23. Select the *Sketch* button ☑.
24. Sketch the section shown in Figure 18.9.
25. Use ☑ and ☑ and dimension to the proper values.
26. Build the feature by pressing ☑.
27. Choose ☑.
28. Enter a depth of 2.00. The plate with the latch is shown in Figure 18.10. Save the part.

FIGURE 18.8 *The model is shown with both ears.*

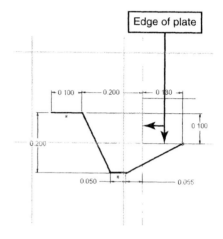

FIGURE 18.9 *Use this geometry to sketch and dimension the latch.*

18.3 THE LIP OPTION

A *lip* is a feature used to interlock two parts, as shown in Figure 18.11. One part contains the protrusion, and the other part, a cut corresponding to the protrusion.

A lip is created by offsetting a given surface called the mating surface. The offset distance is a parameter that must be provided by the user. The lip may have a sloped side at a prescribed angle. The side offset is also a parameter of the lip.

FIGURE 18.10 *The model after the addition of the latch.*

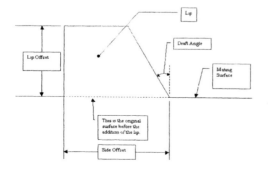

FIGURE 18.11 *A lip is created by offsetting a surface by a certain amount.*

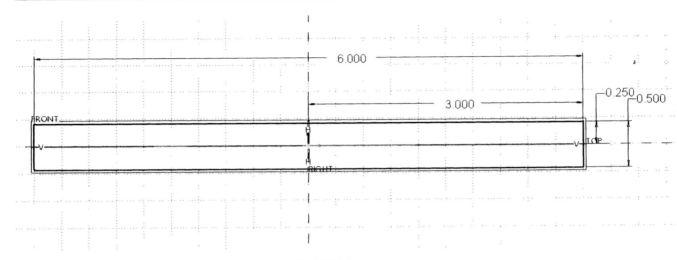

FIGURE 18.12 *Use this geometry to construct the remote control.*

18.3.1 TUTORIAL 18.3: ADDING A LIP TO A PART

This tutorial illustrates:

• Adding a *Lip* to a part

1. Select .
2. Make sure that the *Use default template* option is checked.
3. Enter the name "RemoteControlCoverLower" in the *Name* field.
4. Select *OK*.
5. Select *Insert* and *Extrude* (or simply).
6. Choose the *Sketch* button in the dashboard.
7. With your mouse, choose datum plane Front as the sketching plane.
8. If necessary, use for the *Orientation* field and to "top." Then, click on datum plane Top.
9. Again, the arrow should point into the screen. Use *Flip* if it does not.
10. Select the *Sketch* button .
11. In the Sketcher, draw the section shown in Figure 18.12.
12. Use and to dimension to the values shown in Figure 18.12.
13. Press to build the section.
14. Select . Enter a depth of 2.50". The model at this point is shown in Figure 18.13.

FIGURE 18.13 *This figure shows the model after the creation of the protrusion.*

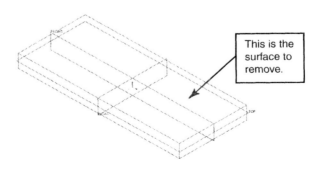

This is the surface to remove.

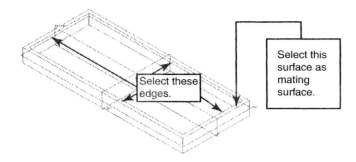

FIGURE 18.14 *The model after the use of the* Shell *option. The* lip *can now be added to the model.*

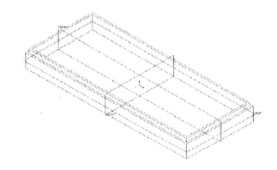

FIGURE 18.15 *A lip is added to the model. Compare this figure with Figure 18.14.*

15. We need to shell the model.
16. Select *Insert* and *Shell* or ▣.
17. Choose the plane shown in Figure 18.13 as the surface to remove.
18. Enter 1/8 for the thickness of the shell. Build the section by pressing ✅.
19. The lip is to be added on the inner portion of the shell thickness using the edges shown in Figure 18.14. Select *Insert, Advanced,* and *Lip*.
20. Then, choose the *Single* option. Select the edges shown in Figure 18.14. Choose *Done*.
21. Choose the surface shown in Figure 18.14 as the mating surface.
22. Enter an offset of 0.125 and a side offset of 0.125/2.
23. Pick the mating surface as the drafting reference plane.
24. The draft angle is 0°. The model with the lip is shown in Figure 18.15. Save this part.

18.4 SURFACE QUILTS

A surface feature may be created by extending one or more curves. In Figure 18.16, we see a practical application of the approach. The shape of the wing is determined by the NACA airfoil section. By extending the datum curve, a three-dimensional surface is created, which models the surface of the wing. The option *Boundary Blend* may be used to generate a surface between two or more curves.

FIGURE 18.16 *By extending a curve, a surface may be constructed. In this case, multiple curves are used and the surface is blended into both sets of curves.*

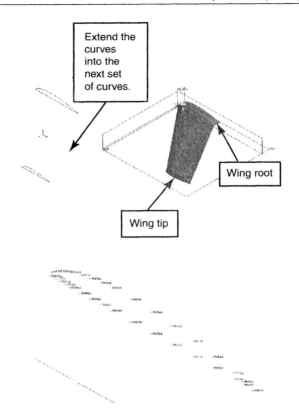

Extend the curves into the next set of curves.

Wing root

Wing tip

FIGURE 18.17 *The datum points from file "CS1.pts" may be used to create the geometry of the wing tip.*

18.4.1 TUTORIAL 18.4: AN AIRCRAFT WING

This tutorial illustrates:

- Using datum curves to generate surfaces
- Using *Boundary Blend* to generate a surface

1. Retrieve the model "Wing." Select ☑ and double-click on the file. Note the datum point in the model (see Figure 18.17).
2. First add additional datum points and curves. Select ▨.
3. Click on the default coordinate system (CS0).
4. Make sure the *Type* is set to *Cartesian*.
5. Press the *Import* button. Find the file CS1.pts. This file is bundled with this book. Double-click on the file to open it.
6. Select *OK*. The software will add the points.
7. Press ▨.
8. From the *CRV OPTIONS* menu, select *Thru Points* and *Done*.
9. Choose *Spline, Single Point,* and *Add Point*.
10. At this point, you will most likely need to zoom in to the model. Now pick the upper points as shown in Figure 18.18. Start with Point 69, then 36, 37, and so on. The last point is Point 52.
11. Choose *Done* and *OK*. Let us call this curve "Curve 3."
12. Repeat the process (steps 7 through 11) for the lower half. This time pick the lower points; start with Point 69 and proceed to

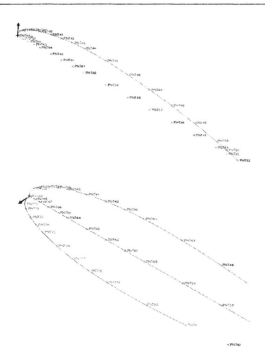

FIGURE 18.18 *Add a datum curve through the upper points by using the options* Spline *and* Single Points.

FIGURE 18.19 *Add a datum curve through the points from file "CS2.pts."*

Point 68, Point 67, and so on. Finish with Point 52. Call this curve "Curve 4."

13. Turn off the display of the datum planes, coordinate system, and points. The curves should appear as in Figure 18.19.

14. We need to add another set of points in order to construct a curve for the end of the wing. Repeat steps 2 through 6. The points are found in the file "cs2.pts."

15. Repeat steps steps 7 through 11. Now pick the lower points; start with Point 70 and proceed to Point 71, Point 72, and so on. Finish with Point 87. Let us call this curve "Curve 5."

16. We need to add two curves offset from Curves 1 and 2 (see Figure 18.20). We will do this using a simple *Copy* and *Move.* Select *Edit, Feature Operations,* and *Copy.*

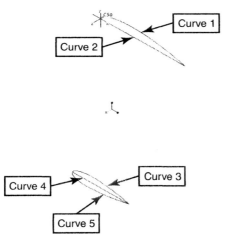

FIGURE 18.20 *The model after the addition of "Curve 5."*

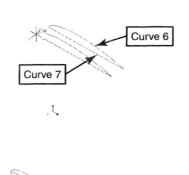

FIGURE 18.21 *The model with the two new curves.*

FIGURE 18.22 *The model with a surface whose boundaries are Curve 2 and Curve 3.*

17. Choose *Move* and *Done*. Pick on Curves 1 and 2 and *Done*.

18. Choose *Translate*. Click on datum plane Front. The arrow should point away from the other datum curves. Use *Flip* to change the direction. Select *OK* when ready.

19. Enter a value of −0.5 and hit *return*. Choose *Done Move, OK,* and *Done*.

20. The software will create a new surface containing Curves 6 and 7, as shown in Figure 18.21.

21. Create the first surface between Curves 1 and 3. Choose *Insert* and *Boundary Blend*.

22. Click on Curve 1. Press your CTRL button and click on Curve 3. Build the feature by pressing ✓. The model at this point is shown in Figure 18.22.

23. Repeat step 22 using Curves 2 and 4.

24. Repeat step 22 using Curves 3 and 5.

25. Repeat step 22 using Curves 4 and 5.

26. Repeat step 22 using Curves 1 and 6.

27. Repeat step 22 using Curves 2 and 7. The model with all the surfaces is shown in Figure 18.23.

28. The remaining half of the wing may be constructed by using *Copy* and *Mirror*. First, we need to create a mirror plane. Choose ▱.

FIGURE 18.23 *The rest of the surfaces may be added to the model by selecting the correct boundaries.*

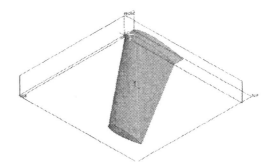

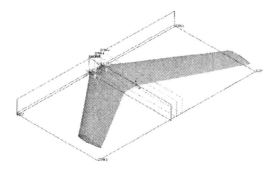

29. Click on datum plane Front and enter an offset of -0.500. Choose *OK*. This will create datum plane DTM1.

30. Choose *Edit, Feature Operations,* and *Copy*.

31. Select *Mirror, All Feat,* and *Done*. Click on datum plane DTM1 and select *Done*.

32. The model is shown in Figure 18.24. Save the part.

18.5 SUMMARY AND STEPS FOR THE OPTIONS CONSIDERED IN THIS CHAPTER

The scope of this chapter is to provide introductory material on surfaces and to examine the options *Draft, Ear,* and *Lip*.

The *Draft* option is used to taper surfaces. The surfaces may be tapered between $\pm 30°$. In general, a *Draft* may be added to a surface by using the following steps:

1. Choose *Insert* and *Draft* or simply select .

2. Click on the surface to be drafted.

3. Click on the *Draft Hinges* cell and choose the appropriate reference.

4. Click on the *Pull Direction* cell and choose the appropriate reference.

5. Enter the draft angle.

6. Build the feature by selecting .

An *ear* is a protrusion that is attached to a base feature and bent at its point of attachment. An ear may be of *Variable* type or *90 deg tab*. In the case of a *Variable* type ear, the user may enter the bend angle. The general sequence of steps for creating an *Ear* is as follows:

1. Select *Insert, Advanced,* and *Ear*.

2. Choose the type of ear—*90 deg tab* or *Variable*.

3. Pick the sketching plane.

4. Select the orientation plane.

5. In the Sketcher, draw the geometry of the ear.

6. Dimension to the proper values.

7. Build the feature.

8. Enter the depth of the ear.

9. If the ear is of type *Variable*, enter the angle of the bend.

A *lip* is a protrusion that is used to interlock multiple parts. The protrusion is constructed on one part and on the mating part, a cut with the same geometry with which the lip is created. A lip may be added to a surface using the following general procedure:

1. Select *Insert, Advanced,* and *Lip*.
2. Choose the edge defining the boundary of the mating surface.
3. Select the mating surface.
4. Enter the offset distance from the mating surface.
5. Enter in the side offset.
6. Select the draft reference plane.
7. Enter the draft angle.

A surface may be created by extending a curve or edge. A surface may be created between one or more boundaries. Surfaces may be constructed by using the same options as for a protrusion, including *Extrude, Revolve, Sweep,* and by choosing ▢ in the dashboard. If the boundaries of the quilt are curves, then the surface may be created by using these steps:

1. *Insert* and *Boundary Blend*.
2. Choose the curves.
3. Build the feature.

18.6 ADDITIONAL EXERCISES

18.1 Using the model of the Howe roof truss from Exercise 13.9 and create a set of trusses 10 meters apart. The result is given in Figure 18.25.

FIGURE 18.25

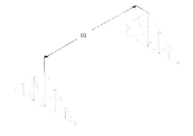

18.2 Use the results from Exercise 18.1 to create surfaces representing the roof. Use Figure 18.26 for reference.

FIGURE 18.26

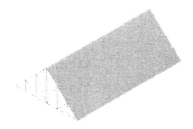

18.3 Add walls to the results from Exercise 18.2 by using *Insert* and *Boundary Blend.* The "walls" are shown in Figure 18.27.

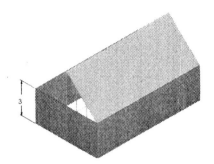

FIGURE 18.27

18.4 Finish off the ends of the house from Exercises 18.1 through 18.3, as shown in Figure 18.28.

FIGURE 18.28

18.5 Create the 14″ × 14″ oil pan illustrated in Figure 18.29. Use a draft angle of 5°.

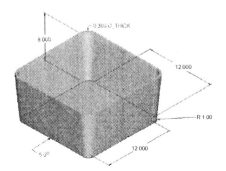

FIGURE 18.29

18.6 Add the second protrusion and lip shown in Figure 18.30 to the oil pan from Exercise 18.5.

FIGURE 18.30

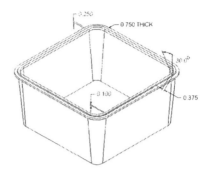

18.7 Create the cover shown in Figure 18.31. Add the ears.

FIGURE 18.31

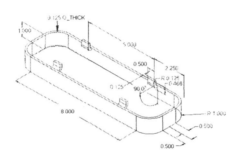

CHAPTER ![19]

ASSEMBLIES AND WORKING DRAWINGS

INTRODUCTION AND OBJECTIVES

Individual parts can be put together to form an assembly. Therefore, complicated machines and structural components can be created by constructing the individual parts (components) independently and then assembling the parts.

In order for an assembly to be created, Pro/E must have access to the original part files. Thus, assembly construction requires large memory storage facilities. Furthermore, because assemblies are quite large, run time for regeneration of an assembly can be considerable.

In Pro/E, assembly datum planes are defined similar to the default datums in the *Part* mode. When using assembly datum planes, the user must first create the datums and then place the components. In general, the names of the assembly datum planes will start with the letter A to distinguish them from datum planes from the part mode. When using the standard template, the names of the assembly datum planes are ASM_FRONT, ASM_RIGHT, and ASM_TOP.

Parts may be constructed in the assembly mode. Creating parts in the assembly mode has two advantages. The first advantage is that the part is automatically placed parametrically in the assembly. The second advantage is that part is constructed in a more straightforward manner, because as the part is sketched, the outline of the other components is visible.

After an assembly has been created, the user may explode the assembly. This is desired to facilitate visualization of how the parts are combined to form the assembly. Furthermore, the user may create working drawings of the assembly containing both the exploded and unexploded states of the assembly.

The objectives of this chapter are to

1. Create an assembly with the use of assembly datum planes
2. Create exploded views of an assembly

3. Construct parts in the assembler
4. Attain proficiency in creating working drawings of an assembly

19.1 CREATING AN ASSEMBLY FILE

The assembly of individual parts is accomplished in the *Assembly* mode. As in the creation of drawing and part files, the Pro/E user creates a file. In this case, the file has the extension '.asm.' The procedure for creating an assembly file is similar to that used for part and drawing files. By using the options *File* and *New* (or the *Create New Object* icon), the user accesses the *New* dialog box shown in Figure 19.1. The assembly file is created by selecting the assembly option, entering a name, and choosing the *OK* button.

After the assembly file is created, the software will load the interface containing the *ASSEMBLY* menu shown in Figure 19.2. Notice the four options *Assemble, Create, Package, Include,* and *Flexible.* These options are used in the following manner:

1. *Assemble.* This option contains the necessary menus with options for constraining components. The constrained components form a *parametric* assembly with their positions specified relative to a base feature.

2. *Create.* The *Create* option is used to construct parts or subassemblies within the assembly mode. The *Component Create* dialog box shown in Figure 19.2 allows the user to choose the desired component to be created.

3. *Package.* By using the *Package* option, components may be placed in an absolute sense; that is, the components are not placed relative to a base part. The components may be repositioned by translating or rotating the component around an axis by using the mouse pointer. The *Package* option is best used when the exact placement of the components is unknown or the user does not want one or all of the components to be located with respect to another component. Components may be reassembled

FIGURE 19.1 *An assembly file may be created by using the* New *dialog box.*

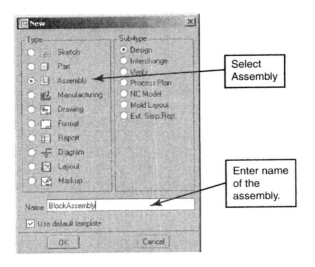

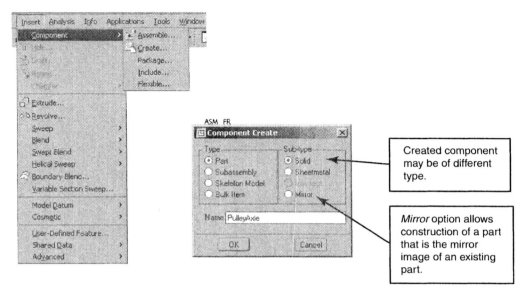

FIGURE 19.2 *A component may be assembled in different modes.*

parametrically after they have been placed with the *Package* option by using the *Assembly* option.

4. *Include.* This option allows a part to be included in the assembly file but is not assembled to the assembly. The part is listed in the assembly model tree.

5. *Flexible.* This option allows the user to place flexible components such as springs. This is convenient because the component may be shown in different assembly states. For example, a spring may be shown in different states of compression or extension.

19.2 CREATING PARAMETRIC ASSEMBLIES

In creating parametric assemblies, the components are positioned with respect to each other. This action can be accomplished by utilizing one or more of the *Constraint Type* options in the *Place* folder illustrated in Figure 19.3. In all, there are eight options for the placement of the components. These options are: *Mate, Align, Insert, Coord Sys, Tangent, Pnt on Line, Pnt on Surface,* and *Edge on Surface.* Usually more than one of these options is needed to fully constrain the component. The first four of these options are probably the most useful. The options are defined as follows:

1. By selecting the *Mate* option, the user is requiring that the selected planes *face each other and be coplanar.* An offset may be prescribed which allows the two surfaces to be mated some distance apart.

2. The *Align* options require that the chosen surfaces face the same direction and are parallel. Again, an offset may be defined which allows the user to place an offset distance between the two faces.

FIGURE 19.3 *The options contained in the* Place *folder are used to fully constrain components for placement in parametric assemblies.*

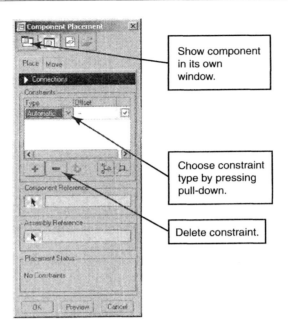

Show component in its own window.

Choose constraint type by pressing pull-down.

Delete constraint.

3. The *Insert* option can be used to place a component "inside" of another component. For example, a model of a bolt can be placed inside of a hole using the *Insert* option.

4. When two or more models contain coordinate systems, aligning the coordinate systems of one model to another by using the *Coord Sys* option may assemble the components. The coordinate systems are chosen by selecting with the mouse, or by selecting the names of the systems from a list.

5. The *Tangent* option is used to place two surfaces in contact. This option is similar to the *Mate* option; however, alignment does not take place.

6. The options *Pnt on Line* and *Pnt on Surface* allow the user to locate a part by constraining a point with a line or surface.

7. The options *Edge on Srf* permit control of an edge with a surface. Thus, the linear edge of a feature can be placed in contact with a defined surface.

In order to get a "feel" for assembling with these options, we have created two parts and bundled the files with this book. These parts are called "AssembleBlockA.prt" and "AssembleBlockB.prt." Make sure that these parts are available for the software to retrieve. Although these models are simple enough that they may be assembled without datum planes, we will use the default template.

19.2.1 TUTORIAL 19.1: ASSEMBLING BY USING PLACE OPTIONS

This tutorial illustrates:

• Generating an assembly using place options

1. Select [ ].

2. Check *Assembly*.

3. Make sure that the *Use default template* option is checked.

4. Enter the name "BlockAssembly" in the *Name* field.

5. Select *OK*.

6. Select *Insert, Component,* and *Assemble* or 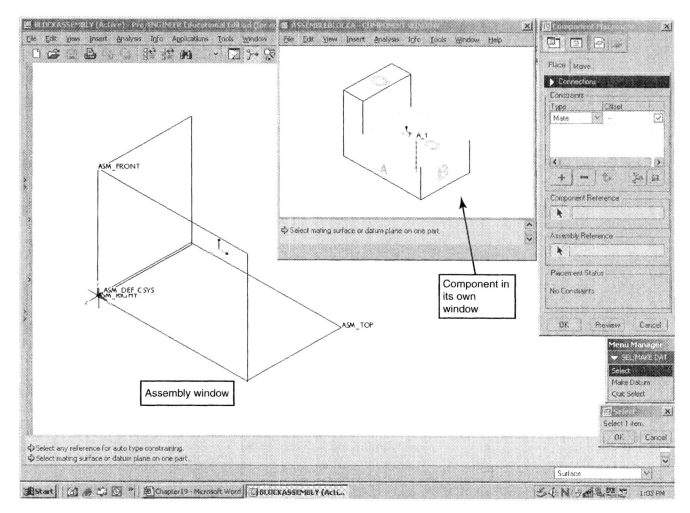.

6. Select *Insert, Component,* and *Assemble* or .

7. Scroll down the list of available parts until you find the part "AssembleBlockA." This part is bundled with the book. Select the part and hit the *Open* button.

8. Show the component in a separate window by picking . Your screen should appear similar to the one shown in Figure 19.4.

9. Select *Mate*.

10. Pick surface G on the component.

11. Pick datum ASM_TOP.

12. Again, select *Mate* and Surface F on the component. Then pick ASM_RIGHT. Note that the model is partially constrained.

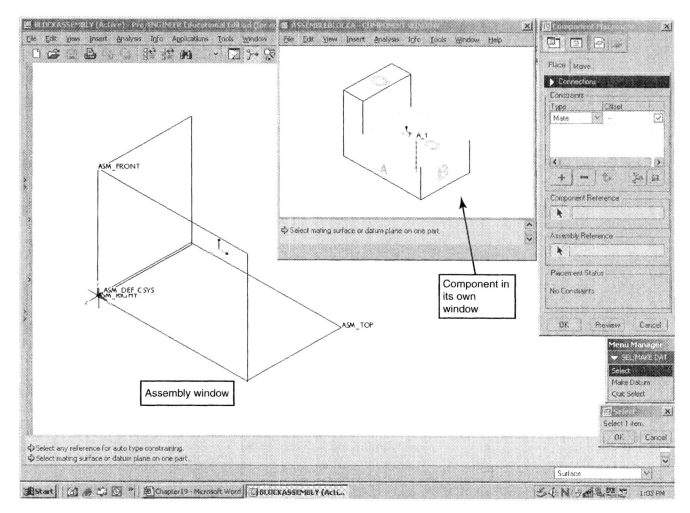

FIGURE 19.4 *The* Component Placement *dialog box keeps track of whether or not a component is fully constrained. It also contains the options for assembling the component to the assembly.*

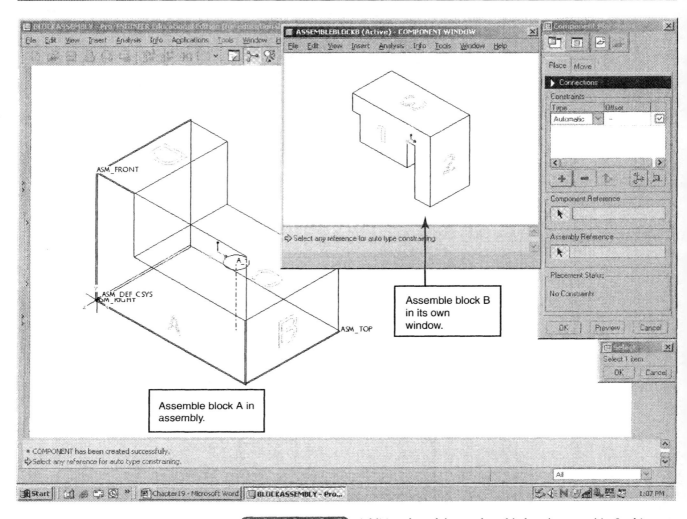

FIGURE 19.5 *Additional models may be added to the assembly. In this case, the component and assembly are shown in different windows.*

13. Select *Align* and pick Surface A. Pick datum ASM_FRONT.

14. The model is now fully constrained. Select *OK*.

15. Add the second block. Select [icon] and find the second part: "AssembleBlockB."

16. Show the new component in a separate window. Note the location of the component and assembly in Figure 19.5.

17. Choose *Mate* from the *Constraint Type* cell. Select Surface 6. Click on Surface C.

18. Now mate Surfaces 8 and B. Choose the *Mate* option. The assembly is still not fully constrained.

19. Finally, align Surfaces 1 and A using the *Align* option. After this constraint is given, the assembly will be fully constrained. Pro/E will notify the user that the component may be placed.

20. Use *Preview* to examine the assembly. The assembled blocks are shown in Figure 19.6. Select *OK*.

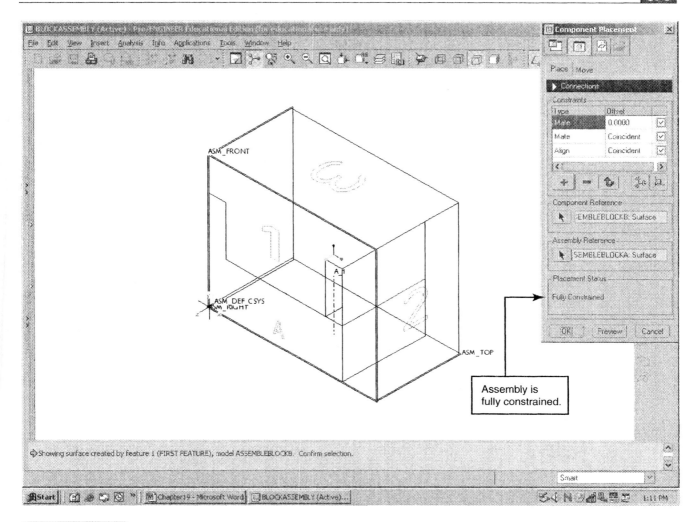

Assembly is
fully constrained.

FIGURE 19.6 *The assembled blocks.*

19.3 ASSEMBLIES WITH MULTIPLE PARTS

The blocks used in the previous section were created to illustrate basic assembling techniques. The blocks are relatively simple to assemble because they have planar orthogonal surfaces. However, most assemblies are complicated and have multiple parts. Datum planes make the assembly process easier for complicated parts.

In creating an assembly with datum planes, the set of datum planes must be created first. Then, the parts can be assembled relative to the datum planes or to each other.

The assembly to be created in the next tutorial makes use of the several parts constructed in previous chapters of this book. These parts are "PlateWithCounterboreHoles" (Chapters 4 and 8), "AngleBracket" (Chapters 2, 4, 7, and 8), and "PulleyWheel" (Chapter 14). In addition, we have:

1. "Half_Inch_1_25Long_Bolt"
2. "Half_Inch_Washer"

3. "Half_Inch_Hex_Nut"

4. "Wheel_Bushing"

These fasteners are bundled with this book. You will need to make sure that you have access to these parts. An additional part will be created in the assembly mode.

We approach the creation of this assembly in several phases. The first phase is to assemble the plate, AngleBracket, and pulley_wheel parts.

19.3.1 TUTORIAL 19.2: A PULLEY ASSEMBLY

This tutorial illustrates:

- Generating an assembly using place options
- Using assembly datum planes

1. Select [image].

2. Check *Assembly*.

3. Make sure that the *Use default template* option is checked.

4. Enter the name "PulleyAssembly" in the *Name* field.

5. Select *OK*.

6. Select [image].

7. Scroll down the list of available parts until you find the part "PlateWithCounterboreHoles."

8. Choose [image].

9. Choose *Mate*. Then, click on datum plane Top in the component. Select datum plane ASM_TOP in the assembly.

10. Select *Align*. Choose datum plane Front and datum plane ASM_FRONT.

11. Select *Mate*. Choose datum plane Right datum plane ASM_RIGHT.

12. The component can now be placed. Select *OK*. Your model should appear as in Figure 19.7.

13. Select [image]. Scroll through the list of available parts and choose "AngleBracket."

14. Choose *Mate*. Pick the two surfaces shown in Figure 19.8.

FIGURE 19.7 *The plate is placed in the assembly. Note the locations of the datum planes.*

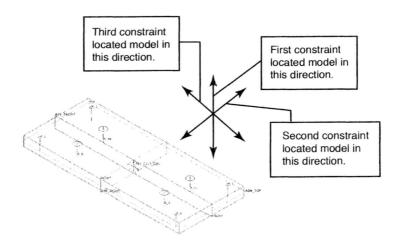

Third constraint located model in this direction.

First constraint located model in this direction.

Second constraint located model in this direction.

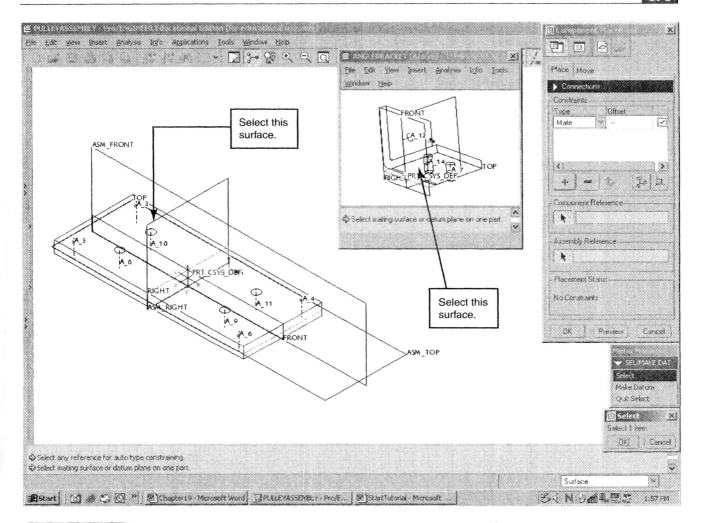

FIGURE 19.8 *Pick the shown surfaces when using the* Mate *option.*

15. Select *Align* and then the two axes A_14 and A_9 shown in Figure 19.9.

16. Because of the orientation and location of the two axes, the bracket will be fully constrained. Choose *OK*. The assembly at this point is shown in Figure 19.10.

17. Select . Scroll through the list of available parts and retrieve "AngleBracket."

18. Choose *Align*. Select axis A_14 and axis A_10 as shown in Figure 19.11.

19. Again, choose *Align*. Pick axis A_7 and A_8 as shown in Figure 19.11.

20. Select *Mate* and choose the surfaces shown in Figure 19.11.

21. The model will be fully constrained, so select *OK*. The assembly at this point is shown in Figure 19.12.

22. Choose . From the list select "PulleyWheel."

23. Choose *Align*. Select datum plane Right and datum plane ASM_RIGHT.

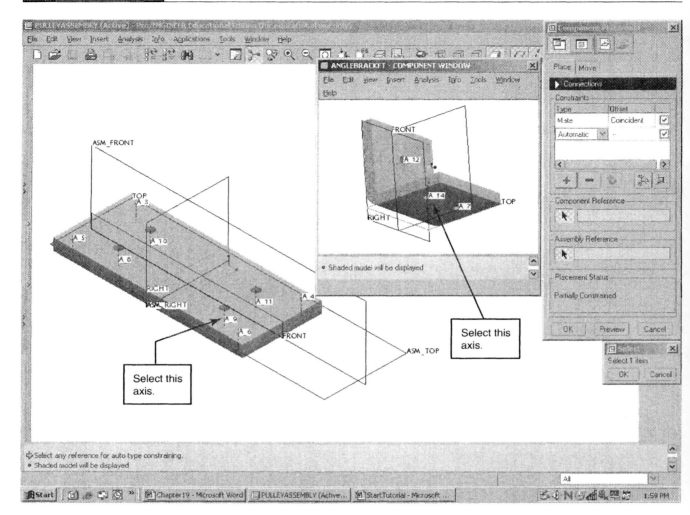

FIGURE 19.9 *Select the shown axes to constrain the model.*

FIGURE 19.10 *The assembly after placing the first bracket.*

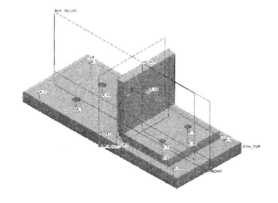

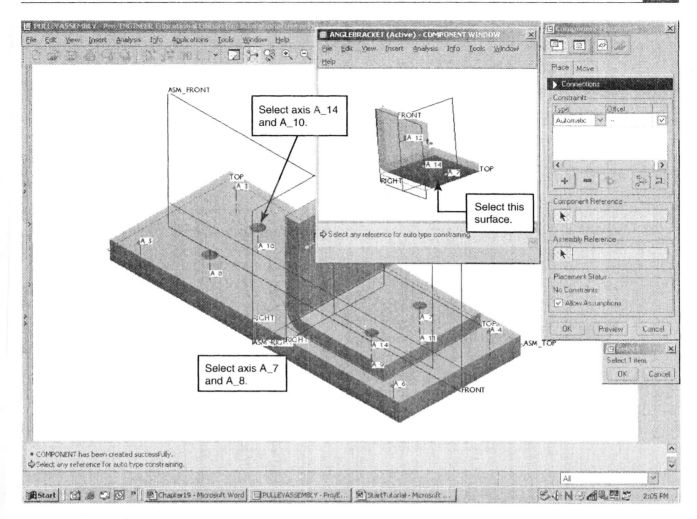

FIGURE 19.11 *Use this figure to select the proper axes.* Mate *the given surfaces.*

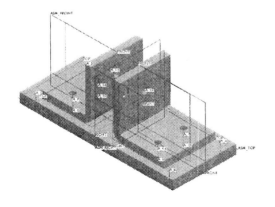

FIGURE 19.12 *The assembly after the addition of the second angle bracket.*

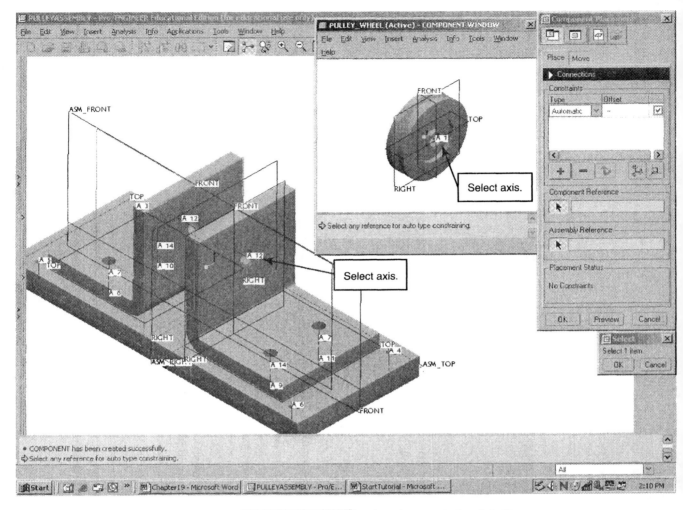

FIGURE 19.13 *Align the axes A_1 and A_12.*

FIGURE 19.13 *Align the axes A_1 and A_12.*

24. Again, choose *Align*. Pick datum plane Front and ASM_FRONT.
25. Choose *Align*. Using Figure 19.13, pick axis A_1 and axis A_12.
26. Choose *OK*. The assembly with the wheel added is shown in Figure 19.14.
27. Save the assembly. Press ▣.

FIGURE 19.14 *With the pulley wheel fully constrained, the model may be added to the assembly.*

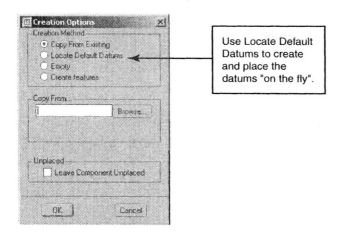

Use Locate Default Datums to create and place the datums "on the fly".

FIGURE 19.15 *The Creation* Options *box is used to initialize the creation of a component in the assembler.*

19.4 CREATING PARTS IN THE ASSEMBLER

Parts created in the assembler may be treated as parts constructed in the part mode. Creating a part in the assembly mode has the advantage that the assembly may be used as a reference for the part. Thus, the size and position of the part, with respect to the assembly, is readily visible. This saves time in the end.

Model construction in the assembler proceeds along a sequence of three steps. First, a base feature is created and assembled. The easiest approach is to load a part containing default datums—that is, a start part. Secondly, the datum planes are assembled. Lastly, additional features may be added to the assembled part.

Figure 19.15 shows the *Creation Options* box. This box contains options for creating the base feature. These options may be used as follows:

1. *Copy From Existing.* Load and assemble an existing start part. The assembly of the component may be redefined. Another advantage of using this option is that if the same start part is used for all the parts in the assembly, then all the parts including the one created in the assembler will have the same standards.
2. *Locate Default Datums.* Create the component datums and assemble. The assembly of the component may be redefined; however, the component may not have the same standards as the rest of the parts.
3. *Empty.* Construct the component without any initial geometry. Geometry may be added from existing parts using the *Copy From* option.
4. *Create Features.* Using existing assembly references, create the geometry. The assembly of the component cannot be redefined. However, the feature may be modified.

In the next tutorial, we will create an axle for the pulley assembly in the assembler. The *Copy From Existing* option is the best way to create such parts. However, for the parts in this assembly, the default standards were used, so it really doesn't matter if a start part is used or not. Furthermore, we can use ASM_FRONT, ASM_RIGHT, and ASM_TOP to create the part. The software will automatically constrain the component to these datums. If we use the option *Copy From Existing,* we would have to constrain the start part to these datum planes. Therefore, in this case, not much time is saved in using a start part.

19.4.1 TUTORIAL 19.3: AN AXLE FOR THE PULLEY ASSEMBLY

This tutorial illustrates:

- Constructing a part in the assembly mode

1. Retrieve the assembly "PulleyAssembly." Select and double-click on the file.
2. Select *Insert, Component,* and *Create* or choose .
3. Choose the *Part* and *Solid* options and enter the name: "Pul-leyAxle." Choose *OK.*
4. Then select *Locate Default Datum, Three Planes,* and *OK.*
5. Select ASM_RIGHT as the sketching plane.
6. Select ASM_TOP as the reference plane.
7. Select ASM_FRONT as the vertical placement.
8. The axle is constructed from two protrusions. For the first pro-trusion, select *Insert, Protrusion,* and *Extrude* or choose .
9. Press . Choose DTM1 as the sketching plane (this datum is generated by Pro/E for the axle).
10. Choose "top" as the orientation and datum plane DTM2. Then, press .
11. In the sketcher, draw a circle that is aligned to the edges of the datum planes ASM_FRONT and datum Top. See Figure 19.16 for the geometry.
12. Dimension to a diameter of 0.75″. Press to build the feature.
13. Choose and extrude to a depth of 1.700. This is the distance between the angle brackets.
14. Press to build the part.
15. Note that the active part is still the axle. We need to add an addi-tional feature to this part. Select .
16. Repeat steps 9 through 11. You can press the *Use Previous* button to use the sketching and reference planes from steps 9 and 10.

FIGURE 19.16 *The cross sec-tion of the axle is a circular section.*

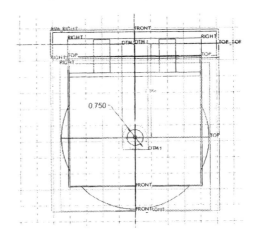

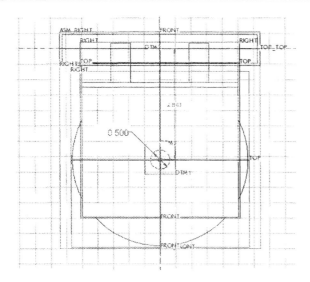

FIGURE 19.17 *The cross section of the second protrusion is also a circular section.*

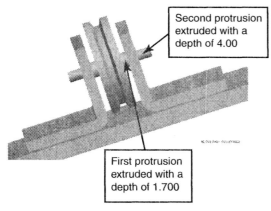

FIGURE 19.18 *The figure shows the axle. In this figure, we have enlarged the area near the axle.*

Second protrusion extruded with a depth of 4.00

First protrusion extruded with a depth of 1.700

Dimension the diameter dimension to 0.500 as shown in Figure 19.17.

17. Extrude *Symmetric* ⊞ to a depth of 4.00. The axle is shown in Figure 19.18. We have enlarged the area around the axle.

18. Reactivate the assembly. In the model tree, select PULLEYASSEMBLY.ASM. Press your right mouse button and choose *Activate*. Note that any of the parts may be modified in the assembly mode by selecting the part in the model tree and pressing *Activate* after pressing the right mouse button.

19. Press ▢ to save the assembly.

19.5 OFFSETS AND ADDITIONAL CONSTRAINTS

The *Mate* option assumes that the chosen surfaces face each other. An offset distance may be prescribed. Likewise, *Align* assumes that the chosen entities face the same direction.

The *Insert* constraint is useful for placing fasteners into an assembly. In general, the threaded surface of the screw, nut, etc. is chosen, along with the threaded surface of the hole.

FIGURE 19.19 *This front view of the assembly shows the bushings to be added.*

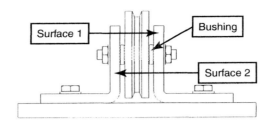

Since both a hole and cylinder have two surfaces, the selection of the appropriate surface may be required. However, because fasteners are symmetric more often than not, it is not very important which surface is chosen.

The pulley assembly may be completed by adding the fasteners. These fasteners are assembled by using *Insert* and the option *Align* or *Mate*.

As shown in Figure 19.19, the bushing is required on either side of the pulley wheel. This prevents the pulley from sliding from side to side along the axle. The available space between the pulley and one of the angle brackets is 0.25″. The bushing is 0.20″. This means that the bushing must be offset from the angle bracket and the pulley by 0.025″ (if equally spaced). The *Mate* option may be used in this case with an offset value of 0.025.

19.5.1 TUTORIAL 19.4: COMPLETING THE PULLEY ASSEMBLY

This tutorial illustrates:

- Adding fasteners to an assembly
- Using *Repeat* to add components to an assembly

1. Retrieve the assembly "PulleyAssembly." Select ☞ and double-click on the file.
2. Select ☒ .
3. Scroll down the list of available parts until you find the part "Wheel_Bushing."
4. Choose ▭ .
5. Select *Align*. Choose axis A_1 on the component and axis A_12 on the assembly. See Figures 19.19 and 19.20 for these entities.

FIGURE 19.20 *Additional fasteners to be added to the assembly are shown.*

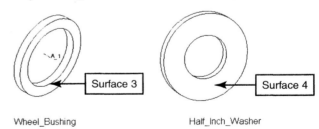

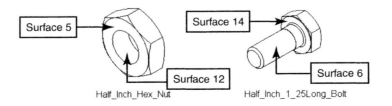

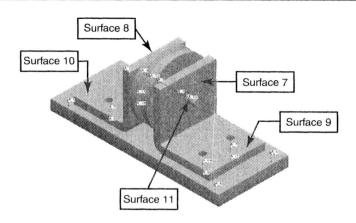

Surface 8

Surface 10

Surface 7

Surface 9

Surface 11

FIGURE 19.21 *The figure illustrates the different surfaces to be used in placing the washers in the assembly.*

6. Select *Mate*. Choose Surface 1 in Figure 19.19. Select Surface 3 in Figure 19.20.

7. Enter the value 0.025 for the offset. Hit *return*.

8. Choose *OK*.

9. Repeat steps 5 through 8 for the second bushing using Surfaces 2 and 3 and the corresponding axes.

10. Choose ▦.

11. Scroll down the list of available parts until you find the part "Half_Inch_Washer."

12. Choose ▦.

13. Choose Align. Select axis A_1 on the washer model in Figure 19.20.

14. Choose axis A_14 in Figure 19.21.

15. Select *Mate*. Choose Surface 4 in Figure 19.20. Select Surface 7 in Figure 19.21.

16. If needed, enter zero for the offset and choose *OK*.

17. Now, select the washer part. Choose *Edit* and *Repeat*.

18. The only difference between the first instance and the next three is that a different axis for the assembly reference should be used. In the *Repeat Component* dialog box (Figure 19.22) choose the *Align* constraint. Then press the *Add* button and click on axis A_7.

19. Click on axis A_7 on the other angle bracket.

20. Click on axis A_14 on the other angle bracket. Choose *OK*.

21. Press *Confirm*.

22. Now, add the hex nuts. Choose ▦.

23. Scroll down the list of available parts until you find the part "Half_Inch_Hex_Nut."

24. Choose ▦.

25. Select *Mate*. Choose Surface 5 in Figure 19.20.

26. Select Surface 15 in Figure 19.22.

27. Enter zero for the offset.

28. Choose *Insert*.

29. Select Surface 12 in Figure 19.20.

30. Choose Surface 11 in Figure 19.21.

31. Enter zero for the offset. Hit *return*.

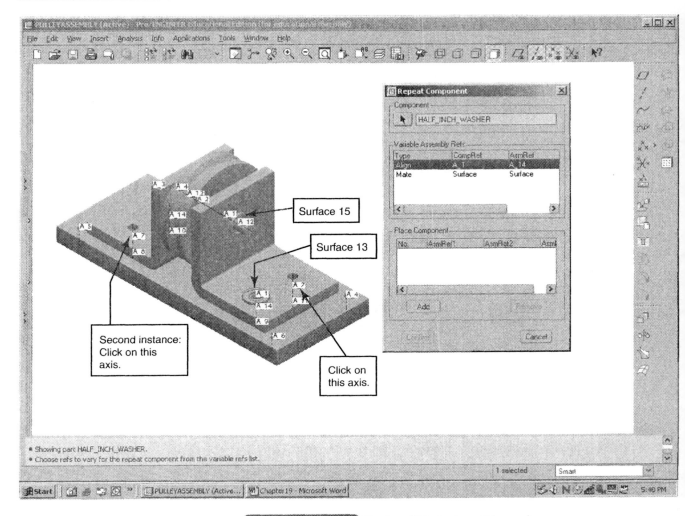

FIGURE 19.22 *Surface 13 is the top of the washer.*

32. *OK*.
33. Repeat steps 22 through 32 for the other side of the axle.
34. Finally, add the cap screws. Choose ⬚.
35. Scroll down the list of available parts until you find the part "Half_Inch_1_25Long_Bolt."
36. Choose ⬚.
37. Select *Mate*.
38. Choose Surface 14 in Figure 19.20 and pick Surface 13 in Figure 19.22. Enter zero for the offset.
39. Choose *Insert*. Select Surface 6 in Figure 19.20. Choose the hole corresponding to each washer.
40. *OK*.
41. Select the fastener. Choose *Edit* and *Repeat*.
42. Select the *Mate* option and press the *Add* button.
43. Now, click on the hole corresponding to each washer.

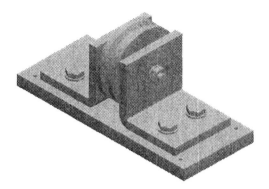

FIGURE 19.23 *The completed assembly is reproduced in this figure.*

44. Choose *OK* and *Confirm*.

45. The completed assembly is given in Figure 19.23. Save the assembly.

19.6 EXPLODED ASSEMBLIES

Assemblies are often created with exploded views; that is, the components are shown as to how they are to be put together. The software automatically generates a default-exploded state. This state may be modified by choosing *View, Explode,* and *Edit Position*. A new exploded state can be created by using the *View Manager* (see Figure 19.24).

Once created, an exploded state may be set as the current state, deleted, redefined, and renamed. Any one of these actions can be easily accomplished by using the *View Manager,* selecting the state in the list, and pressing the right mouse button.

The options in the *Explode Position* dialog box are used to define the exploded state of each component. This dialog box is reproduced in Figure 19.25.

An assembly is exploded by selecting a reference entity along whose direction one or more components is moved. The component(s) may be moved incrementally by a set amount or in a continuous (*Smooth*) manner or by a preset incremental amount. By using the *Translation* option the user

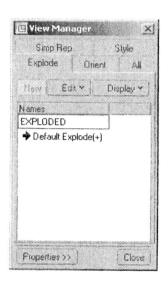

FIGURE 19.24 *The View Manager can be used to define or redefine an exploded state.*

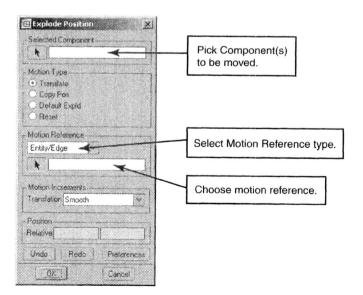

FIGURE 19.25 *The* Explode Position *dialog box menu contains options for defining the direction and amount of translation in an explosion.*

is able to set the way a component is moved. To use the position instruction from one component to another, use the *Copy Pos* command.

The entity that is used as the reference for the move may be selected from the *Explode Position* dialog box using one of the following options:

1. The *View Plane* option uses the viewing plane as a reference to position the component. If a plane other than the viewing plane is to be chosen, then the *Sel Plane* option is helpful. In either case, a plane, parallel to the selected plane, is constructed, which is used to locate the component.

2. The *Entity/Edge* option is used to pick an entity, axis, an edge, or datum plane, which can be used to define the direction of the explosion. The component will be moved so that it is in a line parallel to the feature.

3. The *Plane Normal* option is used to select a plane as the reference plane. The component is placed along a line normal to the plane.

4. The *2 Points* option may be used to define a line between two point or vertices. The component is positioned along a line connecting the two points.

5. By using the *Csys* option, the user may use a coordinate system to reposition a component. The component is placed in the direction of the coordinate system.

A default-exploded state is always created by Pro/E using the *DefaultExpld* option. The exploded positions of the component are determined by default values.

The number of components that are moved is controlled by using the *Preferences* button. The *Move One* option moves one component at a time, while the *Move Many* allows the user to move several components at once. The *Move With Children* option is useful when the move should include the component as well as its children.

The last three options in the *Explode Position* dialog box, *Reset, Undo,* and *Redo,* are used to manipulate the exploded state. The *Reset* option removes all explode instructions and returns the assembly to its original state. The *Undo* command undoes the last performed step, while the *Redo* performs the last step over again.

It is best to create an exploded state in *Isometric* orientation for two reasons. First, most engineers are more familiar with *Isometric* orientation than *Trimetric.* Second, quite often, an entity coincident with one of the isometric axes may be used as the reference for the move.

19.6.1 TUTORIAL 19.5: AN EXPLODED STATE FOR THE PULLEY ASSEMBLY

This tutorial illustrates:

- Creating an exploded state
- Defining the position of components in an exploded state
- Setting an exploded state as the current one
- Viewing an exploded state

1. Retrieve the assembly "Pulley Assembly." Select [icon] and double-click on the file.

2. Then, select *Tools, Environment,* and change the orientation to *Isometric.* Choose *OK.* Repaint the screen.

3. Since the movement of the components will increase the size of the assembly on the screen, minimize the size of the model by zooming out.

4. Select *View, View Manager,* and the *Explode* tab.

5. Press *New* and enter the name "Exploded" as shown in Figure 19.24. Hit *return.*

6. Select "Exploded" in the list, press your right mouse button, and choose *Redefine* from the pop-up menu.

7. Choose *Position.*

8. Select the edge shown in Figure 19.26 as the reference for the move.

9. Press 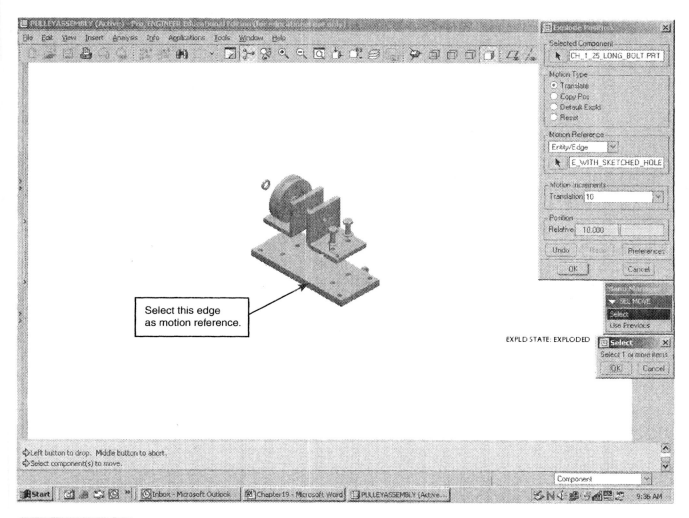 for the *Translate* field. Select an increment value of 10.

10. Press *Preferences* and choose *Move Many.* Then, select *Close.*

FIGURE 19.26 *Select the line forming one edge of the plate as the reference entity.*

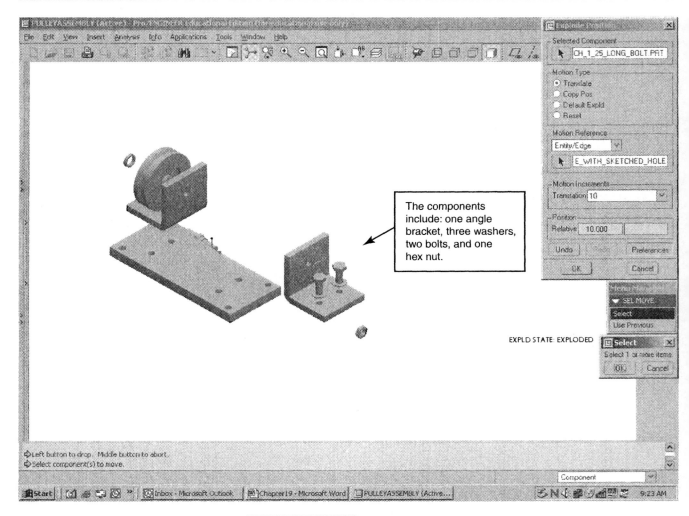

FIGURE 19.27 *Move the components 10 units with respect to the plate.*

11. Now, select one angle bracket, hex nut, washer, and two bolts and their associated washers (see Figure 19.27). Press the CTRL key as you select each component or select the components from the model tree. Note that as each component is selected its edges will be redrawn in red.

12. When you are ready press *OK*. Then click your left mouse button. Drag the mouse to the right. Click the left mouse button to position the components.

13. Repeat steps 11 and 12 for the other angle bracket and associated fasteners (see Figure 19.28).

14. We need to move the axle and the screws up. In order to do this, we need to change the motion reference. Choose ![icon] for the *Motion Reference* and then click on the edge shown in Figure 19.29.

15. Change the *Translation* to 5.

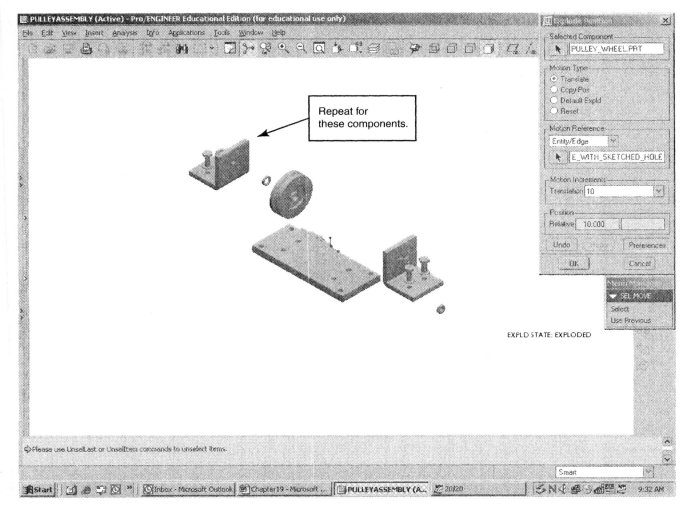

FIGURE 19.28 *Move the components on the other side using the same approach.*

16. Now select all the washers and screws as well as the axle (see Figure 19.30). Choose *OK*. Click on the screen with your left mouse button and drag the mouse up.

17. Change the *Translation* to say 1, and move the washers away (downward) from the screws.

18. Now, use a combination of the two edges used previously as well as the one in Figure 19.30 to move the axle, bushings, and hex nuts so that they are aligned with one another. Your exploded state should be similar to the one shown in Figure 19.31.

19. Select *OK* from the *Explode Position* dialog box, *Done/Return*, and *Close*.

20. Press ⬚ and save the assembly.

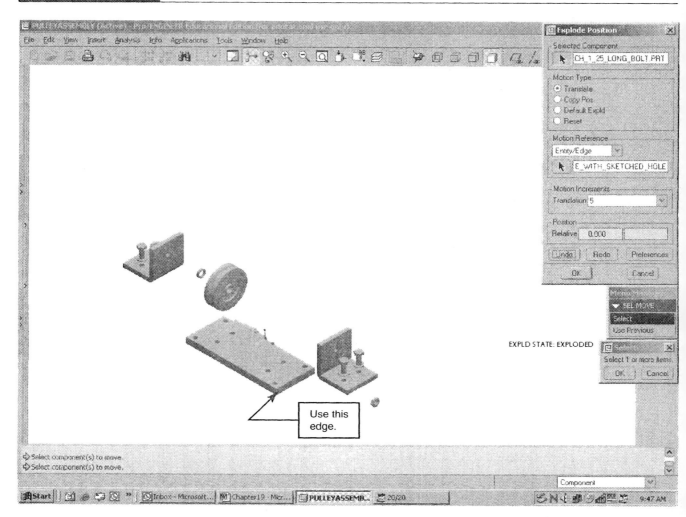

FIGURE 19.29 *Use the given edge.*

19.7 ADDING OFFSET LINES TO ASSEMBLIES

Offset lines are added to exploded assemblies to show how the parts come together. This may be accomplished by using *View, Explode, Offset Lines,* and *Create* or via the *View Manager.*

The offset lines can be *Hidden, Centerline, Phantom line, Cut Plane,* and *Geometry* type. The default is *Hidden.* Use the *Set Def Style* option and the options in the *Line Style* box to set the style of the line to use. The *Line Style* box is shown in Figure 19.32.

Offset lines are drawn between two entities selected by the user. The type of entity can be an *Axis,* normal surface (*Surface Norm* option), or an edge or a curve (*Edge/Curve* option). Select the appropriate option from the menu when drawing the offset lines.

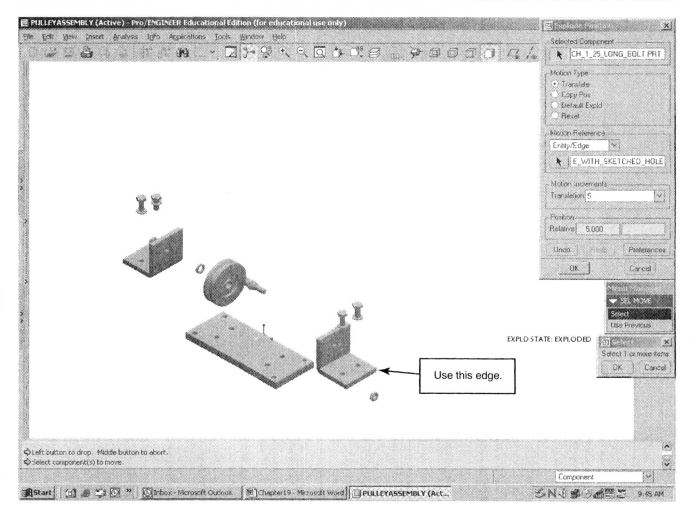

FIGURE 19.30 *Use the edge shown.*

19.7.1 TUTORIAL 19.6: OFFSET LINES FOR THE PULLEY ASSEMBLY

This tutorial illustrates:

- Adding offset lines to an exploded assembly

1. Retrieve the assembly "PulleyAssembly." Select ![icon] and double-click on the file.
2. Turn on the axis display.
3. If you are not in the exploded state use *View* and *View Manager*. Choose *Exploded*. Press your right mouse button and select *Set*.
4. If necessary, the line style may be changed by selecting ![icon] as shown in Figure 19.32.
5. Now, choose the *Properties* button and then ![icon]. Choose *Axis*. Click on the desired axis and then on the axis on the corresponding component. Use Figure 19.33 as a guide.

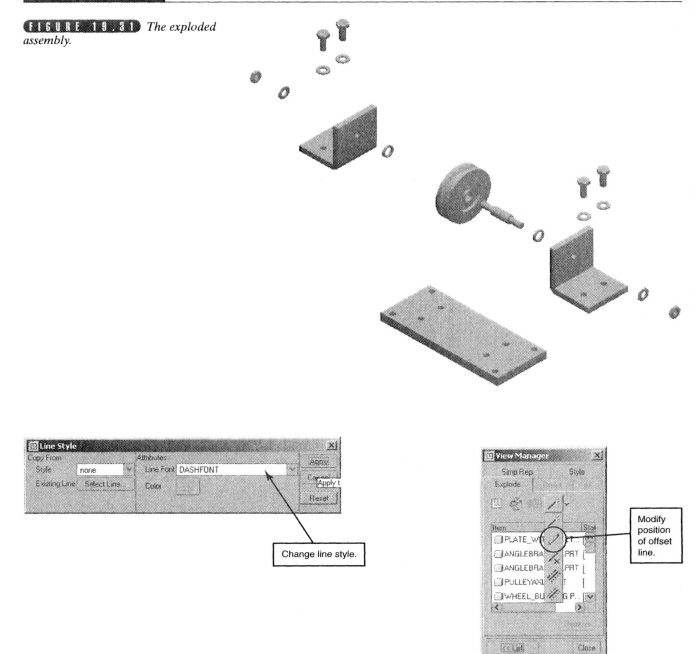

FIGURE 19.32 *The options for creating and defining the style of an offset line.*

6. When all the offset lines have been created choose *Quit*.

7. Note that the position of the offset lines may be modified by selecting ▨. If you want to choose this option, select Move, click on the desired line, and drag the line to a new position. In Figure 19.33,

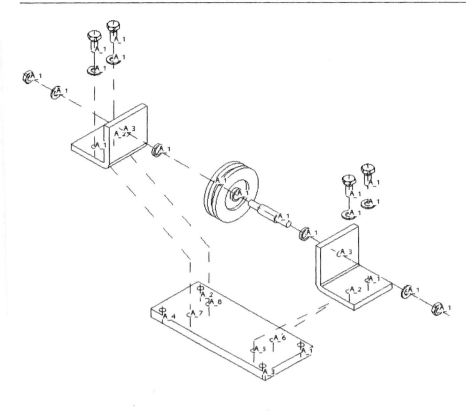

FIGURE 19.33 *The exploded assembly of the pulley with offset lines.*

this method was used to move the lines from the plate to the angle brackets.

8. When you are done choose *Close* in the *View Manager*.

9. Press ⌷ to save the assembly.

19.8 ADDING AN ASSEMBLY TO A DRAWING

An assembly may be added to a drawing by simply specifying the "*.asm" file as the model when adding the view to the drawing. It is advantageous to add both an exploded and unexploded assembly to the drawing. When doing so, the scale of the unexploded assembly should be properly reduced from the scale of the exploded assembly.

Recall that Pro/E identifies the exploded state of the assembly with a name. Thus, when loading the exploded assembly, the name of the assembly must be given. This can be accomplished by choosing the correct exploded state.

19.8.1 TUTORIAL 19.7: A WORKING DRAWING OF THE PULLEY ASSEMBLY

This tutorial illustrates:

• Constructing a drawing with views of an assembly

1. Click on ⌷.

2. Choose the drawing mode. Enter the name: "PulleyAssembly."

3. Click on the *OK* button.

4. Use *Browse* or enter the model name (PulleyAssembly.asm) and then select the *Empty with Format* option.

5. Use *Browse* or enter "AFormat" in the cell provided.

6. Choose *OK*.

7. Change the display to isometric. Choose *Tools, Environment* and change the default orientation to *Isometric*.

8. Select *OK* to save the change.

9. Add the first view. Choose ▣.

10. Choose *Exploded* from the menu and then *Done*.

11. Click on the screen. Choose the *Exploded* state and *Done*.

12. Do not reorient the view by choosing *OK*.

13. Change the scale of the view. Double-click on the scale value and change to say 0.2.

14. Add a second view showing the unexploded assembly. Select ▣.

15. Then, choose *General, Unexploded, No Scale,* and *Done*. The *General* option is chosen so that the new view is independent of the first view. Furthermore, since we have some idea of the available space, we can choose the *No Scale* option.

16. Click somewhere near the lower right-hand side of the screen to locate the new view. After the model is placed, do not change its orientation.

17. Move the views so that your drawing appears as in Figure 19.34. Save the drawing.

FIGURE 19.34 *The drawing with an exploded and unexploded view.*

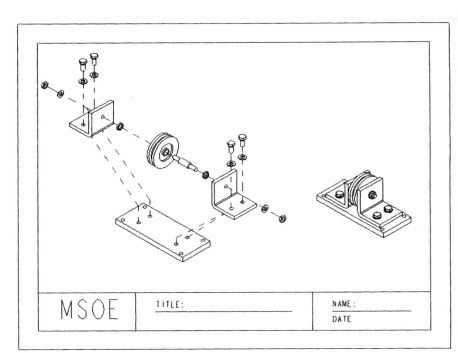

MSOE   TITLE: NAME:

DATE

▮19.9▮ BILL OF MATERIAL (BOM) AND BALLOONS

A working drawing is incomplete without a bill of material (BOM). A BOM may contain the part or component name, the quantity of each component, the material from which it is to be made, and any required finish. If a component is to be obtained from a subcontractor or external supplier, then the supplier is listed, along with the part.

Pro/E has commands for creating a bill of material tailored to the user. A default BOM can be generated by selecting the *Bill of Materials* option from the *Info* option in the toolbar. The information may be sent to the screen, file, or both.

All results sent to a file are saved in a file with the extension .BOM. This file can be loaded into a drawing. By selecting the text and using *Edit* and *Properties* the user may modify the text.

Balloons are used to reference the parts in the assembly to the bill of material. By using the path *Insert, Balloon,* and *Make Note* the user can create a balloon.

19.9.1 TUTORIAL 19.8: ADDING A BOM TO THE WORKING DRAWING

This tutorial illustrates:

- Creating a basic bill of material (BOM)
- Adding the BOM to a drawing
- Editing the BOM
- Adding balloons to a drawing

1. Retrieve the working drawing "PulleyAssembly." Select 🖼 and double-click on the file.
2. Select *Info* from the toolbar and then *Bill of Materials*.
3. Choose the *Top Level* option. Then choose *OK*. The information will be saved to disk and displayed in the browser window.
4. Place the BOM on the drawing by selecting *Insert* and *Note*.
5. For the placement, use the *Default* or *Justify Left;* otherwise, the text may not fit on the screen.
6. Select *File* and *Make Note*. Click somewhere near the upper right-hand side of the screen.
7. Find the file "PulleyAssembly.bom." Open the file by double-clicking on it. Choose *Done/Return*.
8. Edit the note by selecting the text, pressing your right mouse button, and then choosing Properties. Then select *Edit* and *Properties*. Choose the *Editor* button. The software will load the editor.
9. Edit the text. Add the additional text and spaces to produce the BOM shown in Figure 19.35.
10. Choose *OK*.
11. Move the text as necessary.
12. Add the balloons. Select *Insert* and *Balloon*.
13. Choose *Leader, Enter,* and *Make Note*.

FIGURE 19.35 *A working drawing of the pulley assembly.*

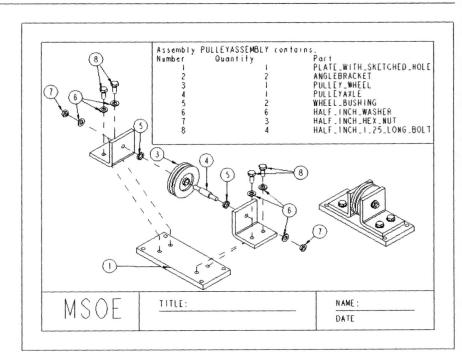

Assembly PULLEYASSEMBLY contains.		
Number	Quantity	Part
1	1	PLATE_WITH_SKETCHED_HOLE
2	2	ANGLEBRACKET
3	1	PULLEY_WHEEL
4	1	PULLEYAXLE
5	2	WHEEL_BUSHING
6	6	HALF_INCH_WASHER
7	3	HALF_INCH_HEX_NUT
8	4	HALF_INCH_1_25_LONG_BOLT

MSOE   TITLE: _____   NAME: _____

DATE

14. Select the entity on the part. If you want the balloon to reference several parts, pick an entity on each part while pressing the CTRL key.

15. Select the type of leader. We suggest the default (*ArrowHead*).

16. *Done.*

17. Pick the location for the note.

18. Enter the text for the balloon.

19. Hit the *return* key twice.

20. Choose *Make Note* to create the next balloon and follow steps 14 through 19.

21. When all the balloons have been created, choose *Done/Return*. Move the balloons as necessary, using your mouse.

22. The arrowheads should be of *Filled* type. If they are not, select *File, Properties,* and *Drawing Options*. Change the arrowhead style (draw_arrow_style) from *Closed* to *Filled*. Choose *Add/Change, Close,* and *Done/Return*. Repaint the screen.

23. If desired add the scale size, your name, title, and date to the format by using *Insert* and *Note*. Save the drawing.

19.10 SUMMARY AND STEPS FOR DEALING WITH ASSEMBLIES

Parts may be put together to form an assembly. Each part in the assembly is called a component. In order for the component to be placed in the assembly, the software must have access to the part.

Assemblies are created as assembly files using the method analogous for part and drawing files. After an assembly file has been created, components may be added to the assembly either parametrically, using the *Assemble* option, or non-parametrically, using the *Package* option. Components placed by using the *Package* option are located in an absolute sense. They are not placed relative to the base part. This is helpful if the exact placement of the part is unknown.

In this chapter, we concentrated on placing the components in a parametric fashion. When placing components using the *Assemble* option, the relative positions of the components must be known. Because the assembly process is parametric, the location of the components may be modified.

An assembly may be created with or without assembly datum planes. If the datum planes are not used, then the first part added to the assembly becomes the base part in the assembly. Assembly datums are useful in locating components that do not have planar surfaces.

The sequence of steps for adding a part to an assembly is:

1. Select *Insert, Component,* and *Assemble* or ▨.
2. Retrieve the desired part.
3. Choose the required constraint.
4. Select the component reference.
5. Choose the assembly reference.
6. Add additional constraints, as needed.
7. When the component is fully constrained, select *OK*.

A part may be constructed in the assembler. In order to ensure that the part has the same standards as the components, it is best to start with a start part containing the standards and the default datum planes. This is the option *Copy From Existing*. Then, the datum planes may be assembled and the new part created using the datum planes as sketching and orientation planes.

If all the parts contain the default standard, the datum planes may be created in the assembler using the option *Locate Default Datums*. The datums are assembled and the part constructed using the assembled datums.

In order to create a part in the *Assembly* mode, use the following sequence of steps:

1. Select *Insert, Component,* and *Create* or choose ▨.
2. Enter a name for the part and select *OK*.
3. Select from *Copy From Existing, Locate Default Datums, Empty,* or *Create First Feature*.
4. Choose *OK*.
5. Assemble the part based on your choice in step 3.
6. Construct the part using the appropriate option and sketching and orientation planes.

After a part has been created in the assembler, it may be modified. Additional features may be added to the part by doing the following:

1. Select the part in the model tree. Press your right mouse button and select *Activate*.
2. Use the appropriate tools to create the desired feature(s).
3. Reactivate the assembly by selecting the assembly in the model tree, pressing your right mouse button and choosing *Activate*.

In order to see how the components come together to form the assembly, an exploded state of the assembly may be created. The exploded state is created by moving the components apart. This can be done in an incremental fashion, or continuously, in a smooth manner. The components are moved with respect to a reference. In general, an exploded state may be created by using the following procedure:

1. Select *View, View Manager*, and the *Explode* tab.
2. Press *New* and enter the name of the exploded state. Hit *return*.
3. Select the state in the list, press your right mouse button, and choose *Redefine* from the pop-up menu.
4. Choose *Position*.
5. Select the reference to be used for the move.
6. If you wish to move the component or components incrementally, select *Preferences* and change the move type.
7. Choose the component or components to move. Press *OK* when done.
8. Click the left mouse button to begin the move.
9. Move the component(s) by moving your mouse.
10. Choose *OK, Done/Return*, and *Close*.

After an exploded state is defined, offset lines may be added to the state. The line type may be changed using the options in the *Line Style* box. The lines are created by defining the feature to which they are attached. This feature may be an axis, normal surface, an edge, or a curve. The sequence of steps to follow for adding offset lines to an assembly is:

1. Select *View* and *View Manager*.
2. Choose the *Explode* tab. Select the desired exploded state.
3. Now, choose the *Properties* button and then ▨.
4. Choose the type of entity.
5. Choose the desired reference on the first part. Select the entity on the next part.
6. After you are finished adding all the offset lines, choose *Close* in the *View Manager*.

Assemblies may be added to a drawing by selecting ▨. Because the assembly may contain an exploded state, choose between the *Exploded* and *Unexploded* option to place the desired state.

A bill of material may be created in the assembly or drawing mode. Select *Info* and *Bill of Material* to do so. Add a bill of material to a drawing by using *Insert* and *Note*. Choose the *File* option and read in the file. This file will have a "bom" extension.

Add balloons to a working drawing by selecting *Insert* and *Balloon*.

19.11 ADDITIONAL EXERCISES

19.1 Using the parts "AssembleBlockA" and "AssembleBlockB," obtain the assembly shown in Figure 19.36.

FIGURE 19.36

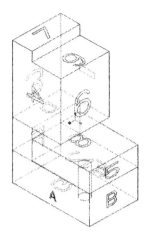

19.2 Assemble the parts "AssembleBlockA" and "AssembleBlockB" to obtain the assembly shown in Figure 19.37.

FIGURE 19.37

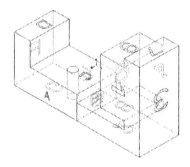

19.3 Use the parts "AssembleBlockA" and "AssembleBlockB" to create the assembly shown in Figure 19.38.

FIGURE 19.38

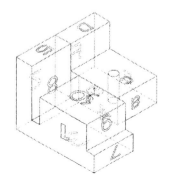

For Exercises 19.4 through 19.11, create the exploded assemblies using the given figures. Some of the parts have already been constructed in the tutorial sections of this book. Additional parts may be required. Consult the BOM associated with each assembly for the parts and the corresponding reference figure, tutorial, or exercise. Create a working drawing if assigned.

19.4 Propeller assembly. Consult Figures 19.39 and 19.40.

FIGURE 19.39

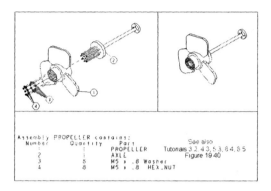

FIGURE 19.40 *Geometry of the axle.*

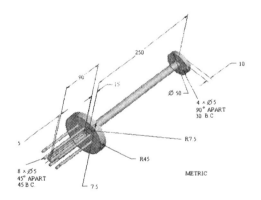

19.5 Linkage assembly. Use Figures 19.41 through 19.45 as a reference.

FIGURE 19.41

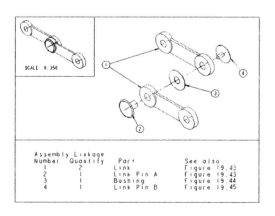

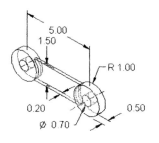

FIGURE 19.42 *Geometry of the part "Link."*

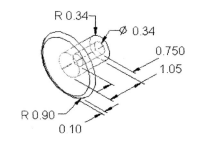

FIGURE 19.43 *Geometry of the "Link Pin A."*

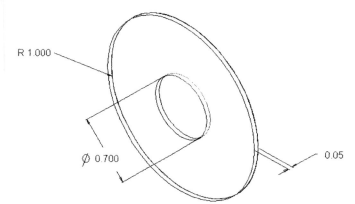

FIGURE 19.44 *Geometry of the part "Bushing."*

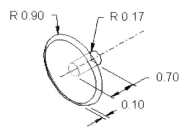

FIGURE 19.45 *Geometry of "Link Pin B."*

19.6 Flood light subassembly. Use Figures 19.46 through 19.48 as a reference.

FIGURE 19.46

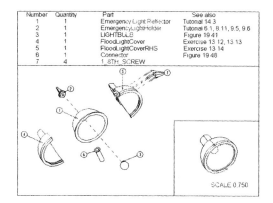

Number	Quantity	Part	See also
1	1	Emergency Light Reflector	Tutorial 14.3
2	1	EmergencyLightHolder	Tutorial 6.1, 8.11, 9.5, 9.6
3	1	LIGHTBULB	Figure 19.41
4	1	FloodLightCover	Exercise 13.12, 13.13
5	1	FloodLightCoverRHS	Exercise 13.14
6	1	Connector	Figure 19.48
7	4	1_8TH_SCREW	

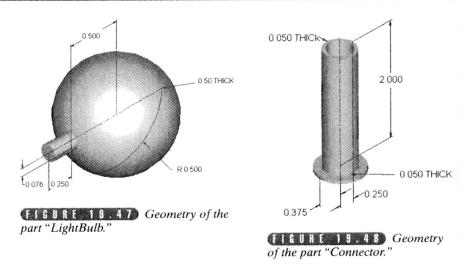

FIGURE 19.47 *Geometry of the part "LightBulb."*

FIGURE 19.48 *Geometry of the part "Connector."*

19.7 Steady rest assembly. Consult Figures 19.49 through 19.51 for the components.

FIGURE 19.49

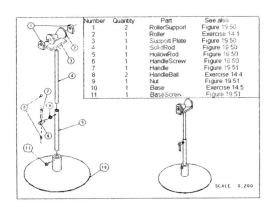

Number	Quantity	Part	See also
1	2	RollerSupport	Figure 19.50
2	1	Roller	Exercise 14.1
3	1	Support Plate	Figure 19.50
4	1	SolidRod	Figure 19.50
5	1	HollowRod	Figure 19.50
6	1	HandleScrew	Figure 19.50
7	1	Handle	Figure 19.51
8	2	HandleBall	Exercise 14.4
9	1	Nut	Figure 19.51
10	1	Base	Exercise 14.5
11	1	BaseScrew	Figure 19.51

SCALE 0.200

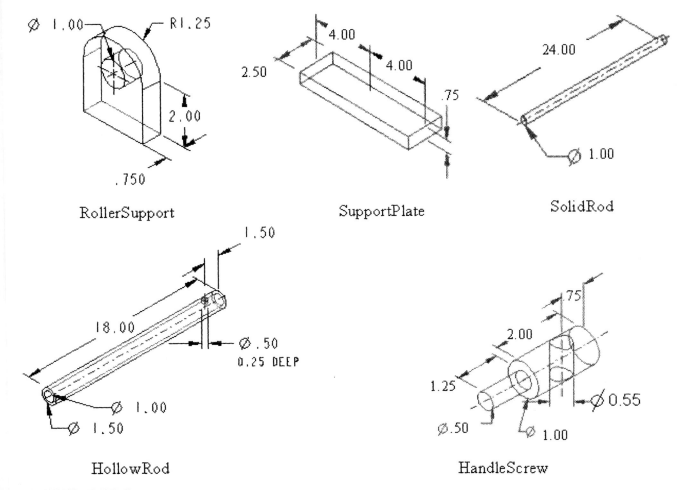

Ø 1.00 — R1.25

2.50
4.00
4.00
.75

24.00
Ø 1.00

2.00
.750

RollerSupport   SupportPlate   SolidRod

1.50

18.00
Ø .50
0.25 DEEP

Ø 1.00
Ø 1.50

75
2.00
1.25
Ø.50
1.00
Ø 0.55

HollowRod   HandleScrew

FIGURE 18.50 *Some parts in the steady rest assembly.*

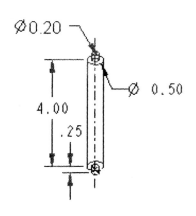

Ø0.20

Ø 0.50

4.00

.25

Handle

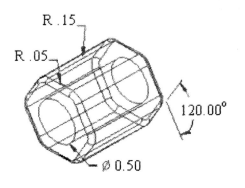

R .15

R .05

120.00°

Ø 0.50

0.866 flat to flat & 1.0 inch long hexagonal nut

Nut

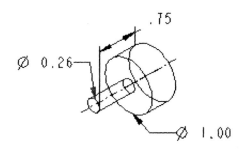

.75

Ø 0.26

Ø 1.00

Base Screw

FIGURE 19.51 *More parts for the steady rest assembly.*

19.8 Blower assembly. Consult Figures 19.52 through 19.56 for the parts in this assembly.

FIGURE 19.52

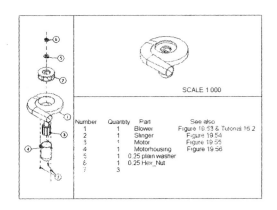

SCALE 1.000

Number	Quantity	Part	See also
1	1	Blower	Figure 19.53 & Tutorial 16.2
2	1	Slinger	Figure 19.54
3	1	Motor	Figure 19.55
4	1	Motorhousing	Figure 19.56
5	1	0.25 plain washer	
6	1	0.25 Hex_Nut	
7	3		

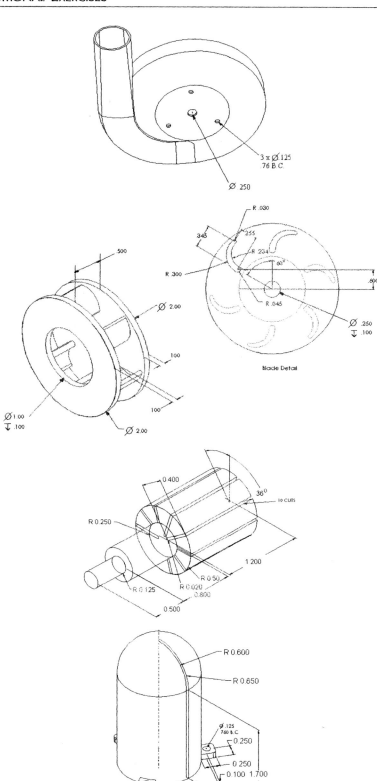

FIGURE 19.53 *Add the holes to the "Blower" from Tutorial 16.2.*

3 x Ø.125
.76 B.C.

Ø .250

FIGURE 19.54 *Part "Slinger."*

R .030

345   255

R .234

R .300   60°

.600

R .045

Ø .250
↧ .100

Blade Detail

.500

Ø 2.00

.100

.100

Ø1.00
↧ .100

Ø 2.00

FIGURE 19.55 *Part "Motor" for the assembly "Blower."*

0.400

36°

10 CUTS

R 0.250

1.200

R 0.50

R 0.125   R 0.020
0.800

0.500

FIGURE 19.56 *Part "Motor-Housing" for the motor in Figure 19.61.*

R 0.600

R 0.650

Ø.125
760 B.C.

0.250

0.250

0.100 1.700

R 0.050

3 Ears 120° APART

19.9 Emergency light assembly. Use Figures 19.57 through 19.59 for this assembly.

FIGURE 19.57

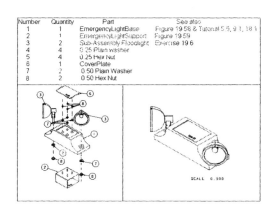

Number	Quantity	Part	See also
1	1	EmergencyLightBase	Figure 19.58 & Tutorial 5.5, 9.1, 18.1
2	1	EmergencyLightSupport	Figure 19.59
3	2	Sub-Assembly Floodlight	Exercise 19.6
4	4	0.25 Plain washer	
5	4	0.25 Hex Nut	
6	1	CoverPlate	
7	2	0.50 Plain Washer	
8	2	0.50 Hex Nut	

FIGURE 19.58 *Additional features for the "EmergencyLightHousing."*

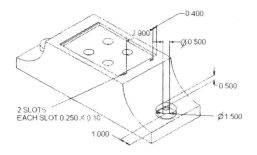

FIGURE 19.59 *Part "EmergencyLightSupport."*

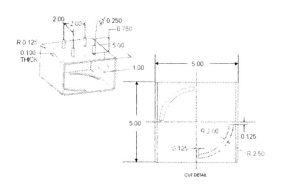

19.10 Butterfly valve assembly. Use Figures 19.60 through 19.62 for this assembly.

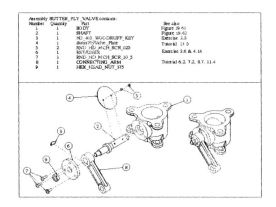

FIGURE 19.60

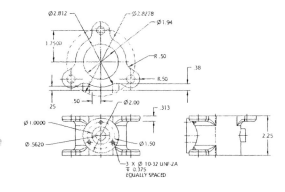

FIGURE 19.61 *Body of the ButterFly Valve.*

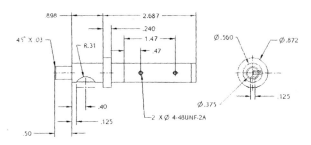

FIGURE 19.62 *Views of the shaft.*

19.11 Hold down assembly. Use Figures 19.63 through 19.66 for this assembly.

FIGURE 19.63

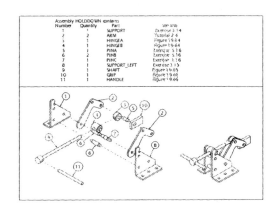

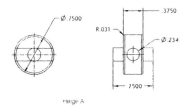

Hinge A

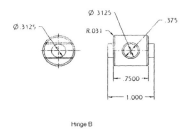

Hinge B

FIGURE 19.64 *Views of the hinges.*

FIGURE 19.65 *Views of shaft.*

FIGURE 19.66 *Multiviews of the parts grip and handle.*

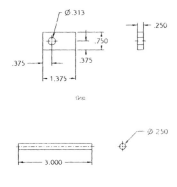

Grip

Handle

19.12 Vise assembly. Use Figures 19.67 through 19.70 for this assembly.

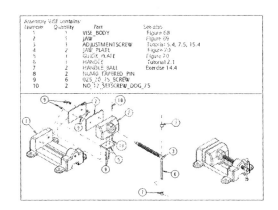

FIGURE 19.67

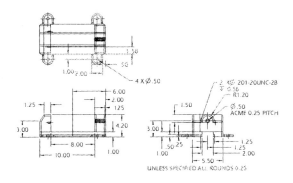

FIGURE 19.68 *Geometry of the vise body.*

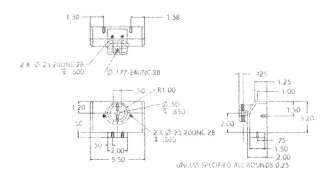

FIGURE 19.69 *Views of the part jaw.*

FIGURE 19.70 *Multiviews of the parts jaw plate and guide plate.*

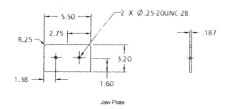

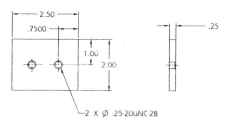

CHAPTER ■20

ENGINEERING
INFORMATION AND
FILE TRANSFER

INTRODUCTION AND OBJECTIVES

Pro/E provides several options for obtaining engineering information, such as mass properties, lengths of a feature, interference, and clearance checks. The options for obtaining such information may be found as suboptions in the *Analysis* command. These include the *Measure* and *Model Analysis* options.

Many of these options require properties of engineering materials. These properties may be entered into the informational database using *Edit* and *Set Up* during model construction. If the properties have not been assigned, then the software will query the user for the required property during the computational phase.

A coordinate system is required for the computation of many of these engineering properties. A coordinate system may be created using the methods outlined in Chapter 13 during model construction, or "on the fly" during calculation of the engineering property.

The Initial Graphics Exchange Specification (IGES) is used to transfer graphics and text to another CADD or CAM (Computer Aided Manufacturing) system. Models and some sections may be imported or exported as IGES files. When exporting a model as an IGES file, the features that the receiving system can receive must be taken into consideration. The proper options must be used in order to achieve the desired results. The software will create two files. The first file is the actual IGES transfer file with the extension ".igs." The second file, with extension " out.log," contains information about the processor, type, and quantity of each entity.

The objectives of this chapter are to

1. Assign material properties to a model
2. Obtain engineering properties and dimensions of a model
3. Find the clearance or interference between two parts
4. Export a part as an IGES file

20.1 DEFINING AND ASSIGNING MATERIAL PROPERTIES

Material properties may be assigned to a model by using the options in the *PART SETUP* menu. This menu is shown in Figure 20.1. The *Materials* option may be used to define and assign material properties. The *Define* option allows the user to enter data about a new material. This material is given a name by the user. After the material is defined, it may be assigned to a part by using the *Assign* option.

When using the *Define* option, material properties are entered in a window. This window is reproduced in Figure 20.2. In order to define a property, the user simply types the value in the appropriate space. Notice that the density may be entered directly by choosing the *Density* option from *PART SETUP* menu. This approach is used when the density is the only material property required.

When entering material properties, enter the values with the correct units. Also, notice that the computation of engineering properties is dependent on the accuracy of the model. The accuracy may be set by using the *Accuracy* option in the *PART SETUP* menu.

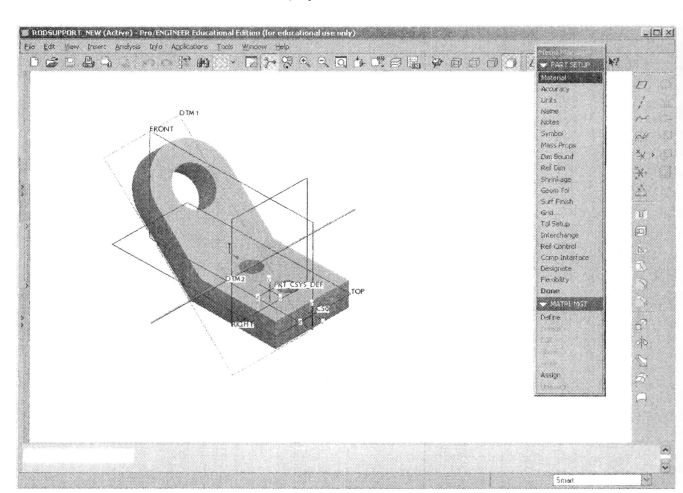

FIGURE 20.1 *Use the* Material *option to define and set material properties for a part.*

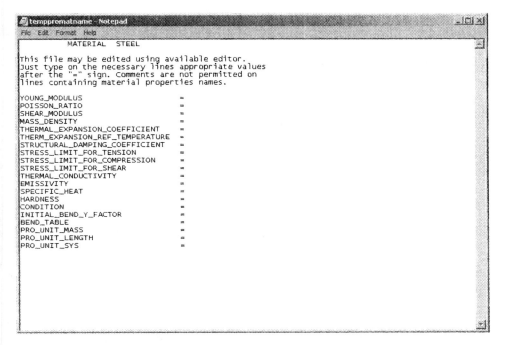

FIGURE 20.2 *The material properties can be defined by using the* Define *option and by entering the values in the appropriate spaces.*

20.1.1 TUTORIAL 20.1: ENTERING THE DENSITY FOR A MODEL

This tutorial illustrates:

- Defining the density of a material

1. Retrieve the model "RodSupport." Select 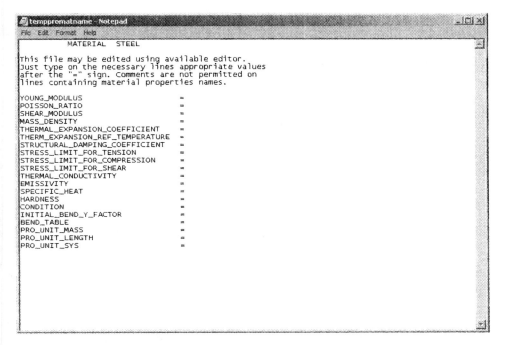 and double-click on the file.
2. For our needs in this chapter, we need to define only the density of the material. Therefore, select *Edit, Set Up,* and *Density.*
3. Select *Material* and *Define.*
4. Enter the name: Aluminum. Hit *return.*
5. The units of the model are inches and the material is aluminum. For aluminum, the density is 170 lb/ft^3 (0.09848 lb/in^3). Click on the MASS_DENSITY cell, and enter the density in lb/in^3.
6. Choose *File, Save,* and *Exit.*

20.2 MODEL ANALYSIS

By using the options *Analysis* and *Model Analysis* the user can obtain an analysis of the model. The *Model Analysis* box, shown in Figure 20.3, will become available. The type of analysis may be one of the following types:

1. Model Mass Properties. The software calculates mass properties of the part or assembly, which include:
   a. The *VOLUME* of the model.
   b. The total *SURFACE AREA* of the model.

FIGURE 20.3 *Use the* Model
Analysis *box to perform an analysis on
the part or assembly.*

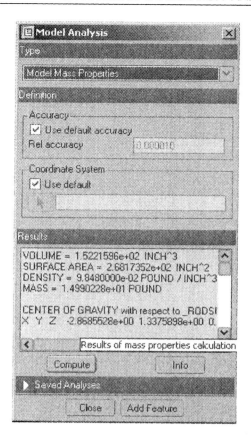

c. The total *MASS* of the model.

d. The *CENTER OF GRAVITY,* with respect to the chosen coordinate system. The center of gravity is shown on the model.

e. The *INERTIA TENSOR* for the specified coordinate system.

f. The *INERTIA TENSOR* at the center of gravity.

g. The *PRINCIPAL MOMENTS OF INERTIA* at the center of gravity and the principal axes. The principal moments are denoted as 1, 2, and 3.

h. The *ROTATION MARIX* for rotation from the chosen coordinate axes to the principal axes. The angle ϕ is the angle of rotation about the axis 1, θ is the angle about axis 2, and ψ is the angle about the axis 3.

i. The *RADII OF GYRATION* at the center of gravity.

j. For an assembly, a summary is provided detailing the mass properties of all the components.

2. *X-Section Mass Properties.* The mass properties are calculated for a chosen cross section.

3. *One-Sided Volume.* The volume on one side of a chosen datum is calculated.

4. *Pairs Clearance.* The software calculates the clearance distance or the amount of interference between any two parts, subassemblies,

entities, or surfaces. If there is interference, the software highlights the interfering parts, subassemblies, entities, or surfaces.

5. *Volume Interference.* The software checks to see if keepin/keepout
   regions have been violated.

6. *Short Edge.* The software highlights all the edges that are shorter
   than a prescribed value.

7. *Edge Type.* For a selected edge, Pro/E lists the edge type.

8. *Thickness.* The thickness of a part is checked to see if it is beyond
   a given bound. The user may select the lower and/or upper bound
   for the thickness.

A few words must be said about the analysis of parts with suppressed or
blanked features. Because suppressed features are removed from the model
generation list, they are not used when performing an analysis. Blanked features are still in the generation list; their visibility has simply been turned
off. Therefore, blanked features are used when carrying out model analysis.

Whether or not two parts interfere with each other is very important in
design. You may recall that in the pulley wheel assembly in Chapter 19, an
axle was created with an RC5 (Running Clearance) fit. This tolerance was
not added to the model. The diameter of the hole and the axle were set the
same. As created, there is no interference, nor a clearance between the two
parts. We can easily check to see if this is the case by using the *Analysis,
Model Analysis,* and *Pairs Clearance* options.

20.2.1 TUTORIAL 20.2: MASS PROPERTIES OF THE ROD SUPPORT

This tutorial illustrates:

- Performing an analysis of a part

1. Retrieve the model "RodSupport." Select ⬜ and double-click on
   the file.

2. Select *Analysis* and *Model Analysis.* Choose *Model Mass Properties.*

3. In Chapter 13, we created a special coordinate axis for this model,
   so uncheck "Use default for the coordinate system" and then
   select coordinate system CS0.

4. The software will also save the results in a file called "rodsup-
   port.m_p." This file is a text file and may be opened and printed.
   Figure 20.4 shows the mass properties for the rod support. Note
   that the results are relative to CS0.

20.2.2 TUTORIAL 20.3: INTERFERENCE BETWEEN TWO PARTS

This tutorial illustrates:

- Checking for interference between two parts

1. Retrieve the model "PulleyAssembly." Select ⬜ and double-
   click on the file.

2. Then, choose *Analysis* and *Model Analysis.* The *Model Analysis*
   box will appear.

3. Select the *Two Pairs* type.

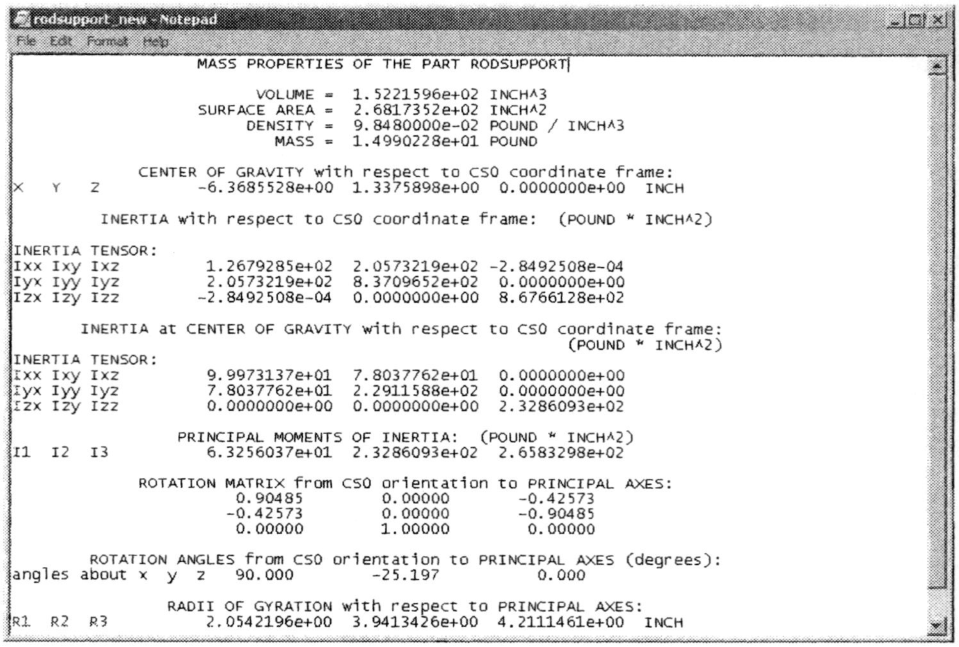

```
rodsupport_new - Notepad                                                    _ |□| x|
File  Edit  Format  Help
                  MASS PROPERTIES OF THE PART RODSUPPORT|

                     VOLUME =  1.5221596e+02  INCH^3
               SURFACE AREA =  2.6817352e+02  INCH^2
                    DENSITY =  9.8480000e-02  POUND / INCH^3
                       MASS =  1.4990228e+01  POUND

            CENTER OF GRAVITY with respect to CS0 coordinate frame:
                         -6.3685528e+00  1.3375898e+00  0.0000000e+00  INCH
 X    Y    Z

           INERTIA with respect to CS0 coordinate frame:  (POUND * INCH^2)

INERTIA TENSOR:
Ixx Ixy Ixz        1.2679285e+02   2.0573219e+02  -2.8492508e-04
Iyx Iyy Iyz        2.0573219e+02   8.3709652e+02   0.0000000e+00
Izx Izy Izz       -2.8492508e-04   0.0000000e+00   8.6766128e+02

          INERTIA at CENTER OF GRAVITY with respect to CS0 coordinate frame:
                                                      (POUND * INCH^2)
INERTIA TENSOR:
Ixx Ixy Ixz        9.9973137e+01   7.8037762e+01   0.0000000e+00
Iyx Iyy Iyz        7.8037762e+01   2.2911588e+02   0.0000000e+00
Izx Izy Izz        0.0000000e+00   0.0000000e+00   2.3286093e+02

            PRINCIPAL MOMENTS OF INERTIA:  (POUND * INCH^2)
                  6.3256037e+01  2.3286093e+02  2.6583298e+02
 I1   I2   I3

            ROTATION MATRIX from CS0 orientation to PRINCIPAL AXES:
                  0.90485        0.00000        -0.42573
                 -0.42573        0.00000        -0.90485
                  0.00000        1.00000         0.00000

         ROTATION ANGLES from CS0 orientation to PRINCIPAL AXES (degrees):
angles about x  y  z   90.000         -25.197         0.000

            RADII OF GYRATION with respect to PRINCIPAL AXES:
                  2.0542196e+00  3.9413426e+00  4.2111461e+00  INCH
 R1   R2   R3
```

FIGURE 20.4 *The mass properties are saved in a text file. This file may be printed.*

4. As shown in Figure 20.5, you need to select the parts in the assembly. Press ![icon] and choose the pulley wheel. Repeat and choose the axle.

5. Then depress the *Compute* button. The software should respond with the statement: "Zero clearance. No interference was detected."

6. Choose the *Close* button. Note that if the upper and lower limits of the respective parts were changed by using the clearance fit, and the analysis redone, the clearance between the two parts would no longer be zero.

20.3 THE MEASURE OPTION

The *Measure* option can be used to analyze the geometry model or assembly. The *Measure* box, shown in Figure 20.6, contains options for carrying out the measurement. The types of measurements that are available are as follows:

1. *Curve Length*. Calculate the length of a curve or edge.
2. *Distance*. Measure the distance between two entities, parts, or subassemblies.
3. *Angle*. Calculate the angle between two entities. The entities may be axes, planar curves, or planar nonlinear curves.
4. *Area*. Obtain the surface area of a chosen entity.
5. *Diameter*. Calculate the diameter of any revolved surface.
6. *Transform*. For two given coordinate systems, the software obtains the transformation matrix.

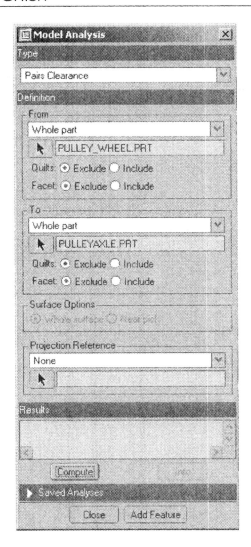

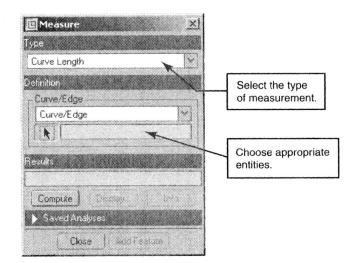

FIGURE 20.5 *Obtain the clearance between the axle and pulley wheel by selecting the two components.*

FIGURE 20.6 *The options in the Measure box may be used to analyze the geometry of a model or assembly.*

Select the type of measurement.

Choose appropriate entities.

FIGURE 20.7 *Reorient the model as shown in this figure. Also, notice the entities to choose when using the* Measure *option.*

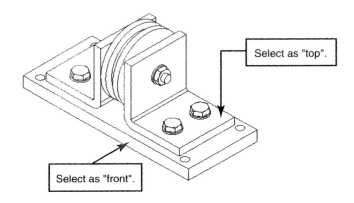

Select as "top".

Select as "front".

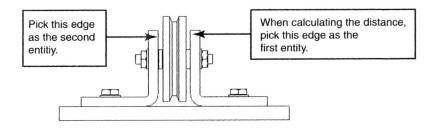

Pick this edge as the second entitiy.

When calculating the distance, pick this edge as the first entity.

20.3.1 TUTORIAL 20.4: USE OF THE MEASURE OPTION

This tutorial illustrates:

- Using *Measure*

1. Retrieve the model "PulleyAssembly." Select 🗁 and double-click on the file.
2. Turn off the display of the datum planes and axes. Reorient the model as shown in Figure 20.7.
3. As an example, let us measure the distance between the two edges shown in Figure 20.7. Choose *Analysis* and *Measure*.
4. Select *Distance* as the measurement type.
5. Pick the two surfaces shown in Figure 20.7.
6. The linear distance between the two lines is 1.700″. Choose *Close*.
7. Now restore the model to its default *Isometric* orientation.
8. Select *Analysis* and *Measure*.
9. For the measurement type, choose *Area*.
10. Pick the "Top" of the plate as the area as shown in Figure 20.7.
11. The surface area is 13.1073 in^2. Choose *Close*.

20.4 IGES FILE TRANSFER

A part or an assembly may be transferred using an IGES format. The creation of IGES files from an assembly is more complicated because of the number of parts.

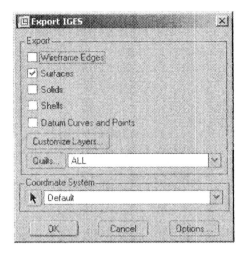

FIGURE 20.8 *The* Export IGES *box is used to export a file.*

The *Export IGES* box shown in Figure 20.8 contains the options for exporting a model in an IGES format. The part may be exported as a wireframe or as a surface model. In a wireframe model, the part is exported with edge information only. The *Surfaces* option, on the other hand, exports the part in a format containing information on the surface as well as the edges. Notice that the *Customize Layers* button may be used to select the appropriate layers.

When exporting an assembly in IGES format, additional information must be provided concerning the treatment of the individual components in the assembly. This information is provided by using the *File Structure* option in Figure 20.9.

The possible choices for the *File Structure* option are as follows:

1. *Flat.* Output the assembly to a single file. The assembly is treated as a single part. When using this option, it is recommended that each component be placed on a separate layer in order to discriminate among the parts.

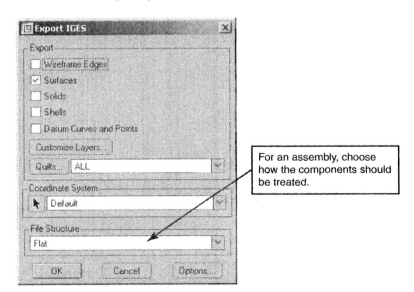

For an assembly, choose how the components should be treated.

FIGURE 20.8 *Additional information must be provided for an IGES transfer of an assembly. Use the* File Structure *option to provide this information.*

2. *One Level.* The assembly is exported to a file with references to the IGES files of the components. The IGES assembly file only contains information regarding the assembly of the components.

3. *All Levels.* The assembly is exported with references to all the components and component IGES files.

4. *All Parts.* Multiple files are created. These files contain information on each part and assembly. If a component is found more than once in the assembly, a file is created for each copy of the component. The format of the file name is "_cpy_#.igs," where # is the numbered copy of the part.

20.4.1 TUTORIAL 20.5: AN IGES FILE OF THE ANGLE BRACKET

This tutorial illustrates:

- Creating an IGES file of a part

1. Retrieve the model "AngleBracket." Select ☞ and double-click on the file.
2. Then, select *File* and *Save A Copy*.
3. Use ∨ for the file type and select the IGES type.
4. Accept the default name of the file by hitting the *return* key.
5. Keep the default settings in the *Export IGES* box.
6. Select *OK*.
7. The software will create two files. The first of these is "AngleBracket.igs" and is the IGES file of the model. The second file, "AngleBracket_out.log," is a log file containing information on the IGES file and processor. Check to see if these files exist in your directory.

20.5   SUMMARY AND THE STEPS FOR THE OPTIONS OF THIS CHAPTER

Parts and assemblies may be analyzed by using various options in the *Analysis* command. These options include *Measure* and *Model Analysis*.

The *Model Analysis* option may be used to obtain mass properties of a model or a cross section. It may also be used to find parts or subassemblies that interfere with one another.

The calculation of mass properties requires the definition of the material of the model. In general, a material may be created, properties assigned, and the material assigned to a part using the following approach:

1. Select *Edit* and *Set Up*.
2. Choose *Material* and *Define*.
3. Enter the name of the material.
4. Select *Assign*.
5. Enter the material properties. *Save* and *Exit*.

The general steps for obtaining an analysis of a model or assembly are as follows:

1. Select *Analysis* and *Model Analysis*.
2. Choose the analysis type.

3. If required, select the desired coordinate system.

4. Enter any additional information required by the type of analysis.

5. Choose *Compute*.

6. Select *Close*.

The *Measure* option may be used to calculate such quantities as distances, surface area, and curve lengths. The following sequence of steps may be used to analyze a part or assembly using the *Measure* option:

1. Select *Analysis* and *Measure*.

2. Choose the measurement type.

3. Select the *Definition* type.

4. Choose the appropriate feature or features.

5. If required, select *Compute*.

6. Choose *Close*.

A part or assembly may be exported in IGES format. The geometry may be exported as a wireframe or surface model. When exporting assemblies, information must be provided as to how to treat the individual components. Options are provided for exporting a single file representing the entire assembly or multiple files containing information on the components and the assembly. In general, a part or assembly may be exported by using the following procedure:

1. Select *File* and *Save A Copy*.

2. Select the IGES type.

3. Enter a name for the IGES file or accept the default name.

4. Pick the export type, coordinate system, and, for an assembly, the *File Structure*. If desired, select the appropriate layer or layers.

5. Choose *OK*.

20.6 ADDITIONAL EXERCISES

For Exercises 20.1 through 20.12, obtain mass properties of the given parts using any desired value for the density. Use the default coordinate system unless instructed otherwise.

20.1 Cement trowel (Exercise 7.1).
20.2 Side guide (Exercise 7.2).
20.3 Control handle (Exercise 7.3).
20.4 External bushing (Exercise 7.4).
20.5 Pipe clamp top (Exercise 7.5).
20.6 Pipe clamp bottom (Exercise 7.6).
20.7 Side support (Exercise 7.7).
20.8 Guide plate (Exercise 7.8).
20.9 Single bearing bracket (Exercise 7.9).
20.10 Double bearing bracket (Exercise 7.10).
20.11 Roller (Exercise 14.1).
20.12 Baseball bat (Exercise 14.2).
20.13 Using the *Distance* option, check the dimensions of the angle bracket. Do these values conform to Figure 2.20?

20.14 Show that in the subassembly "Flood Light" (Exercise 19.6), interference exists between the Parts "FloodLightCover" and "EmergencyLightReflector." Redesign one or more of the parts to eliminate the interference.

20.15 Check the assembly "Steady Rest" (Exercise 19.7) for interference. If interference exists, redesign the conflicting parts.

20.16 Check the assembly "Blower" (Exercise 19.8) for any possible interference. If interference exits, redesign the appropriate parts.

APPENDIX

A.1 THE SKETCHER MODE

The Sketcher mode may be used to create two-dimensional sections. For sections that appear more than once in the same or different parts, it is convenient to create the section in the Sketcher mode and then place the section in a library of sections. The section may be retrieved in the Sketcher when creating a part by using the OPTIONS *Sketch* and *Data from File*.

When a section is loaded into the Sketcher, the user may change the scale as well as the orientation. A dialog box called the *Scale Rotate* box is used to enter the scale and rotation factor.

Sections created in the Sketcher mode are given the extension ".sec" by the software. The mode may be accessed by retrieving or creating a section.

A.1.1 TUTORIAL A.1: CREATING AND USING A SECTION

This tutorial illustrates:

- Generating a section and saving the section to a library
- Retrieving a section from a library into a new part

1. Select ▢.
2. Choose the *Sketch* option and enter the name "C_Section."
3. The software will open the Sketcher.
4. Draw and sketch the section shown in Figure A.1.
5. Dimension to the values shown.
6. Build the section by pressing ✔.
7. Save the section by using *File* and *Save*.
8. Create a new file by selecting ▢.
9. Choose the *Part* mode and give the part a name, "C_Stiffener."
10. Then, select ▢.
11. Choose ▢. The depth is 120.
12. Choose ▢. Enter a thickness of 0.2.
13. Select ▢. Choose datum plane Front as the sketching plane and datum plane Right as the "right" reference plane.
14. Select Sketch.
15. We can now place the section "C_Section." Choose *Sketch* and *Data from File*.
16. From the list, select "C_Section" and then choose the *Open* button.
17. Drag the section to locate it in with respect to the datum planes.

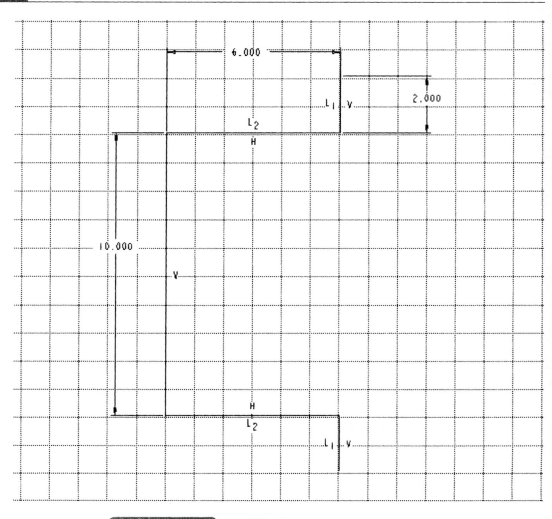

FIGURE A.1 A "C" Section.

18. The dimensions of the section may be changed at this point. Use the same values as in Figure A.2.

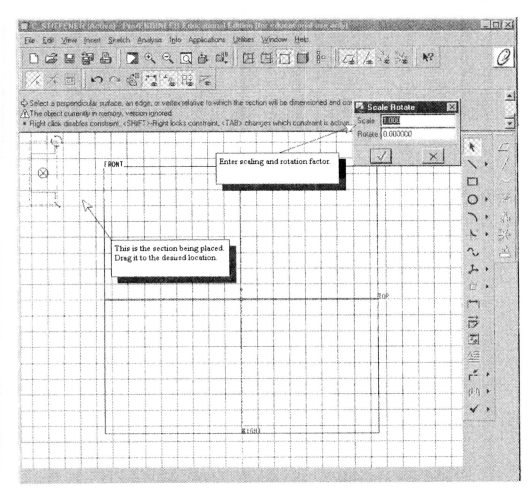

FIGURE A.2 *When the user imports a section in the Sketcher, the section may be scaled and/or rotated.*

19. Fill the thin outward. Build the feature by pressing ☑.
20. The model of the stiffener is shown in Figure A.3.

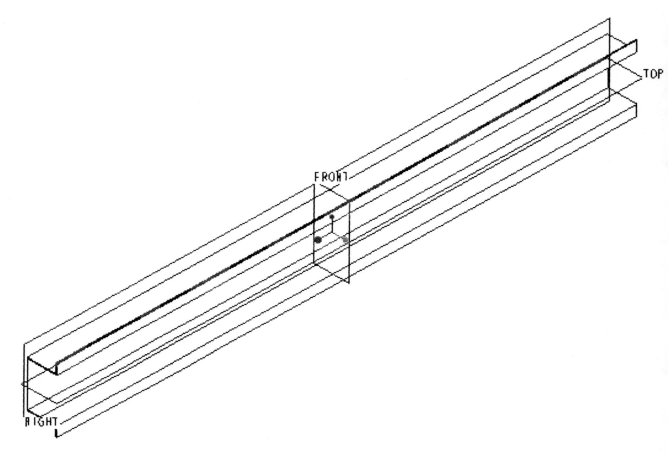

FIGURE A.3 *The stiffener created by placing a preexisting section in the Sketcher.*

BIBLIOGRAPHY

Abbott, I. H., and A. E. Von Doenhoff, 1959. *Theory of Wing Sections*. New York: Dover.

Bertoline, G. R., E. N. Wiebe, C. L. Miller, and J. L. Mohler, 1997. *Technical Graphics Communication,* 2nd ed. Chicago: McGraw-Hill.

Giesecke, F. E., A. Mitchell, H. A. Spencer, I. L. Hill, J. T. Dygdon, J. E. Novak, and S. Lockhart, 1997. *Technical Drawing,* 5th ed., Upper Saddle River, NJ: Prentice Hall.

Parametric Technology Corporation, 1997. *Pro/Engineer Assembly Modeling User's Guide.*

Parametric Technology Corporation, 1997. *Pro/Engineer Drawing User's Guide.*

Parametric Technology Corporation, 1997. *Pro/Engineer Fundamentals.*

Parametric Technology Corporation, 1997. *Pro/Engineer Interface Guide.*

Parametric Corporation, 1997. *Pro/Engineer Part Modeling User's Guide.*

Parametric Technology Corporation, 1998. *Pro/Engineer Release Notes.*

Parametric Technology Corporation, 2000. Pro/Engineer 2001 Online Help.

Parametric Technology Corporation, 2003. Pro/Engineer Wildfire Online Help.

INDEX